ENERGY SAVING COATING MATERIALS

ENERGY SAVING COATING MATERIALS

Design, Process, Implementation and Recent Developments

Edited by

GOUTAM KUMAR DALAPATI

MOHIT SHARMA

ELSEVIER

Elsevier
Radarweg 29, PO Box 211, 1000 AE Amsterdam, Netherlands
The Boulevard, Langford Lane, Kidlington, Oxford OX5 1GB, United Kingdom
50 Hampshire Street, 5th Floor, Cambridge, MA 02139, United States

Library of Congress Cataloging-in-Publication Data
A catalog record for this book is available from the Library of Congress

British Library Cataloguing-in-Publication Data
A catalogue record for this book is available from the British Library

ISBN: 978-0-12-822103-7

For information on all Elsevier publications
visit our website at https://www.elsevier.com/books-and-journals

Publisher: Matthew Deans
Acquisitions Editor: Christina Gifford
Editorial Project Manager: Rafael G. Trombaco
Production Project Manager: Nirmala Arumugam
Designer: Matthew Limbert

Typeset by Thomson Digital

Contents

Contributors

Kamal Alameh
Electron Science Research Institute, School of Science, Edith Cowan University, Joondalup, WA, Australia

Busi Kumar Babu
Department of Chemistry, SRM University, AP – Amaravati, Andhra Pradesh, India

Priyanka Bamola
Department of Physics, School of Physical Sciences, Doon University, Dehradun, Uttarakhand, India

Bikramjit Basu
Materials Research Centre; Interdisciplinary Centre for Energy Research, Indian Institute of Science, Bangalore, Karnataka, India

Sabyasachi Chakrabortty
Department of Chemistry, SRM University, AP-Amaravati, Andhra Pradesh, India

Goutam Kumar Dalapati
Department of Physics, SRM University-AP, Amaravati, Andhra Pradesh, India; School of Engineering & Innovation, The Open University, Walton Hall, Milton Keynes, United Kingdom; Agency for Science, Technology and Research (A*STAR), Institute of Materials Research and Engineering, Singapore

Atasi Dan
Materials Research Centre, Indian Institute of Science, Bangalore, Karnataka, India

Jatindra Kumar Dash
Department of Computer Science and Engineering, SRM University-AP, Amaravati, Andhra Pradesh, India

Charu Dwivedi
Department of Chemistry, School of Physical Sciences, Doon University Dehradun, Dehradun, Uttarakhand, India

Siddhartha Ghosh
Department of Physics, SRM University, AP–Amaravati, Andhra Pradesh, India

Pawan Kumar
Institute of Materials Research and Engineering, Agency for Science Technology and Research, Singapore

Yee-Fun Lim
Institute of Materials Research and Engineering, A*STAR (Agency for Science, Technology and Research), Singapore

Hongfei Liu
Institute of Materials Research and Engineering (IMRE), A*STAR (Agency for Science, Technology and Research), Singapore

Chandreswar Mahata
School of Electronics Engineering, Chungbuk National University, Cheongju, Republic of Korea

Nimai Mishra
Department of Chemistry, SRM University, AP–Amaravati, Neerukonda, Guntur (Dt), Andhra Pradesh, India

Sabyasachi Mukhopadhyay
Department of Physics, SRM University, AP– Amaravati, Guntur, Andhra Pradesh, India

Ashwini Nawade
Department of Physics, SRM University, AP–Amaravati, Guntur, Andhra Pradesh, India

Mohammad Nur-E-Alam
Electron Science Research Institute, School of Science, Edith Cowan University, Joondalup, WA, Australia

Priyanaka
Department of Physics, School of Physical Sciences, Doon University Dehradun, Uttarakhand, India

Kunchanapalli Ramya
Department of Physics, SRM University, AP– Amaravati, Guntur, Andhra Pradesh, India

Ina Rayal
Department of Physics, School of Physical Sciences, Doon University Dehradun, Dehradun, Uttarakhand, India

Durga Venkata Maheswar Repaka
Institute of Materials Research and Engineering, Agency for Science Technology and Research, Singapore

Himani Sharma
Department of Physics, School of Physical Sciences, Doon University, Dehradun, Uttarakhand, India

Mohit Sharma
Agency for Science, Technology and Research (A*STAR), Institute of Materials Research and Engineering, Singapore

Bharti Singh
Department of Applied Physics, Delhi Technological University, Delhi, India

Vishal Singh
Department of Applied Physics, Delhi Technological University, Delhi, India

Ady Suwardi
Institute of Materials Research and Engineering, Agency for Science Technology and Research, Singapore

V.G. Vasavi Dutt
Department of Chemistry, SRM University-AP, Amaravati, Neerukonda, Guntur (Dt), Andhra Pradesh, India

Mikhail Vasiliev
Electron Science Research Institute, School of Science, Edith Cowan University, Joondalup, WA, Australia

CHAPTER 1

Solar radiation and light materials interaction

Vishal Singh[a], Ina Rayal[b], Priyanaka[b], Himani Sharma[b], Charu Dwivedi[c], Bharti Singh[a]
[a]Department of Applied Physics, Delhi Technological University, Delhi, India
[b]Department of Physics, School of Physical Sciences, Doon University Dehradun, Dehradun, Uttarakhand, India
[c]Department of Chemistry, School of Physical Sciences, Doon University Dehradun, Dehradun, Uttarakhand, India

1 Introduction

The supply and demand of energy determine the course of global development in every sphere of human activity. Sufficient supplies of clean energy are intimately linked with global stability, economic prosperity, and quality of life. Finding energy sources to satisfy the world's growing demand is one of society's foremost challenges for the next half-century. It is generally understood that current patterns of energy supply are nonsustainable, and that renewable energy, including solar energy, is bound to play a major role.

1.1 Solar radiation and its characteristics

Solar radiation, often called the solar resource, is a general term for the electromagnetic radiation emitted by the sun *(1,2)*. Solar radiation can be captured and turned into useful forms of energy, such as heat and electricity, using a variety of technologies and materials. The different parts of the electromagnetic spectrum as shown in Fig. 1.1 have very different effects upon interaction with matter and therefore serve different purposes *(3)*. Starting with low frequency radio waves, the matter is quite transparent. (As we can listen to our portable radio inside our home since the waves pass freely through the walls of the house and even through the person beside us.) As we move upward through microwaves and infrared (IR) to visible light (VIS) more, absorption increases. In the lower ultraviolet (UV) range, all the UV from the sun is absorbed in a thin outer layer of our skin. As we move further up into the x-ray region of the spectrum, we become transparent again, because most of the mechanisms for absorption are gone, we then absorb only a small fraction of the radiation, but that absorption involves the more violent ionization events. Each portion of the electromagnetic

Energy Saving Coating Materials
http://dx.doi.org/10.1016/B978-0-12-822103-7.00001-7

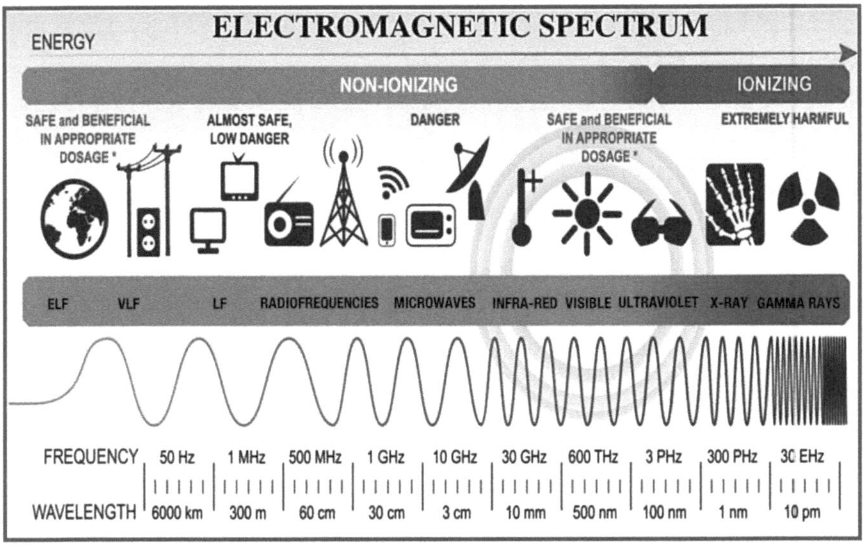

Figure 1.1 ***A diagram showing the electromagnetic spectrum.*** Source: (3) *Copyright.*

spectrum has quantum energies appropriate for the excitation of certain types of physical processes. The energy levels for all physical processes at the atomic and molecular levels are quantized, and if there are no available quantized energy levels with spacings, which match the quantum energy of the incident radiation, then the material will be transparent to that radiation, and it will pass through. If electromagnetic energy is absorbed, but cannot eject electrons from the atoms of the material, then it is classified as nonionizing radiation, and will typically just heat the material.

The spectrum of the incoming solar electromagnetic radiation (250-2500 nm) just outside the earth's atmosphere has a bell shape and defines the sun's temperature (~6000 °C) and is shown in Fig. 1.2 *(4)*. The total amount of energy emitted by the sun and received at the extremity of the Earth's atmosphere is constant-1370 W/m^2/sec. That received per unit area of the Earth's surface is 343 W/m^2/sec. Incoming solar radiations are absorbed by atmospheric gases such as O_2, O_3, CO_2, and H_2O vapor. UV light at wavelengths of <0.18 μm is strongly absorbed by O_2 at altitudes above 100 km (62 mi). Ozone below 60 km (37.2 mi) absorbs most UV between 0.2 and 0.3 μm (200-300 nm). Atmospheric absorption above 40 km (24.8 mi) results in attenuation of ~5% of incoming solar radiation. Under clear-sky conditions, another 10%-15% is absorbed by the lower atmosphere or is scattered back to space, with 80%-85% reaching the ground. It is important to note that the spectrum is limited to $0.25 < \lambda < 3$ μm so that there is no overlap with the spectrum of thermal radiation *(5)*. Hence, it is possible to

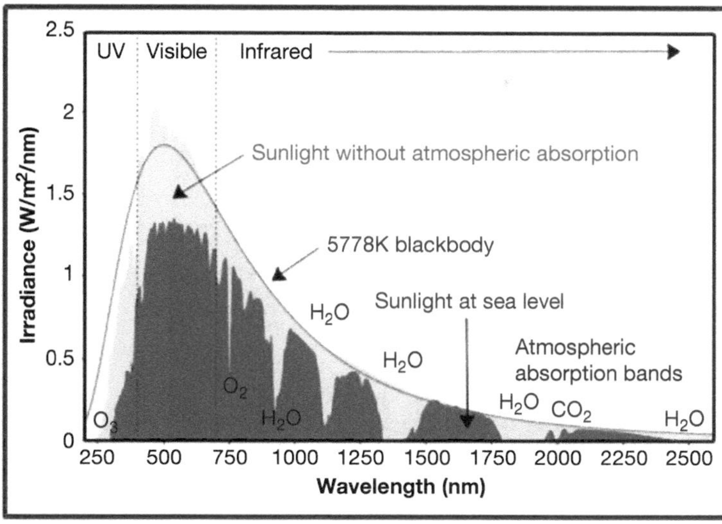

Figure 1.2 ***A diagram showing the solar radiation spectrum.*** Source: (5) *Copyright.*

have surfaces with properties being entirely different with regard to thermal and solar radiation. The balance between the incoming solar radiation and the outgoing thermal radiation is the base for our environmental conditions.

The incoming solar electromagnetic radiation carries energy in tiny packets, which are characterized by their frequency and are called photons. The famous Planck-Einstein relationship between photon frequency (V) and the energy of E.M waves is given by equation

$$E = hv,$$

where h is a fundamental constant and is called as Planck's constant *(6)*. The value of "h;" = 6.626×10^{-34} Js. The wavelength and frequency dependence of photon energy is key in determining how it would interact with matter. Light traveling through a transparent media would travel at speed less than that of light in vacuum, that is, retardation is observed. An example cited earlier where a photon starting at the center of the sun took millions of years to reach the surface is due to this same interaction of photons with matter. Along its path, the photon undergoes a million collisions with the atoms, ions, of the matter, each time gaining or losing energy. But once it reaches the surface, it travels in the vacuum of space taking just 8.3 min to reach the Earth. All these different portions of E.M radiations interact very efficiently with matter. For example, VIS (and near-infrared light) is typically absorbed and emitted by electrons in molecules and atoms that move from one energy level to another. This action allows the chemical mechanisms that underlie human vision and plant photosynthesis. If radiation having a frequency in the visible region of the EM spectrum reflects off an object, and then strikes the eyes, this results in visual perception of the scene. The brain's visual system processes the multitude of reflected frequencies into different shades and hues. This fine-tuned mechanism, a marvel of nature, is what allows us to "see." These types of interaction seem to be also dependent on the type of matter and change with bulk, thin films and nanomaterials. Hence, it becomes important to understand this solar radiation and matter interaction to get the benefit of a natural source of energy.

1.2 Importance of solar radiation

The world's major energy sources are nonrenewable and are faced with ever-increasing demand, and thus are not expected to last long. Besides being nonrenewable, these sources include mainly fossil fuels, contributing tremendously to the perennial problem of global warming because of CO_2 emission. The eminent depletion and pollution problems of the above energy sources make the international community focus attention on alternative sources of energy, especially solar energy appears highly promising. The sun is Earth's natural power source, driving the circulation of global wind and ocean currents, the cycle of water evaporation and condensation that creates rivers and lakes, and the biological cycles of photosynthesis and life. Covering 0.16% of the land on earth with 10% efficient solar conversion systems would provide 20 TW of power, nearly twice the world's consumption rate of fossil energy. These comparisons illustrate the impressive magnitude of the solar resource, providing an energy stream far more potent than present-day human technology can achieve. Solar energy is already being successfully used in residential and industrial settings for cooking, heating, cooling, lighting, space technology, and for communications among other uses. While the reserves of fossil fuels are restricted, there is no limitation to the availability of solar energy.

All routes for utilizing solar energy exploit the functional steps of capture, conversion, and storage. The sun's energy arrives on earth as radiation distributed across the color spectrum from IR to UV. The energy of this radiation must be captured as excited electron hole pairs in a semiconductor, a dye, or a chromophore, or as heat in a thermal storage medium. Excited electrons and holes can be tapped off for immediate conversion to electrical power, or transferred to biological or chemical molecules for conversion to fuel. Natural photosynthesis produces fuel in the form of sugars and other carbohydrates derived from the reduction of CO_2 in the atmosphere and used to power the growth of plants. The plants themselves become available for as biomass for combustion as primary fuels or for conversion in reactors to secondary fuels like liquid ethanol or gaseous carbon monoxide, methane, and hydrogen. We are now learning to mimic the natural photosynthetic process in the laboratory using artificial molecular assemblies, where the excited electrons and holes can drive chemical reactions to produce fuels that link to our existing energy networks. Atmospheric CO_2 can be reduced to ethanol or methane, or water can be split to create hydrogen. These fuels are the storage media for solar energy, bridging the natural day-night, winter-summer, and cloudy-sunny cycles of solar radiation.

In addition to electric and chemical conversion routes, solar radiation can be converted to heat energy. Solar concentrators focus sunlight collected over a large area to a line or spot where heat is collected in an absorber. Temperatures as high as 3000°C can be generated to drive chemical reactions, or heat can be collected at lower temperatures and transferred to a thermal storage medium like water for distributed space heating or steam to drive an engine. Effective storage of solar energy as heat requires developing thermal storage media that accumulate heat efficiently during sunny periods and release heat slowly during dark or cloudy periods. Heat is one of the most versatile forms of energy, the common link in nearly all our energy networks. Solar thermal conversion can replace much of the heat now supplied by fossil fuel. Although many routes use solar energy to produce electricity, fuel, and heat, none are currently competitive with fossil fuels for a combination of cost, reliability, and performance.

2 Different technologies for harnessing solar energy

Technologies and resources are used to convert solar radiation into solar energy to store it or convert it for other applications and can be broadly classified into two groups: (1) photovoltaic (PV) *(7)* and (2) solar thermal *(8)*.

The PV technology converts radiant energy contained in light quanta into electrical energy when light falls upon a semiconductor material by causing electron excitation and strongly enhancing conductivity. Solar thermal technology uses solar heat, which can be used directly for either thermal or heating application or electricity generation. Accordingly, it can be divided into two categories: (1) solar thermal nonelectric and (2) solar thermal electric. The former includes applications such as agricultural drying, solar water heaters, solar air heaters, solar cooling systems, and solar cookers, and the latter refers to the use of solar heat to produce steam for electricity generation, also known as concentrated solar power (CSP). In addition to the above technologies, of utilizing the solar energy spectrum, it is also important to check for the possibility of tailoring the spectral absorptance, emittance, reflectance, and transmittance of a surface to meet different demands in different wavelength intervals, that is, to take advantage of spectral selectivity of the solar grade materials *(9)*.

2.1 Solar PV technology

Solar power can be converted directly into electrical power in PV cells, commonly called solar cells. The sun has a surface temperature of about 6000°C, and its hot gases at this temperature emit light that has a spectrum ranging from the UV, through the visible, into the IR. According to quantum theory, light can behave either as waves or as particles, depending upon the specific interaction of light with matter; this phenomenon is called the wave-particle duality of light. In the particle description, light consists of discrete particle-like packets of energy called photons. Sunlight contains photons with energies that reflect the sun's surface temperature; in energy units of electron volts (eV), the solar photons range in energy (hν) from about 3.5 eV (UV region) to 0.5 eV (IR region). The energy of the visible region ranges from 3.0 eV (violet) to 1.8 eV (red); the peak power of the sun occurs in the yellow region of the visible region, at about 2.5 eV. At high noon on a cloudless day, the surface of the Earth receives 1000 W of solar power per square meter (1 kW/m^2).

PV cells generally consist of a light absorber that will only absorb solar photons above certain minimum photon energy. This minimum threshold energy is called the "energy gap" or "band gap" (Eg); photons with energies below the band gap pass through the absorber, while photons with energies above the band gap are absorbed. The light absorber in PV cells can be either inorganic semiconductors, organic molecular structures, or a combination of both. In inorganic semiconductor materials, such as Si, electrons (e^-) have energies that fall within certain energy ranges, called bands. The energy ranges, or bands, have energy gaps between them. The band containing electrons with the highest energies is called the valence band. The next band of possible electron energies is called the conduction band; the lowest electron energy in the conduction band is separated from the highest energy in the valence band by the band gap. When all the electrons in the absorber are in their lowest energy state, they fill up the valence band, and the conduction band is empty of electrons. This is the usual situation in the dark. When photons are absorbed, they transfer their energy to electrons in the filled valence band and promote these electrons to higher energy states in the empty conduction band. There are no energy states between the valence and conduction bands, which is why this separation is called a band gap and why only photons with energies above the band gap can cause the transfer of electrons from the lower-energy-state valence band into the higher-energy-state conduction band. When photons transfer electrons across the band gap, they create negative charges in the conduction band and leave behind positive charges in the valence band; these positive charges are called holes (h+). Thus, absorbed photons in semiconductors create pairs of negative electrons and positive holes. In a PV cell, the electrons and holes formed upon absorption of light separate and move to opposite sides of the cell structure, where they are collected and pass through wires connected to the cell to produce a current and a voltage thus generating electrical power. The schematic is shown in Fig. 1.3.

2.2 Solar thermal technology

Solar thermal processes encompass a wide variety of technologies, all of which begin by converting the incident sunlight to heat. This heat can be converted to electricity, as is currently done in CSP plants via a Rankine or combined cycle *(10)*, solar thermoelectric generators (STEG) *(11)*, and solar thermophotovoltaics (STPV) [13,14]. Currently, there are 3 GW of installed and underconstructed CSP plants in the world, but STEGs and STPV have yet to be commercialized. Solar thermal energy, or the waste heat from the aforementioned solar-to electricity processes, can also be used for heating or cooling. Nearly 300 GW of solar thermal heat or hot water systems have been installed worldwide *(12)*.

A schematic of a typical solar thermal to electrical energy conversion system is shown in Fig. 1.4. In such solar thermal technologies, the sunlight is absorbed as heat at the absorber. This absorber must strongly absorb the sunlight, while ideally losing little heat to the environment via convection and radiation. An absorber, which has strong solar absorptance and low IR emittance, is called a spectrally selective absorber *(14,15)*.

Absorbers must efficiently convert a solar flux into heat and deliver that heat to the thermal system (Fig. 1.4). The efficiency of the conversion from solar irradiance at the absorber, H_{abs}, to a heat flux delivered to the thermal system, q_h, is defined as the absorber efficiency, η_{abs}:

$$\eta_{abs} = \frac{q_h}{H_{abs}} = \alpha_s - \varepsilon_t \frac{\sigma_{sb}\left(T_h^4 - T_{amb}^4\right)}{H_{abs}} - \frac{h_{conv}\left(T_h - T_{amb}\right)}{H_{abs}}$$

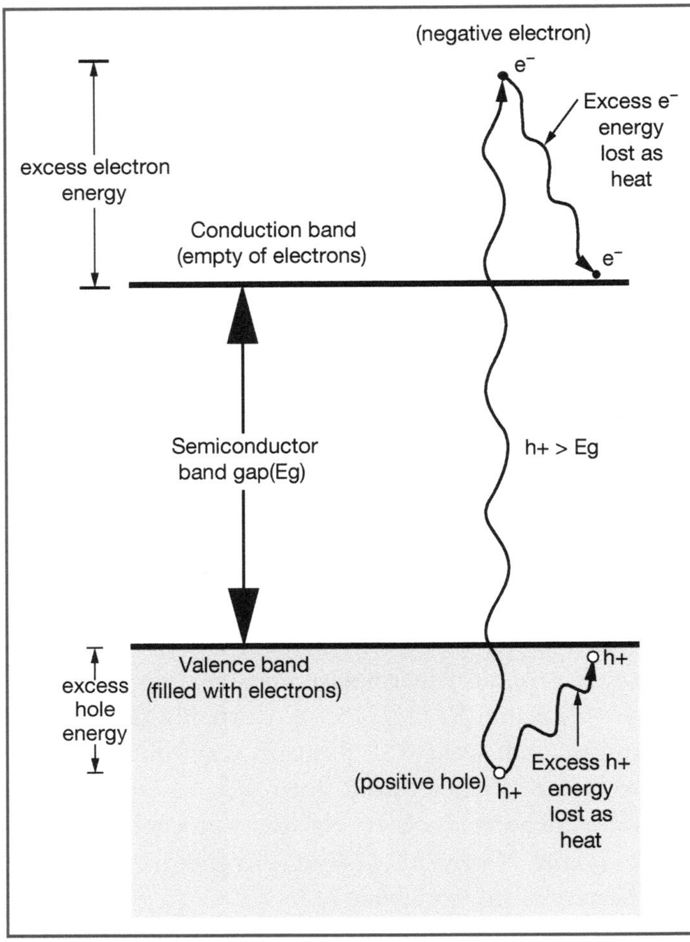

Figure 1.3 ***A diagram showing absorption of photon of energy (h$_\nu$) greater than band gap of semiconducting materials.***

where σ_{sb} is the Stefan-Boltzmann constant, T_h is the absorber temperature, T_{amb} is the temperature of the environment, and h_{conv} is the convective heat transfer coefficient. The solar absorptance, α_s, is defined as:

$$\alpha_s = \frac{\int_0^{\infty}\int_0^{2\pi}\int_0^{\pi/2} \varepsilon'_\lambda(\lambda,\phi,\theta)S(\lambda,\phi,\theta)\cos\theta\,\sin\theta d\lambda\, d\phi\, d\theta}{\int_0^{\infty}\int_0^{2\pi}\int_0^{\pi/2} S(\lambda,\phi,\theta)\cos\theta\,\sin\theta d\lambda\, d\phi\, d\theta}$$

where λ is the wavelength of radiation, ϕ is the azimuthal angle, θ is the polar angle, ε'_λ is the spectral directional emittance at the operational temperature, and S is the incident solar intensity at the absorber. The numerator

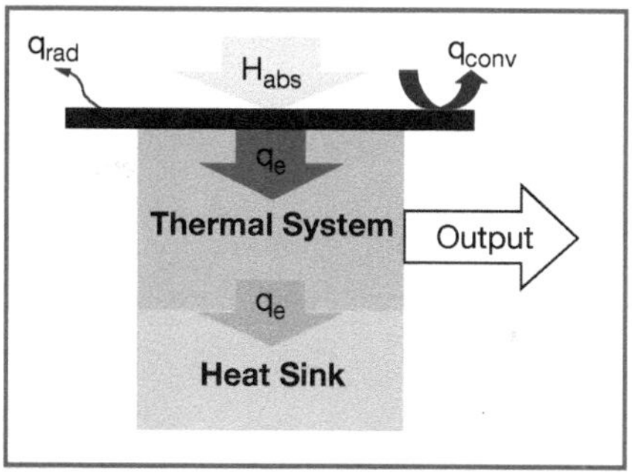

Figure 1.4 ***A typical solar thermal energy conversion system.*** Sunlight incident on the absorber (H_{abs}) is converted to a heat flux (q_h) and delivered to the thermal system, where the desired output (work, electricity, heat, cooling, etc.) is produced. Waste heat (q_c) may be generated in the process and rejected to a heat sink. Radiative (q_{rad}) and convective (q_{conv}) losses at the surface reduce the potential output of the system.

of this equation is the total absorbed solar energy; the denominator is the incident solar flux H_{abs}.

The thermal emittance, ε_t, is defined as:

$$\varepsilon_t = \frac{\int_0^\infty \int_0^{2\pi} \int_0^{\pi/2} \varepsilon'_\lambda(\lambda,\phi,\theta) I_{b\lambda}(\lambda,\phi,\theta)\cos\theta \sin\theta d\lambda \, d\phi \, d\theta}{\int_0^\infty \int_0^{2\pi} \int_0^{\pi/2} I_{b\lambda}(\lambda,\phi,\theta)\cos\theta \sin\theta d\lambda \, d\phi \, d\theta}$$

where $I_{b\lambda}$ is the blackbody intensity given by Planck's formula.

One of the efficient spectrally selective absorbers is a mixture of ceramics and metals (cermet). When integrated with an antireflection coating (ARC) and an IR-reflective base layer (either an intrinsically IR reflecting substrate or an IR-reflective coating on a substrate), cermet-based coatings (Fig. 1.5) can be very effective spectrally selective absorbers.

By having a large emittance in the solar wavelengths and a small emittance in the IR, absorbers can have high solar absorptance and low thermal emittance, thus resulting in high absorber efficiency. To maximize the absorber efficiency, multiple absorption mechanisms should be employed. One strategy is to explore new materials, including metal inclusions and dielectrics, which have high thermal stability and excellent optical properties for high temperature applications.

3 Interaction of matter with solar radiation

Rapid developments in nanoscience have created an unusual opportunity for advances in basic solar energy research. Because the individual microscopic steps of solar energy use take place on the nanometer scale, the ability

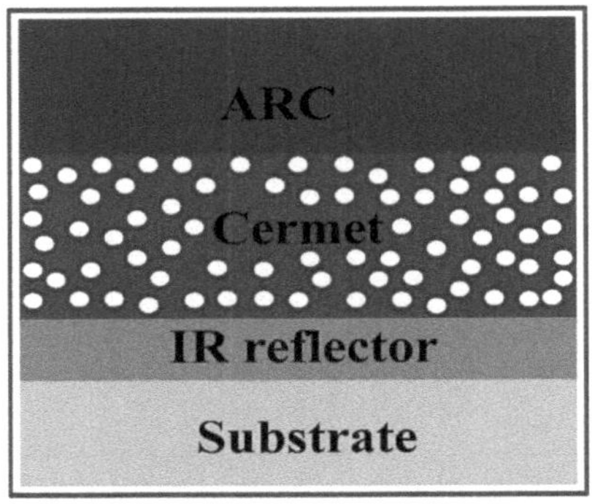

Figure 1.5 ***Schematic representation of a cermet-based spectrally selective absorber.*** The antireflection coating (ARC) reduces solar reflection off the surface; the cermet provides selective absorption; and the IR reflector (typically Cu, Al, Mo, or other metal with low intrinsic emissivity) reduces radiation losses. The substrate is usually metal (to conduct heat well) or glass (for lower cost). The cermet comprises a dielectric host (*dark red*) with metal particle inclusions (*circles*).

to pattern and control matter on this length scale presents unusual opportunities for researchers to create new materials for solar energy conversion and use. The emergence of new fundamental physical properties on this length scale could potentially lead to far more efficient technologies for conversion of solar energy to electricity and fuels. To take advantage of these new materials and processes, researchers must create complex arrangements of nanoscale components.

The most popular term in the nanoscale world is quantum confinement effect, which is essentially due to changes in the atomic structure as a result of direct influence of ultra-small length scale on the energy band structure *(16)*. The length scale corresponds to the regime of quantum confinement ranges from 1 to 25 nm for typical semiconductor groups of IV, III-V, and II-VI, in which the spatial extent of the electronic wave function is comparable with the particle size. As a result of these "geometrical" constraints, electrons "feel" the presence of the particle boundaries and respond to changes in particle size by adjusting their energy. This phenomenon is known as the quantum-size effect. Quantization effects become more important when the particle dimension of a semiconductor is near or below the bulk semiconductor Bohr exciton radius, which makes material's properties size dependent. In general, the Bohr radius of a particle is defined as *(17)*

$$r_b = \varepsilon \frac{m}{m^\star} a_0$$

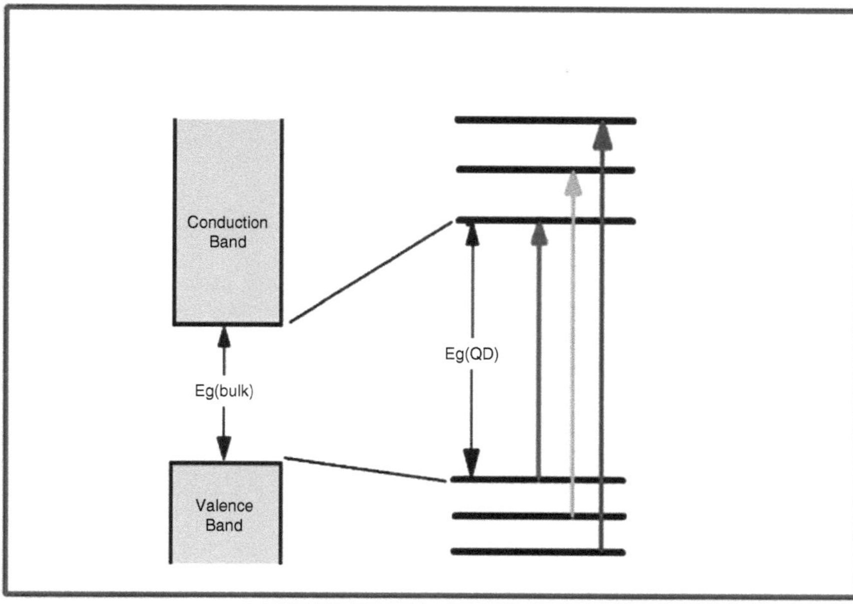

Figure 1.6 ***A schematic of the discrete energy level of a semiconductor.***

where ε is the dielectric constant of the material, m^* is the effective mass of the particle, m is the rest mass of the electron, and a_o is the Bohr radius of the hydrogen atom. When the particle size approaches Bohr exciton radius, the quantum confinement effect causes increasing of the excitonic transition energy and blue shift in the absorption and luminescence band gap energy *(17)*. For example, 4.8 nm diameter PbSe nano crystals (NCs) show an effective band gap of approximately 0.82 eV, exhibiting a strong confinement induced blue shift of >500 meV compared to the bulk PbSe band gap of 0.28 eV (the Bohr exciton radius in PbSe is 46 nm) *(18)*. In addition, quantum confinement leads to a collapse of the continuous energy bands of a bulk material into discrete, atomic like energy levels. The discrete structure of energy states leads to a discrete absorption spectrum, which is in contrast to the continuous absorption spectrum of a bulk semiconductor as shown in Fig. 1.6. A quantum confined structure is one in which the motion of the carriers (electron and hole) are confined in one or more directions by potential barriers. Based on the confinement direction, a quantum confined structure will be classified into three categories as quantum well, quantum wire, and quantum dots (QD) or nanocrystals. The basic type of quantum confined structure is given in Table 1.1.

Table 1.1 Classification of quantum confined structures.

Structure	Quantum confinement	Number of free dimension
Bulk	0	3
Quantum well/superlattices	1	2
Quantum wire	2	1
Quantum dot/nanocrystals	3	0

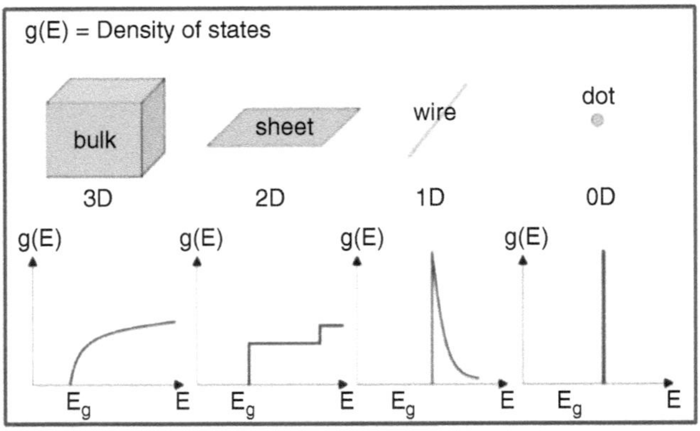

Figure 1.7 ***Density of electron states of a semiconductor as a function of dimension. The optical absorption spectrum is roughly proportional to the density of states.***

In QDs, the charge carriers are confined in all three dimensions which the electrons exhibit a discrete atomic-like energy spectrum. Quantum wires are formed when two dimensions of the system are confined. In quantum well, charge carriers (electrons and holes) are confined to move in a plane and are free to move in a two-dimensional. Also, the energy level of one of the quantum numbers changes from continuous to discrete. Compared with bulk semiconductors, the quantum well has a higher density of electronic states near the edges of the conduction and valence bands, and therefore a higher concentration of carriers can contribute to the band-edge emission. As more number of the dimension is confined, more discrete energy levels can be found; in other words, carrier movement is strongly confined in a given dimension. Density of electron states in bulk, 2D, 1D, and 0D semiconductor structures, are shown in Fig. 1.7. 0D structures have very well-defined and quantized energy levels. The quantum confinement effect corresponding to the size of the nanostructure can be estimated via a simple effective-mass approximation model.

This method can predict the confined energy levels of nanostructures by solving Schrodinger equation assuming the barriers have an infinite confining

potential. The "effective mass" solutions of the Schrödinger equation for electrons confined in a quantum dot or NCs, quantum wire and quantum well are:
Quantum dot or nanocrystal:

$$E_{n,m,l} = \frac{n^2\hbar^2}{2m^*}\left(\frac{n^2}{L_z^2} + \frac{m^2}{L_y^2} + \frac{l^2}{L_x^2}\right), = \phi(z)\phi(y)\phi(x)$$

Quantum wire:

$$E_{n,m}(k_x) = \frac{n^2\hbar^2}{2m^*}\left(\frac{n^2}{L_z^2} + \frac{m^2}{L_y^2}\right) + \frac{\hbar^2 k_x^2}{2m^*}, = \phi(z)\phi(y)exp(ik_x x)$$

Quantum well:

$$E_{n,m,l} = \frac{n^2\hbar^2 n^2}{2m^* L_z^2} + \frac{\hbar^2}{2m^*}(k_x^2 + k_y^2), = \phi(z)exp(ik_x x + ik_y y)$$

where $n, m, 1 = 1, 2 \ldots$ the quantum confinement numbers; L_x, L_y and L_z are the confining dimensions; $\exp(ik_x x + ik_y y)$ is the wave function describing the electronic motion in x and y directions, similar to free electron wave functions.

In addition to the quantum confinement effect of nanostructures, their small dimensions, results into extremely large surface area to volume ratio, which makes a large number of surface or interfacial atoms, are resulting in more surface dependent properties. Nanomaterials can be classified depending on the dimensions such as (1) 0D spheres and clusters, (2) 1D nanofibers, nanowires, and nanorods, (3) 2D films, plates, and networks, (4) 3D nanomaterials and is shown in Fig. 1.8. While most of micro structured materials have similar properties to the corresponding bulk materials, the

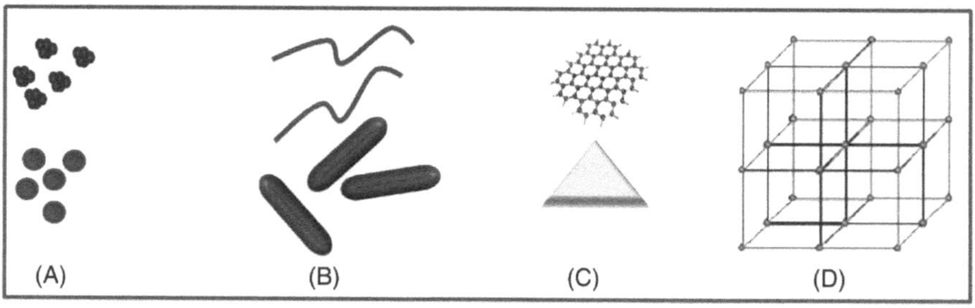

Figure 1.8 Classification of nanomaterials (A) 0D spheres and clusters, (B) 1D nanofibers, wires, and rods, (C) 2D films and networks, (D) 3D nanomaterials.

properties of materials with nanometer dimensions are significantly different from those of atoms and bulk materials. This is mainly due to the nanometer size of the materials, which render them: (1) large fraction of surface atoms, (2) high surface energy, (3) spatial confinement, and (4) reduced imperfections cannot be seen exist in the corresponding bulk materials *(19)*. Reduced imperfections are also an important factor in the determination of the properties of the nanomaterials.

4 Effect of nanostructuring on the solar energy harvesting

Due to these changes in the properties of nanoparticles as compared to their bulk counterparts, Nanostructure-based solar energy is attracting significant attention as a possible candidate for achieving drastic improvement in the field of solar energy conversion to some useful applications *(20)*. To improve the efficiency of solar energy conversion technologies, we have to increase the absorbance of the materials; this can be achieved by taking that material into the nano scale range because due to change in size, the band gap and density of states also change. So, we can use nanostructured layers instead of traditional thin-film solar cells. Nanostructured layers in thin-film solar cells offer three important advantages. First, due to multiple reflections, the effective optical path for absorption is much larger than the actual film thickness. Second, light-generated electrons and holes need to travel over a much shorter path and thus recombination losses are greatly reduced. As a result, the absorber layer thickness in nanostructured solar cells can be as thin as 150 nm instead of several micrometers in the traditional thin film solar cells. Third, the energy band gap of various layers can be made to the desired design value by varying the size of nanoparticles. This allows for more design flexibility in the absorber of solar cells. Examples to this type of layers are: polycrystalline thin-film solar cells such as $CuInSe_2$ (CIS), $Cu(In,Ga)Se_2$ (CIGS), and CdTe compound semiconductors *(21,22)*. Because of the high absorption coefficient ($\sim 10^5$ cm^{-1}), a thin layer of ~2 mm is sufficient to absorb the useful part of the spectrum. Fig. 1.9 shows the schematic structure of CIGS-based solar cells obtained by low-temperature pulsed electron deposition, with the experimentally achieved efficiency of 15% *(23)*.

Moreover, nanomaterials are smaller in size and also have lightweight; due to this, lightweight and flexible solar cells can yield a high specific power (W/kg) and open numerous possibilities for a variety of applications. Interaction of solar radiations with matter at nanoscale has two advantages;

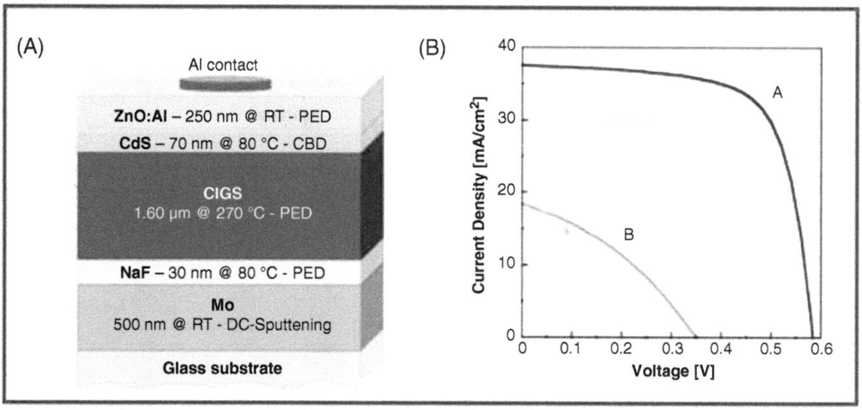

Figure 1.9 (A) Schematic diagram of CIGS photovoltaic (PV) structure. (B) ($I - V$) Characteristic for the CIGS solar cell. Source: (23) *Copyright.*

it improves light absorption and reduces the amount of material needed. Another advantage of nanoscale is that some of the indirect band gap semiconductors can be transferred to strongly absorbing direct band gap semiconductors and can be used as a thin film solar cells such as silicon, which has mostly dominated the PV industry.

As we know, electromagnetic radiation (primarily in the visible and near-infrared regions of the spectrum) is emitted from the sun and absorbed by the solar cell. A photon will then excite a negatively charged electron from the valence band (low energy state) to the conduction band (a higher energy state) leaving behind a positively charged vacancy, called a hole. For this energy transfer to create any usable energy, the photon must have energy greater than the bandgap of the material, or else the electron will immediately relax down and recombine with the hole and the energy will be lost as heat. Upon excitation above the bandgap, the photon creates an electron and a hole, which are now free to move throughout the semiconductor crystal. These act as charge carriers, which transport the energy to the electrical contacts, which results in a measurable external current. Current solar cells cannot convert all the incoming light into usable energy because some of the light can escape back out of the cell into the air. Additionally, sunlight comes in a variety of colors and the cell might be more efficient at converting bluish light while being less efficient at converting reddish light. Higher energy light does excite electrons to the conduction band, but any energy beyond the band gap energy is lost as heat. All these problems can be resolved by using the QDs instead of the bulk materials *(24)*. The access energy, which is lost in the form of heat after excitation of

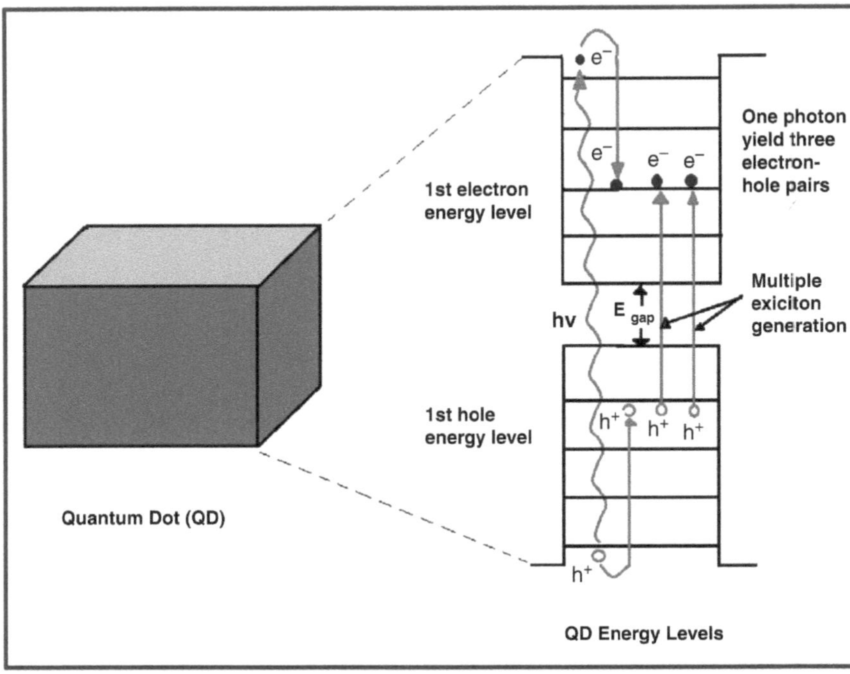

Figure 1.10 ***Multiple exciton generation in quantum dots.*** Because of quantum confinement in the small nanoscale semiconductor QD particle, the energy levels for electrons and holes are discrete. This slows hot exciton cooling and enhances multiple exciton formation. A single absorbed photon that has energy at least 3 times the energy difference between the first energy levels for electron and holes in the QD can create 3 excitons. The bandgap of the bulk semiconductor is indicated as Eg.

electron, is further used in QDs for multiple exciton generation as shown in the schematic given in Fig. 1.10 *(25,26)*.

In bulk semiconductors, this process is known as electron hole pair multiplication (EHMP) and occurs when more than one electron hole pair is produced by absorption of one photon of energy at least twice the band gap. Nozik et al. reported that production of these electron-hole pair increases by ~2 in PbSe QDs compared to bulk PbSe. EHMP is more efficient in PbSe QDs than in bulk PbSe *(29)*. It is reported that in QDs there are at least three fundamental properties that are modified due to quantum confinement and affect the EHPM process. (A) Crystal momentum is no longer a good quantum number. There are three factors that can affect $h\nu_{th}$ (threshold energy) related to crystal momentum: (1) absorption selection rules are modified; (2) conservation of crystal translational momentum is relaxed, allowing $h\nu_{th}$ to be less than that required by momentum conservation. In fact, we find $h\nu_{th}$ in QDs of PbSe is lower than that allowed by

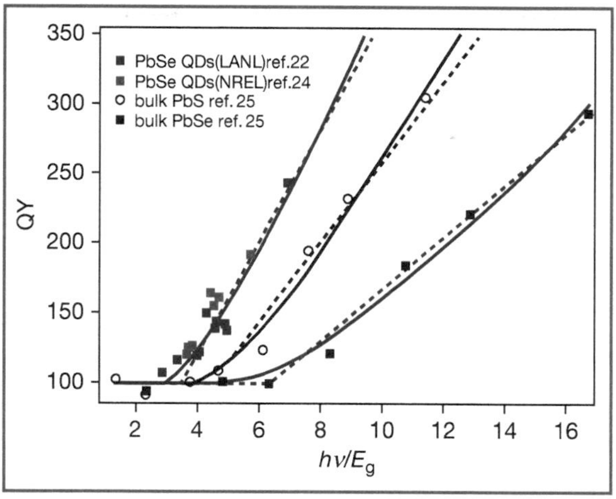

Figure 1.11 ***QY vs. hν/Eg for PbSe QD, PbSe bulk, and PbS bulk samples.*** Source: *[29–31] Copyright.*

momentum conservation in bulk PbSe, and (3) single- and multiexcitonic states can be coupled through the Coulomb operator to form a superposition of states. Such coupling is not possible in bulk systems with well-defined momentum. The best way to compare EHPM processes between semiconductor QDs and bulk semiconductors is to plot the quantum yield (QY) vs. $h\nu$/Eg (threshold energy needed to produce extra electron-hole pair) as this provides a direct determination of the EHPM efficiency, η_{EHPM}, and allows for determination of the relative contributions of EHPM and competing relaxation channels and is shown in Fig. 1.11 for PbSe QD and PbSe and PbS bulk samples. (B) The discrete structure of semiconductor QD energy bands, due to quantum confinement and intimate control over surface states and surface ligands, can be used to modify carrier relaxation rates. (C) Increased Columbic coupling between excitons in QDs increases Auger-related processes like multiple-exciton-generation (MEG).

Due to these properties of QDs, solar cells produced from QDs can have much higher power conversion efficiencies than their bulk counterparts and are given in Fig. 1.12. The results are based on Shockley-Queisser (SQ) detailed balance calculations given [I]. In Fig. 1.12, curve 6 (black solid line) is the conventional SQ (Shockley-Queisser) calculation with just 1 EHP created per photon at the band gap; curve 1 (solid red curve) assumes the maximum multiplication energetically allowed and is based on M_{max}. Curve 2 (solid green line) is based on $h\nu_{th}$ = 2Eg followed by creation of one extra exciton created per Eg (defined as the L2 characteristic); curve 3 (solid blue

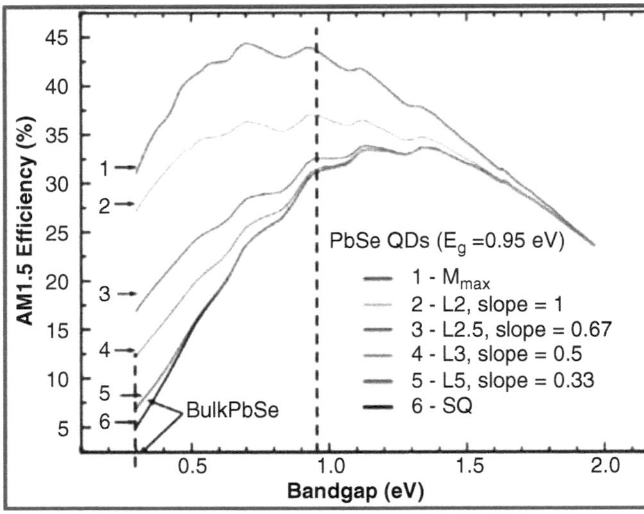

Figure 1.12 ***PV power conversion efficiencies at AM1.5 vs. band gap for various characteristics of MEG QY.*** Source: (29) *Copyright.*

line) is based on a threshold of 2.5 Eg, curve 4 is based on a threshold of 3Eg (defined as the L3 characteristic); curve 5 is based on a threshold of 4.5Eg with η_{EHPM} 0.19 (defined as the L5 characteristic and is the experimental bulk characteristic for PbSe *(29)*.

As seen in this Fig. 1.12, the maximum thermodynamic conversion efficiency of ~5% for bulk PbSe is only marginally enhanced when the experimentally measured EHPM is included. Thus, EHPM in bulk PbSe cannot produce a significant enhancement of conversion efficiency. In contrast, for PbSe QDs with a quantized band gap of 0.95 eV, the maximum thermodynamic conversion efficiency is 31% for the SQ calculation, 37% for the L2 characteristic (solid green line), 32% for the L3 characteristic (solid purple line), and 42% for the M_{max} characteristic (solid red line). These calculations show that PbSe QDs will always have a much higher theoretical conversion efficiency compared to bulk PbSe (by factors ranging from 2.7 (L3 characteristic) to 3.5 (M_{max} characteristic) *(29,30)*.

5 Importance of solar thermal technology

Another aspect of using the incoming solar photon is photo thermal energy production, and to look for new technologies to efficient conversion of solar energy into electricity, it has become much important to look for several ways of minimizing energy utilization, as conservation of energy is the most effective way of reducing CO_2 concentration, which has risen from ~315 ppm at the end of the 1950s so that it now exceeds ~400 ppm, without any sight for this reduction. In view of this, it becomes important to reduce the usage of fossil fuels and also the electricity usage, which reduces the harmful effect of CO_2 emission, that is, global warming, by making energy efficient building, windows, and vehicle. In this direction, it becomes very important to study a unique class of materials called as solar energy materials, for thermal applications and have optical properties that make them well adapted for utilizing solar energy and for reaching energy efficiency, especially in the built environment *(31)*. As we know that electricity utilization in-house and industrial is must. For example, in the United States, air conditioning-based in-house cooling consumes ~15% of the primary energy used by buildings.

Solar energy materials have properties that are tailored to the characteristics of the electromagnetic radiation in our natural surroundings, especially its spectral distribution, angle of incidence, and intensity. This tailoring can be made with regard to solar irradiation, thermal emission, atmospheric absorption, VIS, photosynthetic efficiency, and more. Solar energy materials can be of many kinds, for example, metallic, semiconducting, dielectric, glassy, polymeric, gaseous, etc. In particular, thin surface coatings of solar energy materials may exhibit the desired properties in their own right or may yield such properties when backed by an appropriate substrate. By using these smart solar energy materials for strategically cooling or heating, which keeps indoor comfort without electricity requirement, will therefore impact the global energy consumption *(32)*. Energy-efficient window is therefore considered to be the first step for reducing energy consumption through the reduction of heat transfer between indoor and outside environments. Windows are the key components to any building's design and provide the strategy to improve energy efficiency of the building. The conventional glass windows without any coatings films allow the passage of visible transmittance as well as solar heat that in turn increase the heating as well as cooling of the building structure. However, a functional coating on the glass eliminates solar heat and reduces the energy consumption. The functional materials and energy efficient windows have become a subject of intensive study toward energy conservation that has been discussed aplenty *(33–53)*.

There are a variety of coating technologies such as thermo-chromic, low-emission (low-e) coating, aerogel, photochromic, electrochromic, gasochromic, suspended particle, and liquid crystal that are demonstrated for energy saving windows application. Fig. 1.13 shows the passive IR regulating coating for smart windows application. As shown in Fig. 1.13 above,

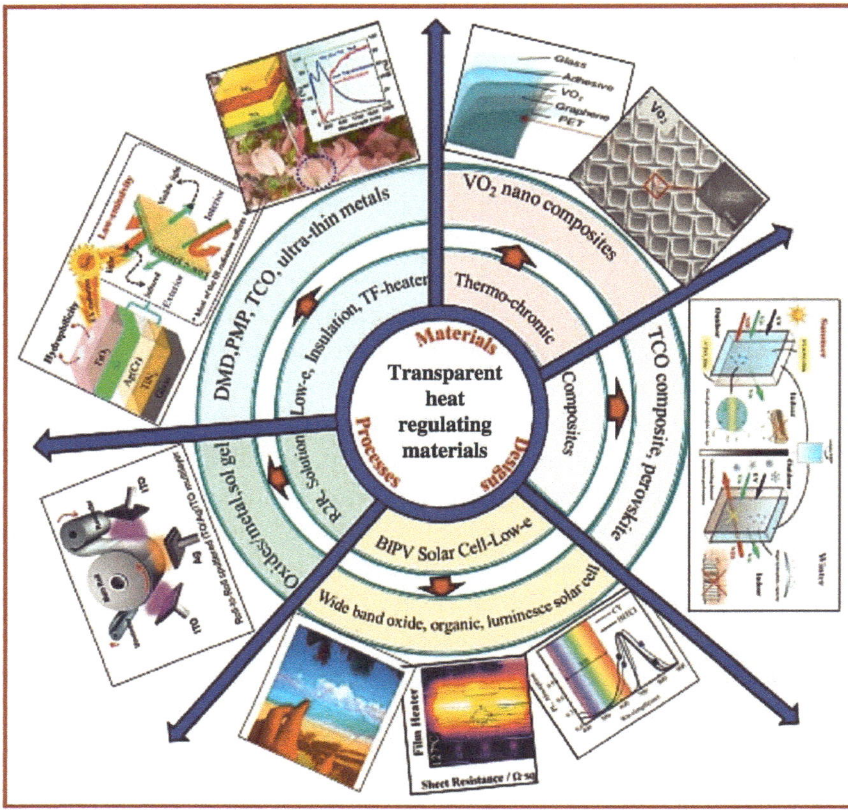

Figure 1.13 ***Schematic diagram of transparent heat regulating (THR) materials, multi-layer coating, and applications.*** Source: *[56–63] Copyright.*

the thermochromic coating of metal oxides on the window glass of the building can be used for reducing the energy loss by controlling the heat flow across them. Thermo-chromic materials are capable of changing their optical properties when exposed to heat. The transmittance and reflectance can be significantly altered due to phase transition; in common language, there is a change in the color of the materials coating due to influence of thermal energy. The performance of thermo-chromic materials can be evaluated by phase transition temperature (T_c), luminous transmittance (T_{lum}), visible transmittance, and modulation capability of solar energy (ΔT_{sol}). The transition temperature represents the critical value of temperature beyond which the thermo-chromic materials change its phase from semiconducting to metallic or shows change in color *(35)*. The energy saving efficiency of thermo-chromic coating can be estimated by the change in transmittance of the coating at 2500 nm before and after metal to semiconductor transition temperature; that is called switching efficiency *(62)*. The ratio of

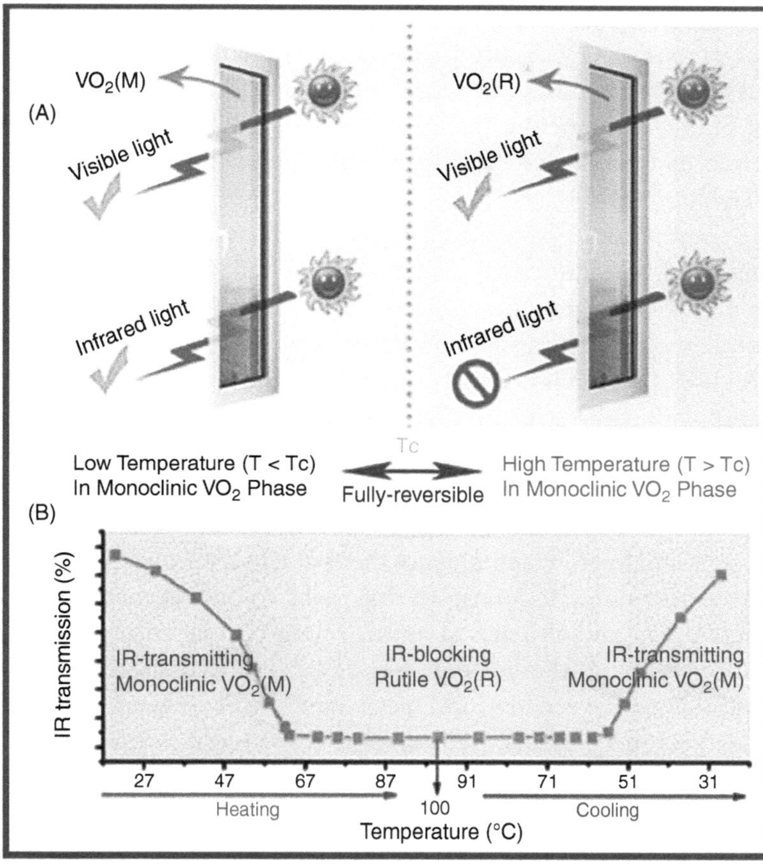

Figure 1.14 (A) Schematic of visible and infrared light transmission from VO2 film at low and high temperature corresponds to insulator to metallic phases. (B) IR transmittance profile of VO2 film during heating and cooling. Source: (61) *Copyright*.

transmitted light to incident light is defined as luminous transmittance and presented as % of light transmitted. The transmittance of solar energy before and after transition is measured by the solar modulation efficiency (ΔT_{sol}) *(35)*. Among the various transition metal oxides (lower oxides of vanadium, titanium, iron, and niobium), thermo-chromic properties of VO_2 are extensively investigated due to lower transition temperature and sharp transition features. Thermo-chromic properties in transition metal oxide appear due to d-d transitions with shift in Fermi level under thermal excitation and lattice expansion and temperature-dependent electron-phonon coupling *(63)*. Insulator-to-metal transition at 68°C in vanadium dioxide (VO_2) is responsible for electrochromic behavior. Fig. 1.14 shows the schematic diagram for insulator to metal transition of VO_2 and its impact in optical properties. Numerous efforts have been reported to improve the thermochromic properties of VO_2 materials especially to reduce the T_c to make the technology disruptive for commercial utilization. It has been found that the doping of VO_2 by numerous foreign cations (W^{6+}, Nb^{5+}, Ti^{4+}, Cr^{3+}, or Al^{3+}) is reported, and it is noticed that doping changes the transition temperature considerably *(64)*.

The conventional glass emits 84% of the radiant heat falling upon it by absorbing and transmitting IR radiation while reflecting only 16% of IR radiation *(65)*. Transparent heat reflector reflects IR of wavelength ($k > 700$ nm) and allows the transmission of VIS with 400 nm $< k <$ 700 nm *(33,85)*. Harmful UV radiation is normally absorbed by the glass substrate or dielectric layers within the structure. The dielectric/metal *(66)* and dielectric/metal/dielectric-based multilayers *(33,38–41,42,44,50,54,67,68,69–73)* with low emission (low-e) materials act as heat reflecting coatings. Low-thermal emittance (low-e) coatings effectively reduce IR radiation transfer through windows, which reduces thermal leakage from indoors to the outdoor surroundings. It consists of thin metal to protect metal layer from the environment and also acts as antireflecting coating to increase visible transmission through the coating layer *(54,74)*. The undercoat layer improves adhesion between glass and metal layer *(48,68)*. Design of dielectric layer based on metal oxides and sulfide is very crucial as it enhances visible transmission without reducing IR reflection and protects coating from the harsh ambient *(43,52,54,75–78)*. Microstructural properties of thin metal layer and dielectric passivation layers have significant roles on the IR reflectance and visible transmittance *(45–47,79)*. Growth mechanism and choice of interfacial layer enhance the stability and durability of the coating *(39,45,48,50,77,80–83)*. Apart from presenting the desirable optical performance, commercial transparent heat reflector coating is designed to be chemically, thermally, and mechanically stable.

Low-e coating can be divided into two categories; high solar gain low-e and low solar gain low-e, as shown in Fig. 1.15. The high solar gain low-e coating is more suitable for cold weather, as it allows visible transmission and near infra-red (NIR) into the building, whereas, low solar gain low-e is suitable for hot weather since the latter reflects NIR into the atmosphere and keeps the building cool. By choosing appropriate materials and multilayer structure, optical performance of the coating can be tuned. Thickness of dielectric and metal plays important role to develop low solar gain low-e and high solar gain low-e *(46,49,52,54,74,84–86)*. Furthermore, optical properties (visible transmittance and IR reflectance) of the multilayer

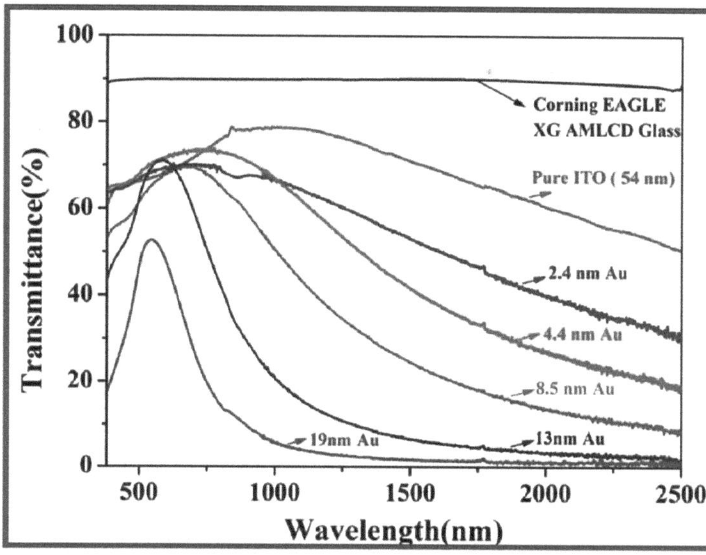

Figure 1.15 ***Direct spectral transmittance of trilayered samples (ITO/Au/ITO) with different gold spacer thicknesses.*** Source: *(91) Copyright.*

depend on the synthesis process, microstructural property, interface engineering, and design of the multilayers *(43,51–53,75,72,76-78,87–90)*.

6 Novel materials for regulating IR radiation

Besides low-e and thermo-chromic materials, transparent solar cells on glass/plastic substrates are emerging rapidly for building-integrated photovoltaics (BIPV) for energy-harvesting windows. The transparent solar cell is capable of generating electricity while simultaneously reducing heating and cooling demands for indoor comfort. Transparent solar cells with heat mirror can empower the solar energy harvesting and reduces the electricity consumption required for indoor comfort *(92)*. Inorganic and organic semiconductors have been employed for transparent/semitransparent solar cells *(92–94)*. Semitransparent perovskite solar cells (PSCs) are not only highly efficient but also very effective in thermal-mirror operation. The average power conversion efficiency of semitransparent PSCs is 13.3% and the outstanding NIR rejection is ~85.5%. This demonstrates their great potential for ideal "energy generating and heat-rejecting" solar window applications making smart use of solar energy. In addition, dye-sensitized solar cells (DSSCs) and organic lead halide PSCs have attracted significant attention as next-generation low-cost photo-voltaics because of the potential

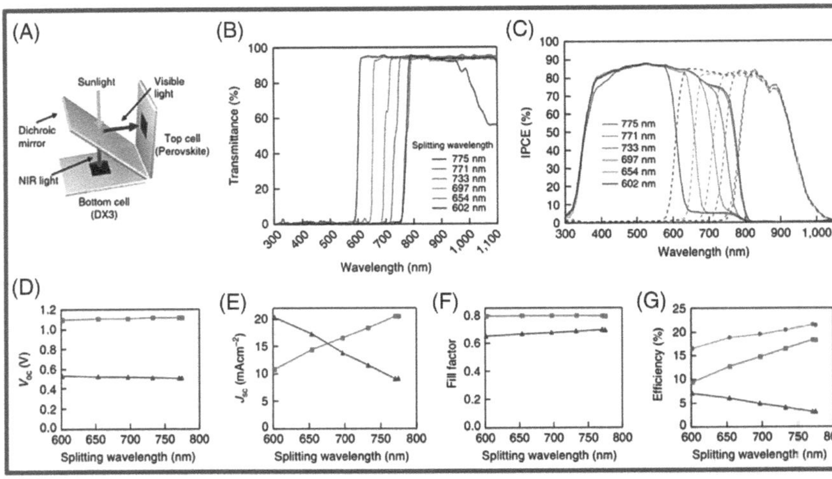

Figure 1.16 ***Device structure, IPCE spectra, and J-V characteristics of the hybridized cell.*** (A) Device structure of the hybridized cell using a dichroic spectrum splitter. (B) Transmittance spectra of dichroic mirrors under measurement conditions such as incident of a 45° angle. (C) IPCE spectra of the individual components of the hybridized cell with various dichroic mirrors: visible absorbing cell, PSC (*solid line*); near-IR absorbing cell, DX3-based DSSC (*dashed line*). (D-G) PV parameter and total power conversion efficiency dependence on the splitting wavelength of the dichroic mirrors: PSC (*red*), DX3-based DSSC (*blue*), and total power conversion efficiency of the hybridized cell (*green*). Source: (93) *Copyright.*

widespread applications. Fig. 1.16 shows below a hybridized solar cell employing a spectral splitting system, which was constructed using dichroic mirrors at an angle of 45° with splitting edge wavelengths of 602, 654, 697, 733, 771, and 775 nm. It can be seen that by using series connection, the light absorption band can be adjusted so as to equal the photocurrent of each component under standard sunlight. In this, a high-voltage PSC with a lead bromide perovskite will show the high-energy bandgap (B 2.3 eV), and therefore the series-connected tandem cell would achieve up to 22% PCE.

Transparent polymer solar cells selectively absorb solar spectrum and convert into electricity *(94–98)*. All solution-based transparent solar cells with novel materials are ideal to regulate NIR for smart windows application *(92,93,98)*. Luminescent materials-based fully transparent solar cells not only reduce IR transmittance into the building but also convert NIR into electricity *(61,99)*.The near-infrared transparent luminescent solar cells (TLSCs) based on organic salts provide an alternative strategy for transparent solar harvesting systems that can ultimately enhance the overall system efficiency of combined UV and NIR TLSCs. As shown in Fig. 1.17. Lunt et al. have designed and fabricated the first visibly transparent luminescent

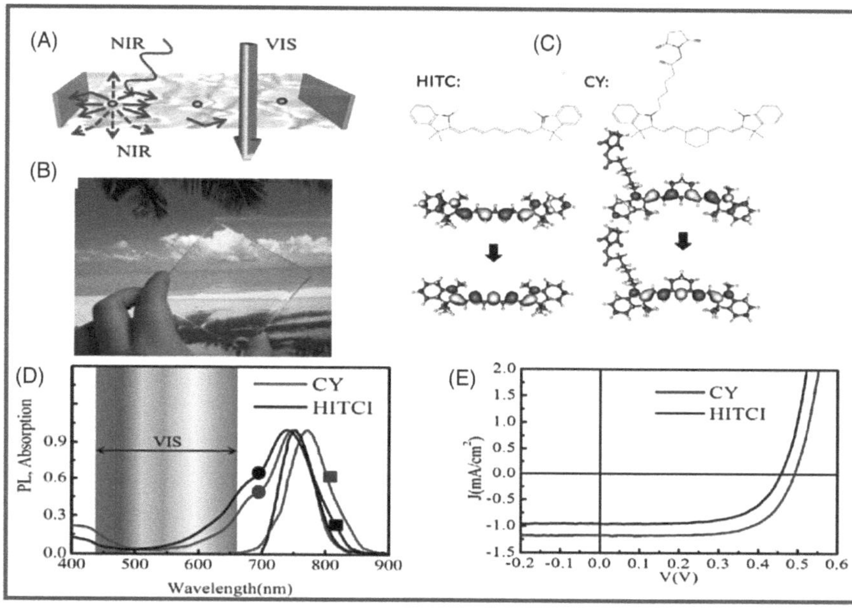

Figure 1.17 (A) Schematic of the transparent luminescent solar concentrator. (B) Photograph of the transparent LSC system incorporating CY luminophore. (C) CY and HITC molecular cation structures (*top*); the natural excited-state transition orbital pairs for HITC (*left*) and CY (*right*). The hole orbitals are shown on the top of the excited electron orbitals. Two cyanine derivatives: 2-[7-(1,3-dihydro-1,3,3-trimethyl-2H-indol-2-ylidene)-1,3,5-heptatrienyl]-1,3,3-trimethyl3H-indolium (HITC) iodide (HITCI) and 1-(6-(2,5-dioxopyrrolidin-1-yloxy)-6-oxohexyl)-3,3-dimethyl-2-((E)-2-((E)-3-((E)-2-(1,3, trimethylindolin-2ylidene)ethylidene) cyclohex-1-enyl)vinyl)-3H-indolium chloride (CY) have been used for the TLSCs. (D) Normalized absorption (*circle symbols*) and emission (*square symbols*) spectra of the NIR-absorbing luminophores CY (*blue line*) and HITCI (*black line*) films. (E) Current density as a function of voltage for the fully assembled TLSC systems with two of the luminophores. Source: (61) *Copyright.*

solar concentrator devices, which selectively harvest NIR photons based on fluorescent organic salts. These transparent TLSCs exhibit a nontinted transparency of 86 ± 1% in the visible spectrum combined with an efficiency of 0.4 ± 0.03% and have the potential for efficiencies up to 10% due to the large fraction of photon flux in the near-infrared *(61)*. These transparent NIR LSCs provide a new route to transparent light-harvesting systems with tremendous potential for high defect tolerances and processability. Vasiliev et al. demonstrated a new class of solar window system ready for industrial application *(77)*. UV and IR radiation energy are converted and/or deflected geometrically toward the panel edge for collection by copper-indium selenide ($CuInSe_2$)-based solar cells. The power conversion efficiencies 3.04%

in 10 cm × 10 cm vertically placed clear glass panels facing direct sunlight, and up to 2.08% in 50 cm × 50 cm installation ready framed window systems were demonstrated. Davy et al. demonstrated near-ultraviolet solar cells with electro-chromic windows for smart management of the solar spectrum *(99)*. Single-junction organic solar cells integrating with a low-cost, polymer-based electro-chromic window enables intelligent management of the solar spectrum *(99)*. Hence, we can say that for the economic commercial utilities, multifunctional performance of coatings is essential to enhance the energy conversion from the incoming solar radiation.

7 Conclusions

This chapter gives us the insight of technologies available for harnessing the huge amount of incoming solar radiation for its successful conversion into energy and hence can be treated as alternative to traditional fossil fuel-based sources of energy. It was discussed that using nanostructured materials having specific physical and chemical properties will lead to improvement in the efficiency of existing solar PV cell industry. In addition, the outcome from the chapter comes out to be that the use of spectrally selective coating over the windows/glass of the building can be used for regulating the heating and cooling of the inside environment. This type of functional coating on the glass eliminates solar heat and reduces the energy consumption, which will lead to energy saving, and will further tackle the CO_2 emission globally. Further, the plastic/glass-based transparent solar cells integrated with dielectric/metal/dielectric-based transparent conductor have been shown to empower the heat mirror and solar electricity for smart windows applications. This chapter, in turn, describes how incorporation of nanotechnology can improve the efficiency of solar energy conservation. One thing should be clear from this chapter: that solar energy and related materials research represents a dynamic field, whose importance and industrial viability are bound to increase in the future.

References

1. Bitran, M.; Murray, W. Infrared, Visible, and Ultraviolet Radiation. In *Patty's Toxicology*. John Wiley & Sons, Inc.: New York, 2001.
2. Poole, C. P., Jr.; Owens, F. J. *Introduction to Nanotechnology* John Wiley & Sons, Inc.: Hoboken, New Jersey, 2003.
3. Elliott, D. A.; Holland, J. M. Oral, Head, Neck Oncology, Reconstructive Surgery. In *Electromagnetic waves*; Zhurbenko, V., Ed.; BoD–Books on Demand: Denmark, 2018.
4. Granqvist, C. G., Ed. *Materials Science for Solar Energy Conversion Systems*; Pergamon: Oxford, UK, 1991.

5. Granqvist, CG Transparent Conductors as Solar Energy Materials: A Panoramic Review. *Sol. Energy Mater. Sol. Cells* **2007**, *91*, 1529–1598.
6. Hans, Kangro, Ed. *Planck's Original Papers in Quantum Physics*Taylor & Francis Ltd: London, 1972.
7. Archer, M. D., Hill, R., Eds. *Clean Electricity from Photovoltaics: Volume 1 of Series on Photoconversion of Solar Energy*Imperial College Press: London, UK, 2003.
8. Sangadevan, S. Recent Trends on Nanostructures Based Solar Energy Applications a Review. *Adv. Mater. Sci.* **2013**, *34*, 44–61.
9. Granqvist, C. G. Radiative heating and cooling with spectrally selective surfaces. *Appl. Opt., Optic. Soc. Am.* **1981**, *20* (15), 2606–2615.
10. Mills, D. Advances in solar thermal electricity technology. *Sol Energy* **2004**, *76*, 19–31.
11. Kraemer, D.; Poudel, B.; Feng, H. P.; Caylor, J. C.; Yu, B.; Yan, X., et al. High-performance flat-panel solar thermoelectric generators with high thermal concentration. *Nat. Mater.* **2011**, *10*, 532–538.
12. Mauthner, F.; Weiss, W. *Markets and Contribution to the Energy Supply 2011* Solar Heating & Cooling Programme, International Energy Agency: Paris, 2013.
13. Wang, N.; Han, L.; He, H. C.; Park, N. H.; Koumoto, K. A novel high-performance photovoltaic-thermoelectric hybrid device. *Energ Environ. Sci.* **2011**, *4*, 3676–3679.
14. Kennedy, C. E. *Review of Mid- to High-Temperature Solar Selective Absorber Materials* National Renewable Energy Laboratory: United States, 2002.
15. Niklasson, G. A.; Granqvist, C. G. Surfaces for selective absorption of solar energy: an annotated bibliography 1955-1981. *J. Mater. Sci.* **1983**, *18*, 3475–3534.
16. Wise, F. W. Lead salt quantum dots: The limit of strong quantum confinement. *Acc Chem Res* **2000**, *33* (11), 773–780.
17. YOFFE, A. D. Low-Dimensional Systems: Quantum Size Effects and Electronic Properties of Semiconductor Microcrystallites (Zero-Dimensional Systems) and Some Quasi-Two-Dimensional Systems. *Adv. Phys.* **1993**, *42* (2), 173–266.
18. Wise, F. W. Department of Applied Physics, Cornell University; Lead Salt Quantum Dots: the Limit of Strong Quantum Confinement. *Acc. Chem. Res.* **2000**, *33*, 773–780.
19. Corkish, R.; Kettemann, S.; Nelson, J. Low-Dimensional Systems and Nanostructures. *Phys. E* **April 2002**, *14* (1-2).
20. Corkish, R.; Kettemann, S.; Nelson, J. Proceedings of the Workshop on Nanostructures in Photovoltaics. Max Planck Institute for the Physics of Complex Systems, Dresden, Germany, July 28-4 August, 2001.
21. Ramanathan, K.; Contreras, M. A.; Perkins, C. L., et al. Properties of 19.2% efficiency ZnO/CdS/CuInGaSe2 thin-film solar cells. *Prog Photovolt: Res. Appl.* **1999**, *11*, 225.
22. Wu, X.; Kane, J.C.; Dhere, R.G.; DeHart, C.; Albin, D.S.; Duda, A.; Gessert, T.A.; Asher, S.; Levi, D.H.; Sheldon, P. In Proceedings of the 17th European Photovoltaic Solar Energy Conference and Exhibition, Munich, 2002, p. 995.
23. Rampino, S.; Armani, N.; Bissoli, F., et al. 15% Efficient Cu(In,Ga)Se2 Solar Cells Obtained by Low-Temperature Pulsed Electron Deposition. *Appl. Phys. Lett.* **2012**, *101* (132107).
24. Yoon, W.; Boercker, J. E.; Lumb, M. P.; Placencia, D.; Foos, E. E.; Tischler, J. G. Enhanced Open-Circuit Voltage of PbS Nanocrystal Quantum Dot Solar Cells. *Sci Rep* **2013**, *3*, 2225.
25. Yun, D.; Feng, W.; Wu, H.; Yoshino, K.; Efficient Conjugated Polymer-ZnSe and -PbSe Nanocrystals Hybrid Photovoltaic Cells Through Full Solar Spectrum Utilization. *Sol Energy Mater Sol Cells* **n.d.,** *93*, 1208-1213.
26. Schaller, Richard D., Melissa A. Petruska, and Victor I. Klimov. Effect of electronic structure on carrier multiplication efficiency: Comparative study of PbSe and CdSe nanocrystals. *Appl. Phys. Lett.* **87** (25), 2005, 253102.
27. Midgett, A. G.; Hillhouse, H. W.; Hughes, B. K.; Nozik, A. J.; Beard, M. C. Flowing versus static conditions for measuring multiple exciton generation in PbSe quantum dots. *J Phys Chem C* **2010**, *114* (41), 17486–17500.

28. Pijpers, J. J. H.; Ulbricht, R.; Tielrooij, K. J.; Osherov, A.; Golan, Y.; Delerue, C., et al. Assessment of Carrier-Multiplication Efficiency in Bulk PbSe and PbS. *Nat. Phys.* **2009**, *5*, 811–814.
29. Beard, M. C.; Midgett, A. G.; Hanna, M. C.; Luther, J. M.; Hughes, B. K.; Nozik, A. J. Comparing Multiple Exciton Generation in Quantum Dots to Impact Ionization in Bulk Semiconductors: Implications for Enhancement of Solar Energy Conversion. *Nano Lett* **2010**, *10*, 3019–3027.
30. Klimov, V. I. Detailed-Balance Power Conversion Limits of Nanocrystal-Quantum-Dot Solar Cells in the Presence of Carrier Multiplication. *Appl. Phys. Lett.* **2006**, *89*, 123118.
31. Granqvist, C. G.; Wittwer, V. Materials for Solar Energy Conversion: An Overview. *Sol. Energy Mater. Sol. Cells* **1998**, *54* (1-4), 39–48.
32. Raman, A. P.; Anoma, M. A.; Zhu, L.; Rephaeli, E.; Fan, S. Passive Radiative Cooling Below Ambient Air Temperature Under Direct Sunlight. *Nature* **2014**, *515*, 540–544.
33. Granqvist, C. G. Transparent Conductors as Solar Energy Materials: A Panoramic Review. *Sol. Energy Mater. Sol. Cells* **2007**, *91*, 1529–1598.
34. Yang, Y. -S.; Zhou, Y.; Chiang, F. B. Y.; Long, Y. Tungsten Doped VO2/Microgels Hybrid Thermochromic Material and Its Smart Window Application. *RSC Adv.* **2017**, 7, 7758–7762.
35. Rezaei, S. D.; Shannigrahi, S.; Ramakrishna, S. A Review of Conventional, Advanced, and Smart Glazing Technologies and Materials for Improving Indoor Environment. *Sol. Energy Mater. Sol. Cells* **2017**, *159*, 26–51.
36. Zhou, J.; Gao, Y.; Zhang, Z.; Luo, H.; Cao, C.; Chen, Z., et al. VO2 Thermochromic Smart Window for Energy Savings and Generation. *Sci. Rep.* **2013**, *3*, 3029.
37. Zhao, Y.; Xu, R.; Zhang, X.; Hu, X.; Knize, R. J.; Lu, Y. Simulation of Smart Windows in the ZnO/VO2/ZnS Sandwiched Structure with Improved Thermochromic Properties. *Energy Build* **2013**, *66*, 545–552.
38. Wu, C. -C.; Shih, W. -C. Development of a Highly Transparent, Low-Resistance Lithium-Doped Nickel Oxide Triple-Layer Film Deposited by Magnetron Sputtering. *Chem. Commun.* **2017**, *53*, 1634–1637.
39. Sharma, V.; Kumar, P.; Kumar, A.; Surbhi; Asokan, K.; Sachdev, K. High-Performance Radiation Stable ZnO/Ag/ZnO Multilayer Transparent Conductive Electrode. *Sol. Energy Mater. Sol. Cells* **2017**, *169*, 122–131.
40. Lin, T. -C.; Huang, W. -C.; Tsai, F. -C. Hydrogen Plasma Effect Toward the AZO/CuCr/AZO Transparent Conductive Electrode. *Microelectron Eng.* **2017**, *167*, 85–89.
41. Kim, D. -H.; Cho, K. -S.; Kim, H. -K. Thermally Evaporated Indium-Free, Transparent, Flexible SnO_2/AgPdCu/SnO_2 Electrodes for Flexible and Transparent Thin Film Heaters. *Sci. Rep.* **2017**, 7, 2550.
42. Chen, K. -N.; Yang, C. -F.; Wu, C. -C.; Chen, Y. -H. Development of the a-IGZO/Ag/a-IGZO Triple-Layer Structure Films for the Application of Transparent Electrode. *Materials* **2017**, *10*, 226.
43. Peres, L.; Bou, A.; Barakel, D.; Torchio, P. ZnS|Ag|TiO2 Multilayer Electrodes with Broadband Transparency for Thin Film Solar Cells. *RSC Adv.* **2016**, *6*, 61057–61063.
44. Meng-Qi, Z.; Han-Dong, J.; Peng-Qing, B.; Fu-Jian, Z.; Jin, M.; Xiao-Tao, H. Performance Improvement of TiO2/Ag/TiO2 Multilayer Transparent Conducting Electrode Films for Application on Photodetectors. *J Phys D Appl Phys* **2016**, *49*,115108. Dalapati, G.K. et al. *Prog Mater Sci* **2018**, *95*, 42-131, 115.
45. Kim, J. -H.; Kim, D. -H.; Kim, S. -K.; Bae, D.; Yoo, Y. -Z.; Seong, T. -Y. Control of Refractive Index by Annealing to Achieve High Figure of Merit for TiO2/Ag/TiO_2 Multilayer Films. *Ceram. Int.* **2016**, *42*, 14071–14076.
46. Kim, J. H.; Lee, H. -K.; Na, J. -Y.; Kim, S. -K.; Yoo, Y. -Z.; Seong, T. -Y. Dependence of Optical and Electrical Properties on Ag Thickness in TiO2/Ag/TiO_2 Multilayer Films for Photovoltaic Devices. *Ceram. Int.* **2015**, *41*, 8059–8063.

47. Kulczyk-Malecka, J.; Kelly, P. J.; West, G.; Clarke, G. C. B.; Ridealgh, J. A.; Almtoft, K. P., et al. Investigation of Silver Diffusion in TiO2/Ag/TiO2 Coatings. *Acta Mater.* **2014**, *66*, 396–404.
48. Alvarez, R.; González, J. C.; Espinós, J. P.; González-Elipe, A. R.; Cueva, A.; Villuendas, F. Growth of Silver on ZnO and SnO2 Thin Films Intended for Low Emissivity Applications. *Appl. Surf. Sci.* **2013**, *268*, 507–515.
49. Wu, H. -W.; Yang, R. -Y.; Hsiung, C. -M.; Chu, C. -H. Influence of Ag Thickness of Aluminum-Doped ZnO/Ag/Aluminum-Doped ZnO Thin Films. *Thin Solid Films* **2012**, *520*, 7147–7152.
50. Al-Kuhaili, M. F.; Al-Aswad, A. H.; Durrani, S. M. A.; Bakhtiari, I. A. Energy-Saving Transparent Heat Mirrors Based on Tungsten Oxide-Gold WO3/Au/WO3 Multilayer Structures. *Sol. Energy* **2012**, *86*, 3183–3189.
51. Neghabi, M.; Behjat, A.; Ghorashi, S. M. B.; Salehi, S. M. A. The Effect of Annealing on Structural, Electrical and Optical Properties of Nanostructured ZnS/Ag/ZnS Films. *Thin Solid Films* **2011**, *519*, 5662–5666.
52. Leng, J.; Yu, Z.; Xue, W.; Zhang, T.; Jiang, Y.; Zhang, J., et al. Influence of Ag Thickness on Structural, Optical, and Electrical Properties of ZnS/Ag/ZnS Multilayers Prepared by Ion Beam asslsted Deposition. *J. Appl. Phys.* **2010**, *108*, 073109.
53. Dobrikov, G. H.; Rassovska, M. M.; Andreev, N. M.; Boyadzhiev, S. I.; Gesheva, K. A.; Ivanova, T. M., et al. Development of Transparent Heat Mirrors Based on Metal Oxide Thin Film Structures. *Thin Solid Films* **2009**, *518*, 1091–1094.
54. Dalapati, G. K.; Masudy-Panah, S.; Chua, S. T.; Sharma, M.; Wong, T. I.; Tan, H. R., et al. Color Tunable Low Cost Transparent Heat Reflector Using Copper and Titanium Oxide for Energy Saving Application. *Sci. Rep.* **2016**, *6*, 20182.
55. Kim, H.; Kim, Y.; Kim, K. S.; Jeong, H. Y.; Jang, A. R.; Han, S. H., et al. Flexible Thermochromic Window Based on Hybridized VO_2/Graphene. *ACS Nano* **2013**, *7*, 5769–5776.
56. Liu, T.; Liu, B.; Wang, J.; Yang, L.; Ma, X.; Li, H., et al. Smart Window Coating Based on F-TiO2-KxWO3 Nanocomposites with Heat Shielding, Ultraviolet Isolating, Hydrophilic and Photocatalytic Performance. *Sci. Rep.* **2016**, *6*, 27373.
57. Lu, Q.; Liu, C.; Wang, N.; Magdassi, S.; Mandler, D.; Long, Y. Periodic Micro-Patterned VO_2 Thermochromic Films by Mesh Printing. *J. Mater. Chem. C* **2016**, *4*, 8385–8391.
58. Loka, C.; Park, K. R.; Lee, K. -S. Multi-Functional TiO2/Si/Ag(Cr)/TiNx Coatings for Low-Emissivity and Hydrophilic Applications. *Appl. Surf. Sci.* **2016**, *363*, 439–444.
59. Ko, E. -H.; Kim, H. -J.; Lee, S. -J.; Lee, J. -H.; Kim, H. -K. Nano-Sized Ag Inserted into ITO Films Prepared by Continuous Roll-to-Roll Sputtering for High-Performance, Flexible, Transparent Film Heaters. *RSC Adv.* **2016**, *6*, 46634–46642.
60. Kang, T. -W.; Kim, S. H.; Kim, C. H.; Lee, S. -M.; Kim, H. -K.; Park, J. S., et al. Flexible Polymer/Metal/Polymer and Polymer/Metal/Inorganic Trilayer Transparent Conducting Thin Film Heaters with Highly Hydrophobic Surface. *ACS Appl. Mater. Interfaces* **2017**, *9*, 33129–33136.
61. Zhao, Y.; Meek, G. A.; Levine, B. G.; Lunt, R. R. Near-Infrared Harvesting Transparent Luminescent Solar Concentrators. *Adv. Opt. Mater.* **2014**, *2*, 606–611.
62. Kamalisarvestani, M.; Saidur, R.; Mekhilef, S.; Javadi, F. S. Performance, Materials and Coating Technologies of Thermochromic Thin Films on Smart Windows. *Renew. Sustain. Energy Rev.* **2013**, *26*, 353–364.
63. Wang, Y.; Runnerstrom, E. L.; Milliron, D. J. Switchable Materials for Smart Windows. *Annu. Rev. Chem. Biomol. Eng.* **2016**, *7*, 283–304.
64. Béteille, F.; Livage, J. Optical Switching in VO_2 Thin Films. *J. Sol-Gel Sci. Technol.* **1998**, *13*, 915–921.
65. Karlsson, J.; Roos, A. Annual Energy Window Performance vs. Glazing Thermal Emittance-The Relevance of Very Low Emittance Values. *Thin Solid Films* **2001**, *392*, 345–348.

66. Ahmad, S. H. A.; Al-Kuhaili, M. F.; Durrani, S. M. A.; Faiz, M. M.; Ul-Hamid, A. Bi-Layered Energy Efficient Coatings as Transparent Heat Mirrors Based on Vanadium Oxide Thin Films. *Sol. Energy Mater. Sol. Cells* **2017**, *169*, 258–263.
67. Fan, J. C. C.; Bachner, F. J.; Foley, G. H.; Zavracky, P. M. Transparent Heat-Mirror Films of TiO_2/Ag/TiO_2 for Solar Energy Collection and Radiation Insulation. *Appl. Phys. Lett.* **1974**, *25*, 693–695.
68. Grosse, P.; Hertling, R.; Müggenburg, T. Design of Low Emissivity Systems Based on a Three-Layer Coating. *J. Non-Cryst. Solids* **1997**, *218*, 38–43.
69. Cheng, J.; Cao, G.; Zong, H.; Kang, C.; Jia, E.; Zhang, B., et al. Highly Transparent Conductive AZO/Zr50Cu50/AZO Films in Wide Range of Visible and Near Infrared Wavelength Grown by Pulsed Laser Deposition. *Results Phys.* **2017**, 7, 910–913.
70. Fang, Y.; He, J.; Zhang, K.; Xiao, C.; Zhang, B.; Shen, J., et al. Ar Plasma Irradiation Improved Optical and Electrical Properties of TiO_2/Ag/TiO_2 Multilayer Thin Film. *Opt. Lett.* **2015**, *40*, 5455–5458.
71. Chiu, P. -K.; Lee, C. -T.; Chiang, D.; Cho, W. -H.; Hsiao, C. -N.; Chen, Y. -Y., et al. Conductive and Transparent Multilayer Films for Low-Temperature TiO(2)/Ag/SiO(2) Electrodes by E-Beam Evaporation with IAD. *Nanoscale Res. Lett.* **2014**, *9*, 35.
72. Gong, L.; Lu, J.; Ye, Z. Conductive Ga Doped ZnO/Cu/Ga Doped ZnO Thin Films Prepared by Magnetron Sputtering at Room Temperature for Flexible Electronics. *Thin Solid Films* **2011**, *519*, 3870–3874.
73. Yang, X.; Gao, P.; Yang, Z.; Zhu, J.; Huang, F.; Ye, J. Optimizing Ultrathin Ag Films for High Performance Oxide-Metal-Oxide Flexible Transparent Electrodes Through Surface Energy Modulation and Template-Stripping Procedures. *Sci Rep* **2017**, 7, 44576. 116 Dalapati, G.K. et al. *Progr Mater Sci* **2018,** *95*, 42-131.
74. Dhar, A.; Alford, T. L. Optimization of TiO_2/Cu/TiO_2 Multilayer as Transparent Composite Electrode (TCE) Deposited on Flexible Substrate at room Temperature. *ECS Solid State Lett.* **2014**, *3*, N33–N36.
75. Leftheriotis, G.; Papaefthimiou, S.; Yianoulis, P. Integrated Low-Emittance-Electrochromic Devices Incorporating ZnS/Ag/ZnS Coatings as Transparent Conductors. *Sol. Energy Mater. Sol. Cells* **2000**, *61*, 107–112.
76. Cho, H.; Yun, C.; Park, J. -W.; Yoo, S. Highly Flexible Organic Light-Emitting Diodes Based on ZnS/Ag/WO3 Multilayer Transparent Electrodes. *Org. Electron.* **2009**, *10*, 1163–1169.
77. Leftheriotis, G.; Yianoulis, P.; Patrikios, D. Deposition and Optical Properties of Optimised ZnS/Ag/ZnS Thin Films for Energy Saving Applications. *Thin Solid Films* **1997**, *306*, 92–99.
78. Liu, X.; Cai, X.; Qiao, J.; Mao, J.; Jiang, N. The Design of ZnS/Ag/ZnS Transparent Conductive Multilayer Films. *Thin Solid Films* **2003**, *441*, 200–206.
79. Wu, C. -C.; Chen, P. S.; Peng, C. -H.; Wang, C. -C. TiO_x/Ag/TiO_x Multilayer for Application as a Transparent Conductive Electrode and Heat Mirror. *J. Mater. Sci. Mater. Electron* **2013**, *24*, 2461–2468.
80. Leftheriotis, G.; Yianoulis, P. Characterisation and Stability of Low-Emittance Multiple Coatings for Glazing Applications. *Sol. Energy Mater. Sol. Cells* **1999**, *58*, 185–197.
81. Ando, E.; Miyazaki, M. Durability of Doped Zinc Oxide/Silver/Doped Zinc Oxide Low Emissivity Coatings in Humid Environment. *Thin Solid Films* **2008**, *516*, 4574–4577.
82. Durrani, S. M. A.; Khawaja, E. E.; Al-Shukri, A. M.; Al-Kuhaili, M. F. Dielectric/Ag/Dielectric Coated Energy-Efficient Glass Windows for Warm Climates. *Energy Build* **2004**, *36*, 891–898.
83. Sivaramakrishnan, K.; Alford, T. L. Metallic Conductivity and the Role of Copper in ZnO/Cu/ZnO Thin Films for Flexible Electronics. *Appl. Phys. Lett.* **2009**, *94*, 052104.

84. Lampert, C. M. Coatings for Enhanced Photothermal Energy Collection I. Selective Absorbers. *Solar Energy Mater.* **1979**, *1*, 319–341.
85. Lampert, C. M. Heat Mirror Coatings for Energy Conserving Windows. *Solar Energy Mater.* **1981**, *6*, 1–41.
86. Ghasemi Varnamkhasti, M.; Fallah, H. R.; Mostajaboddavati, M.; Hassanzadeh, A. Influence of Ag Thickness on Electrical, Optical and Structural Properties of Nanocrystalline MoO_3/Ag/ITO Multilayer for Optoelectronic Applications. *Vacuum* **2012**, *86*, 1318–1322.
87. Ando, E.; Miyazaki, M. Moisture Resistance of the Low-Emissivity Coatings with a Layer Structure of Al-doped ZnO/Ag/Al-Doped ZnO. *Thin Solid Films* **2001**, *392*, 289–293.
88. Boiadjiev, S. I.; Dobrikov, G. H.; Rassovska, M. M. M. Preparation and Properties of RF Sputtered Indium-Tin Oxide Thin Films for Applications as Heat Mirrors in Photothermal Solar Energy Conversion. *Thin Solid Films* **2007**, *515*, 8465–8468.
89. Dhar, A.; Alford, T. L. High Quality Transparent TiO_2/Ag/TiO_2 Composite Electrode Films Deposited on Flexible Substrate at Room Temperature by Sputtering. *Appl. Mater.* **2013**, *1*, 012102.
90. Fan, J. C. C. Sputtered Films for Wavelength-Selective Applications. *Thin Solid Films* **1981**, *80*, 125–136.
91. Fang, X.; Mak, C. L.; Dai, J.; Li, K.; Ye, H.; Leung, C.W. ITO/Au/ITO Sandwich Structure for Near-Infrared Plasmonics. *ACS Appl. Mater. Interfaces* **2014**, *6*, 15743–15752.
92. Kim, H.; Kim, H. -S.; Ha, J.; Park, N. -G.; Yoo, S. Empowering Semi-Transparent Solar Cells with Thermal-Mirror Functionality. *Adv. Energy Mater.* **2016**, 6 1502466-n/a.
93. Kinoshita, T.; Nonomura, K. .; Jeon, N. J.; Giordano, F.; Abate, A.; Uchida, S.; Kubo, T.; Seok, S. L.; Nazeeruddin, M. K.; Hagfeldt, A.; Gratzel, M.; Segawa, H. Spectral Splitting Photovoltaics Using Perovskite and Wideband Dye-Sensitized Solar Cells. *Nat. Commun* **2015**, *6*, 8834.
94. Betancur, R.; Romero-Gomez, P.; Martinez-Otero, A.; Elias, X.; Maymo, M.; Martorell, J. Transparent Polymer Solar Cells Employing a Layered Light-Trapping Architecture. *Nat. Photon* **2013**, 7, 995–1000.
95. Zhou, E.; Nakano, M.; Izawa, S.; Cong, J.; Osaka, I.; Takimiya, K., et al. All-Polymer Solar Cell with High Near-Infrared Response Based on a Naphthodithiophene Diimide (NDTI) Copolymer. *ACS Macro Lett* **2014**, *3*, 872–875.
96. Za Tan; Li, S.; Wang, F.; Qian, D.; Lin, J.; Hou, J., et al. High Performance Polymer Solar Cells with As-Prepared Zirconium Acetylacetonate Film as Cathode Buffer Layer. *Sci. Rep.* **2014**, *4*, 4691.
97. Xia, X.; Wang, S.; Jia, Y.; Bian, Z.; Wu, D.; Zhang, L., et al. Infrared-Transparent Polymer Solar Cells. *J. Mater. Chem.* **2010**, *20*, 8478–8482.
98. Chen, C. -C.; Dou, L.; Zhu, R.; Chung, C. -H.; Song, T. -B.; Zheng, Y. B., et al. Visibly Transparent Polymer Solar Cells Produced by Solution Processing. *ACS Nano* **2012**, *6*, 7185–7190.
99. Huang, X.; Han, S.; Huang, W.; Liu, X. Enhancing Solar Cell Efficiency: The Search for Luminescent Materials as Spectral Converters. *Chem. Soc. Rev.* **2013**, *42*, 173–201.

Further readings

1. Cao, F.; Kenneth, M.; Gang, C.; Zhifeng, R. A Review of Cermet-Based Spectrally Selective Solar Absorbers. *Energy Environ. Sci.* **2014**, 7 (5), 1615.
2. McGuire, J. A.; Joo, J.; Pietryga, J. M.; Schaller, R. D.; Klimov, V. I. New Aspects of Carrier Multiplication in Semiconductor Nanocrystals. *Acc. Chem. Res.* **2008**, *41* (12), 1810–1819.

CHAPTER 2

Infrared radiation and materials interaction: Active, passive, transparent, and opaque coatings

Charu Dwivedi[a], Priyanka Bamola[b], Bharti Singh[c], Himani Sharma[b]
[a]Department of Chemistry, School of Physical Sciences, Doon University, Dehradun, Uttarakhand, India
[b]Department of Physics, School of Physical Sciences, Doon University, Dehradun, Uttarakhand, India
[c]Department of Applied Physics, Delhi Technological University, Delhi, India

1 Introduction

A broad spectrum of wavelengths ranging from 295 to 2500 nm reaches in our environment as sun light, scientifically in forms of visible, UV (ultra violet), &IR (Infrared) *(1–4)*. In IR regime, mainly three bands are considered: (1) short wave IR (SWIR); (2) mid-wave IR (MWIR) and (3) long wavelength IR (LWIR), depending on the wavelength, as shown in Fig. 2.1.

The origin of life on earth is mainly generated by the infrared radiation, and half of the sun's energy within the star spectrum falls on the IR radiation. IR, generally termed as thermal radiation, is the most important spectrum, which lies between visible and UV spectra. It has wavelengths from 0.74 to 100 μm.

William Herschel discovered IR radiation in 1800, when he found a set of an even warmer temperature measurements with the colors of visible spectrum (blue to red), using various thermometers. But Herschel could not explain the complete nature of IR radiation *(5,6)*. During the second half of the 19th century, J.C. Maxwell contributed his four equations on electromagnetism, which unified the fundamental laws of electromagnetism *(7)*. Later, Hertz proved these equations experimentally. By then, it was deducted that light and IR radiation have common nature and properties *(7)*. Clearly, the idea of electromagnetism changed the whole perceptions of IR radiation. It supported further analysis and progressive approaches toward IR radiations. In 19th century, W. De. W. Abney and E.R. Festing mapped IR radiation in solar spectrum by their photographic emulsions and found a unique pattern. This laid a foundation of analyzing IR spectroscopy in modern science.

Energy Saving Coating Materials
http://dx.doi.org/10.1016/B978-0-12-822103-7.00002-9

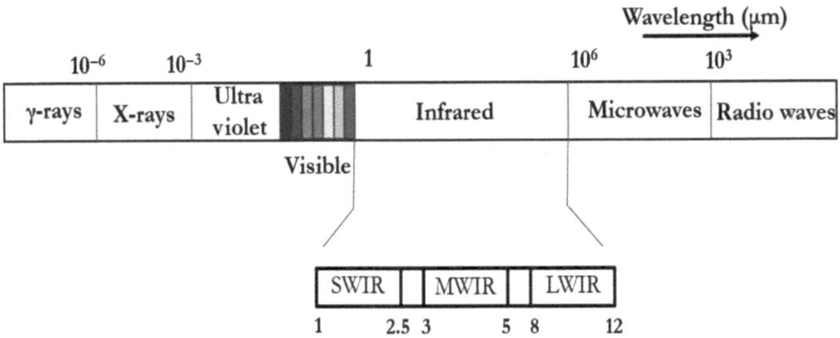

Figure 2.1 ***A diagram of electro magnetic spectrum and IR region.***

In recent times, metals and nanostructures are being used in the study of IR radiation, which is important for its advancement in the modern science and technology, as well as in the overall development of mankind *(4,8,9)*. IR radiation's silent presence in our surroundings does not affect its necessity for our existence on earth. Remote controlling devices, wireless connections, security systems, sensor development, optical fiber, thermal imaging, and various apparatus in hospitals daily use the technology of IR radiation (10–19). Analyzing the transferring and receiving patterns of IR radiation is the basic for these technologies. By using the absorption and emission spectra, we can determine the molecular structures. It is also helpful in determining the complex chemical mixture of molecules by both qualitative and quantitative analyses. Photovoltaic IR photo detectors based on thin-film materials (such as HgCdTe, InSb, InGaAs) have been widely used for earth observation, environment monitoring, and remote sensing *(4)*.

In modern times, where the energy demand is increasing rapidly in our daily lives and industries, cost-effective and efficient medium is required, which also favors the environment. Solar energy is very important in this regard and it consists of energy in terms of UV, IR, visible and microwave regions.

The principle of absorption is applied generally to convert sunrays into various forms of usable energy, which are renewable. For example, in the absorption of UV radiation (43.0 eV), molecular ionization is used for the ejection of electrons beyond their molecular energy or transition from the ground state to upper electronic state within atoms in metal oxides (CuO, Cu_2O, TiO_2, etc.) *(8,13,19)*. Semiconductors are generally used for the absorption of visible light (1.7–3.0 eV) to generate electricity by photovoltaic devices *(4,8)*. The absorption excites electron from the valence bond to conduction band due to heat and impurities of the semiconductor *(8,15,20,21)*. Now, the external circuit can produce electricity or can also be transferred to the photochemical cells for inducing reduction/oxidation *(20,21)*. Normally, the absorption of IR radiation requires transitions from the ground state to excited states of molecular vibrations, which finally produce energy in the form of heat *(21)*. Metals with high thermal conductivity can absorb or reflect IR radiation quickly as the quantum states of their vibrations lie within this region, although most metals reflect almost all IR radiation *(18,21)*. Highly efficient thermal detection devices can be realized by increasing the IR absorbing capability *(18)*.

2 Light reflection mechanism of IR radiations with metals and nanostructures

When a metal sample is exposed in direct sunlight, three phenomena can occur, namely reflection, transmission, and absorption *(22)*. Reflection is of two types: mirror like reflection' (specular refection) and diffuse reflection. Mirror-like reflection is more significant for optically smooth and highly absorbing sample surfaces. On the other end, diffuse reflection occurs when incident radiation penetrates into the powder and is further mirrored by the grain boundaries of the particle. In the diffuse reflections, particle size is most important, as the particle size decreases, the number of reflections at the grain boundaries will increase *(8)*. As a result, the depth of penetration of incident light decreases, which leads to a decrease in absorption and an increase in reflectance.

Absorption of light is the excitation of an electron from one bonding state to higher bonding states *(9)*. An interesting fact is that all types of light wavelengths cannot generate energy transitions (due to energy differences), and a specific wavelength is required for the process of absorption of light. For example, the wavelength between 400 and 700 nm is necessary to motivate electronic transitions *(23,24)*. Anything higher than this wavelength (>700 nm) is not enough to generate power. Sun also has a wavelength of 1500 nm, which is ineffective for causing any electronic transition. Thus, the wavelength of 1500 nm beam will not be observed and when further being mirrored and scattered, it leads to the diffused reflection of NIR light *(9)*. Although, there is no methodology for predicting the IR reflectivity of chemical and inorganic compounds *(9)*.

Metals are opaque at IR wavelengths. This is because metals are characterized by their large concentrations of free electrons, which is the property that gives them their highly reflective nature *(25)*. Reflective metals can be pure metals (such as Au, Ti, Cu, Al, and Ag), metals with surface coatings (AgS on Ag and AlO(OH) on Al), and multiple-layered structures: TiO_2/Au, silicon powder, metal-coated particles, etc. *(8,9)*. Coated metals have a decreased reflection factor corresponding to visual light. For example, the coating outside of the metal, which can be considered as a metal oxide, can absorb visual light, whereas the IR light of a longer wavelength passes through the coating and can be reflected by the underlying metal *(26)*. These multilayered structures are designed to minimize the reflection of visual light whereas giving a high reflection of IR light *(27)*.

High reflective pure metals such as Ag, Au, Cu, and Al are continuously being used in the visible and infrared spectral region *(28)*. Material's reflectivity can be derived from the basic optics.

$$R = |1 - \hat{n}/1 + \hat{n}|^2 = (1-n)^2 + k^2/(1+n)^2 + k^2 = 1 - 4n/(1+n)^2 + k^2 \tag{2.1}$$

where R, reflectivity of Light at normal incidence; $\hat{n}$, Complex refractive index; n, real part of refractive index; k, extinction coefficient.

In the near infrared region if $n \ll k$, we can derive the following expressions:

$$k \approx \sqrt{\mu c^2 \tau / \rho(1+\omega^2\tau^2)} \tag{2.2}$$

$$n \approx 1/2\omega\sqrt{\mu c^2 / \rho\tau(1+\omega^2\tau^2)} \tag{2.3}$$

and

$$R \approx 1 - 4n/k^2 \tag{2.4}$$

where μ, magnetic permeability; c, speed of light; ρ, electrical resistivity; ω, angular frequency; τ, mean free time between charge carrier's collisions.

Reflectivity is then obtained by inserting Eqs. (2.2) and (2.3) into Eq. (2.4).

$$R \approx 1 - \sqrt{4/\mu} \quad c^{-1}\tau^{-1/2}\rho^{1/2} \tag{2.5}$$

Table 2.1 Comparison of the electrical resistivity and light reflectivity of pure Ag, Cu, Al, Mg crystalline films, several AgMgAl, AgCuMgAl, AgCuMg, and ZrCu thin-film coatings on glass substrate (all compositions are given in atomic percent) *(28)*.

	Electric resistivity *ρ*, *n**m*	**Light reflectivity** *R*, **at 1000 nm**
Ag	16	0.98
Cu	17	0.97
Al	27	0.94
Mg	45	0.92
$Ag_{35}Cu_{35}Mg_{30}$ (crystalline)	189	0.98
$Ag_{76}Mg_{17}Al_{7}$ (crystalline)	230	0.95
$Ag_{73}Mg_{17}Al_{10}$ (crystalline)	286	0.92
$Ag_{60}Mg_{27}Al_{13}$ (crystalline)	304	0.91
$Ag_{30}Mg_{45}Al_{25}$(amorphous)	1032	0.79
$Ag_{30}Cu_{30}Mg_{15}Al_{25}$(amorphous)	1120	0.77
$Ag_{45}Mg_{37}Al_{18}$(amorphous)	1178	0.77
$Zr_{50}Cu_{50}$ (amorphous)	2975	0.60

Magnetic permeability has negligible effect on reflectivity because it is essentially constant for ferromagnetic metals. It is observed in Eq. (2.5) that a material with low ρ and high τ would exhibit high reflectivity.

Hu et al. investigated electrical resistivity and optical reflectivity, where they prepared a series of multicomponent AgMgAl alloys on glass substrate by sputtering technique *(28)*. Table 2.1 shows that reflectivity is a function of resistivity for multi component AgMgAl alloys and it scales with the square root of electrical resistivity, which can be explained by the classical reflection theory given by Eq. (2.5).

3 Factors affecting infrared reflectivity

IR reflectivity basically depends on the relative ratio of the particles, neighboring medium, and distribution of particles within the coating, concentration, particle's size, and wavelength of the incident light *(29)*. Coating thickness is an important criterion that affects film reflection factor. Higher coating thickness ends up with higher reflection factor due to the upper range of nanoparticles on the substrate for reflective *(29)*. Recently, the optical properties of nanocrystalline and connected materials are becoming theoretical and experimental interest. Nanocrystalline materials possess fascinating optical properties like increased band gap and increased light intensity. Among the nanocrystalline materials, metal compound nanoparticles belong to a stimulating category of compounds *(30)*. Metal compound nanocrystals will be ready in distinctive shapes and they exhibit remarkably totally different chemical properties compared to macrocrystals. The particle size of a compound will have an effect on its color that successively will have an effect on its reflection factor properties. Moreover, metal compound nanoparticles possess high surface areas, which make them very helpful for a range of applications as well as coatings *(31)*. The present oxides of the metallic element at air pressure exhibit three phases: anatase, brookite, and mineral *(8)*. The mineral part is thermodynamically the foremost stable and possesses the best density with a compact atomic structure. TiO_2 within the mineral part is essentially used in reflective coatings attributable to its effective lightweight scattering properties. Recently, it has been found that the mineral part nanoparticles have higher absorption within the visible region compared to those of the anatase part *(32,33)*. Coatings containing even bit of mineral part TiO_2 particles replicate nearly the complete visible spectra. Reflective coatings of TiO_2 are applied in business optical merchandise like optical fibers, integration spheres, luminaries, reflective panels, and optical device cavity mirrors *(9)*. Studies on the reflection factor of TiO_2 particles are the topic of serious analysis for several decades as their applications need high photo stability and high reflection efficiency. The improvement in the photo stability of mineral part of TiO_2 while maintaining reflection factor at an equivalent time has been a difficult subject.

Doping TiO_2 with Al, Li, and K is one among the better ways to enhance the reflection factor of TiO_2 particles *(34)*. It creates defects within the TiO_2 space lattice and introduces traps for electrons and holes. This will stop the migration of electrons and holes toward the surface of TiO_2. The impact of Al doping on the reflective properties of TiO_2 nanoparticles and synthesized exploitation of the sol–gel technique has been investigated in 2013 by Kumar et al. *(34)* It has been shown that with Al doping, the transition temperature for anatase to mineral part was increased; but no change in morphology was observed. TiO_2 nanoparticles doped with 0.1% Al showed high photo stability with no amendment in reflection factor *(34)*. The coating with Al-doped TiO_2 nanoparticles has been applied on a plastic substrate with totally different coating thicknesses to design lightweight reflectors. These reflectors are found to possess diffuse reflection factor of 98.17%–98.29% for 0.25-mm-thick coatings. In 2010, A. Ranade were synthesized niobium (Nb)-doped TiO_2 films using several different techniques *(35)*. Nb was originally chosen because of the dopant material attributable to its similarity in atomic radius to Ti^{4+} and also the valence state it prefers *(35)*. Zinc

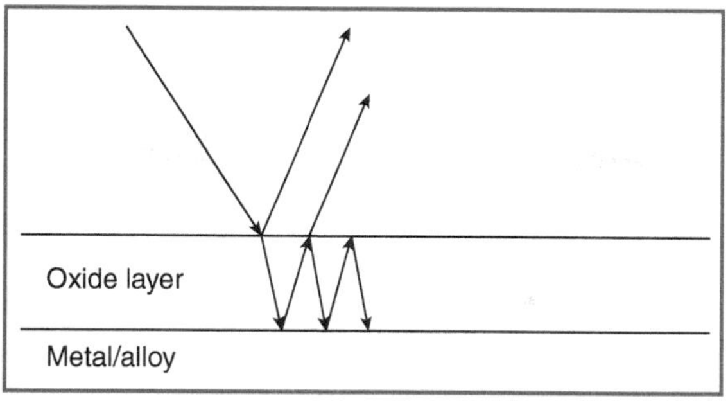

Figure 2.2 ***An oxidized surface capturing the light inside the oxide layer, which may further enhance absorption (25).***

oxide nanoparticles coated with metallic element increased the reflective performance. In comparison to the uncoated material, the coated nanoparticles possess ten percent more reflection factor in 400–800 nm *(36)*. Fe_2O_3, Cr_2O_3, Sb_2O_3, and ZrO_2 are other metal oxides having high refractive indices; to achieve high reflectance over a wide spectral range, more than one type of metal oxides can be used *(37)*. Metals also naturally have a layer (or several layers) of oxides on the surface. The chemical and optical properties of the oxides can often be very different from the properties of the metal or alloy underneath *(25)*. In Fig. 2.2, the oxide layer enhances the absorption by capturing the light *(25)*.

Cadmium stannate (Cd_2SnO_4) is one among the foremost clear, heat reflective semiconductors. Cadmium stannate films on an oxide plate replicate infrared at 1.5 μm film thickness *(38)*. These films provide eighty percent reflectivity at 2 μm film thickness and 90% reflectivity at 6 μm thickness. These properties build metallic element stannate films extremely appropriate for greenhouse window applications *(38)*. Another study describes the employment of colored metallic pigments like mica flakes and aluminum for better infrared reflectivity *(26)*. According to the study, color has been incorporated on the metallic surfaces in such a way that it does not interfere with the ability of metallic pigments to control infrared reflectivity *(26)*.

4 Regulating infrared radiation using inorganic and organic reflective materials

4.1 IR regulating inorganic materials

The IR reflective material either reflects in IR region or transmits the light. The reflectance of any material is usually due to specular reflectance and diffuse reflectance. Specular reflectance is associated with the visual brightness or mirror quality of a surface and it depends on the surface smoothness *(39)*. As the roughness of surface increases, the multiple reflections occur in the material and the component of diffuse reflectance increases and surface appears dull or matte finished. Diffuse reflectance of a material is determined by the particle shape, size, and distribution of the material *(40–42)*. The total reflectivity of reflective coating is also decided by the chemical composition and refractive index of the pigment and the binder material *(43)*. If the refractive index of the pigment is similar to that of the binder refractive index in the IR region, it will be transparent to the IR radiation *(43)*. An ideal reflective material should have not only high refection but also low emissivity and good corrosion resistance.

Over the years, different inorganic and organic materials are being investigated for their applicability as IR reflective materials. Inorganic IR regulating materials are the most commercialized ones due to their ease of processing and high stability. Inorganic IR reflecting materials include transition metals such as Ag, Au, Cu, Al, Ti, and Rh, the nitrides, boride and carbides of Ti, Zr, Hf, V, Nb, Ta, Cr, Mo, and W, metal oxides of Cr, Fe, Ti, Ce, and Mg *(44–47)*.

Transition metal-based IR reflective materials have reported to have highest reflectivity in the IR band range and variable absorption in the visible region. The advantage of metallic coating lies in their broadband reflectivity and high tolerance to thickness variations. Reflectivity of silver is highest among all other materials and it makes Ag the most popular choice for commercial applications *(48–50)*. Silver coating can be used as a transparent window for visible light and reflective surface for IR radiation at the same time. The composite formation with silver and nanocrystalline silver can be processed into composite material by using metal oxide as a base and top layer *(51)*. This process reduces the metallic glare of the silver without significant reduction in its reflectivity. The metal oxide layer also provides scratch and corrosion resistance to the reflective layer. The reflectivity of other coinage metals such as copper and gold is also comparable to silver, but oxidizing tendency of copper and high cost of gold makes them less popular choice of reflective materials *(52,53)*. Dalpati et al. have demonstrated fabrication and application of TiO_2/Cu/TiO_2 heat reflector coating for use in heat regulating smart windows. They observed that on annealing the reflector film, the crystal quality of the TiO_2 improved and its IR reflectance is also increased up to 85% at 1200 nm (Fig. 2.3) *(54)*. Aluminum is another transition metal, which is used extensively as reflecting material,

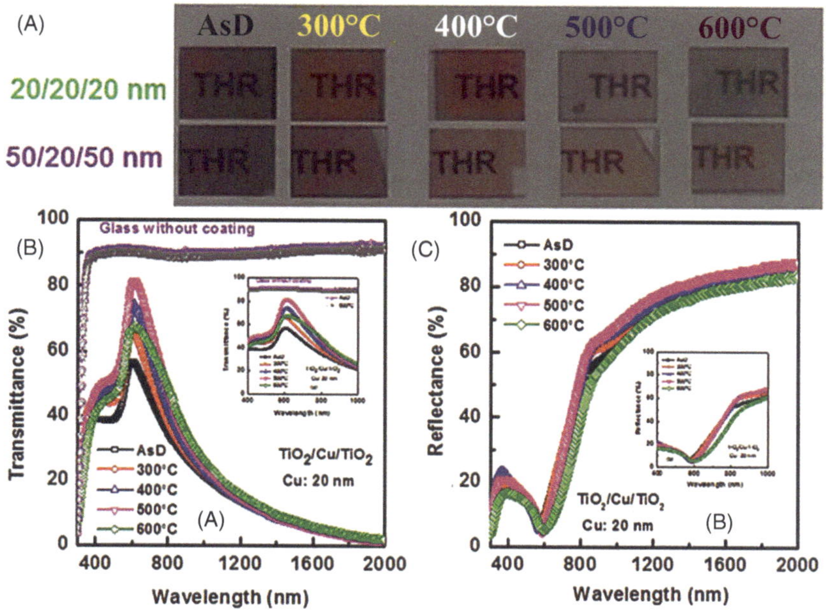

Figure 2.3 (A) Demonstration of $TiO_2/Cu/TiO_2$ heat reflector on glass substrate after thermal treatment (B) transmittance spectra and (C) reflectance spectra of TiO_2/Cu/TiO_2 thermal heat reflection with Cu layer thickness of 20 nm top TiO_2 layer thickness of 25 nm (C) *inset* of (B, C) show clear variation of hat reflector properties in the visible range and NIR region. Transmittance spectra of plain glass with and without thermal at 600 are also compared.

especially in the applications, which require broad spectral range of reflection *(55)*. Aluminum is used as reflective surface in telescope mirrors and as parabolic or curved mirror in flat solar concentrators, but it is prone to weathering, which decreases its specular reflectance. The high IR reflectivity of metal-based coatings is due to the abundance of free electrons. But this also makes them susceptible toward easy wear and tear and corrosion.

The durability of the metallic coating can be enhanced by making its hybrid with a transparent protective coating. The coating materials are usually other transition metals or metal oxides such as TiO_2 and SiO_2. These materials are applied to protect the front surface of the reflective metal by dip coating, sol–gel thin coating, PVD, CVD or sputtering, electro or electroless plating *(56)*. The efficiency of the processes various with the choice of protective coating used and the method of coating. The optical properties of metal-based coating depend upon nucleation and growth phenomena and the impurities. The optical properties of metal and metal oxide coatings are also significantly directed by the chemical composition, stoichiometry, and defects along with crystallinity and crystal structure. Therefore, various techniques for metal and metal oxide film deposition techniques are being developed globally. It has been observed that even the slight variation in the deposition technique or variation in the composition of the chemicals used can alter the optical and physical properties of the reflective coating *(57,58)*. CVD, PVD, ion, and magnetron sputtering and electroplating are the most commonly employed deposition techniques *(59,60)*. CVD and PVD techniques result in uniform and thin reflective layer but the setup for these techniques is expensive and processing on large surface area with these techniques is a challenge. There are not many options available for finding thermally stable precursors for use in CVD techniques. At higher temperatures, most of the metals tend to form their respective oxides, which create defects and impurities in the reflective coatings. The major drawback of electroplating is that it results in the films of thickness 50 nm or higher; additionally, the substrate is also required to be conducting surface.

Metal oxides and sulfides-based IR reflective materials are mostly used in paints, pigments, and camouflage materials *(42,45)*. These materials show some absorption in visible band and emit rest of the visible radiation. The radiations from the visible band, which are emitted by these pigments, correspond to their color. For use in painting materials, these reflective materials are synthesized in the form of fine crystalline powder, which not only impart the paint its color and reflective property but also help in regulating viscosity of the entire formulation. The industrial production of metal oxide and sulfide nanocrystalline is relatively a cost-effective process. The oxides and sulfides of Ti, Fe, and Mn are usually doped with metals such as Fe, Cd, Ni, Sb, Zn, Cr, Bi, Cu, and Y to adjust the response of the host material towards radiations *(61–63)*. Kalbunde et al. have investigated the effect of the size of the nanocrystalline metal oxide on the percentage reflectivity of various metal oxides. It is observed that the reflectivity of the metal oxide decreases with the increase in the crystalline size (Table 2.2) *(45)*.

The IR radiations can also be regulated by doping the reflective coating with other transition metal ions. In this process, the amount of the dopant is required to be controlled as excess dopant could decrease the band gap of the native coating material and make it more absorptive in the IR region. Shi and coworkers investigated the effect of dopant on the reflectivity of the aluminum phosphate reflective coating. They observed that on Fe^{3+} doping, there was a significant decrease in the band gap from 5.98 to 3.60 eV resulting in the absorption of more NIR and decrease in the reflectance *(64)*. The

Table 2.2 NIR reflectance properties of nanocrystalline metal oxides and macrocrystalline metal oxides.

Metal oxide	*R*% at 810 nm	Mean particle aggregate size (μm)	Average crystalline size (nm)	Surface area (m^2/g)
NC-CeO_2	110	61.6	≤7	≥50
MC-CeO_2	92	95	55	6
NC-Al_2O_3	102	2.6	Amorphous	≥275
MC-Al_2O_3	92	2.6	70	68
NC-TiO_2	108	14.5	Amorphous	≥500
MC-TiO_2	98	16.4	79	8
NC-MgO-I	109	4.2	8	≥230
N-MgO-II	105	7.8	≤4	≥600
MC-MgO	103	2.4	23	45

MC, macrocrystalline; *NC*, nanocrystalline; *R%*, the relative reflectivity was calculated compared to bulk materials.

reflectivity of the surface could also be increased using a combination of dopants. Yang and coworkers have synthesized Fe/N-doped $MgTiO_3$ as IR reflective pigment by sol gel method. The doping with Fe/N not only made the $MgTiO_3$ pigment more IR reflective but also imparted it a dark red blue color *(65)*. In another work, George and coworkers have synthesized yttrium molybdenum oxide-based IR reflective material using silicone and praseodymium as dopants. The authors have reported that the color, hue, and the IR reflectivity of the pigment are regulated by change in the stoichiometric addition of the dopant material *(66)*. The mechanism by which the dopant affects the IR reflectivity is different in the case of Si and Pr. In silicon-doped material, an addition phase α-$Y_2Si_2O_7$ is formed, which increases the apparent concentration of Mo^{6+} ions in the crystal lattice; this results in decrease in the band gap from 2.60 to 2.45 eV. Whereas in the case of Pr, an additional energy level is created between O^{2-} and Mo^{6+} conduction bands due to the $4f^1$ electrons. This change in the energy levels causes the red shift in the absorption edge and decrease the band gap *(66)*.

Other class of inorganic reflective material is metal carbides. Among these, silicon carbide has generated profound interest in aerospace industry due to its tunable IR reflectivity, high strength, moderate density, smaller thermal expansion coefficient, higher thermal conductivity, higher specific stiffness, and lower thermal deformation *(67,68)*. Metal nitrides and metal oxynitrides are being explored as IR regulating materials for solar applications. The advantages of these materials such as AlN, TiN, and TiAlN and their complex nitrides lie in the fact that they are highly heat resistant, abrasive, and oxidation resistant, and have broad spectral selectivity *(69,70)*. The production of these films/coatings can be achieved by DC magnetron sputtering or PVD techniques *(25)*. The quasi-covalent bond formed due to hybridization between p and d orbitals makes the nitrides, borides, and carbides chemical-resistant, hard, and abrasion-resistant, and the unbound d electron in the transition metal imparts them metal like electrical properties. Therefore, a number of reflective and antireflective coatings are being developed for solar and electro optical applications.

4.2 NIR regulating organic materials

IR regulating organic materials are rare. The difference in the refractive indices of the organic materials is generally too low and a multilayer system is required to achieve significant IR reflection. The durability of these materials is not as good as that of the inorganic counterparts. However, the organic materials have found wide spread applications as a base material or one of the coatings in multilayered IR regulating surface. Organic pigments are also used in textile printing, inks, and paints. Examples of organic origin pigments include halogenated copper phthalocyanine, azo compounds, and a few perylene derivatives *(71–73)*. Chlorophyll is naturally occurring IR reflective compound but due to its poor chemical stability, it cannot be used in commercial application. IR regulation is also reported by using liquid crystal (LC)-based switchable devices *(74–77)*. These devices work on the fact that the orientation of the liquid crystals can be altered by applying an electric field, which also affects their transmittance. Liquid crystals are mainly of six types such as nematic, smectic, twisted nematic, cholesteric, guest–host, and ferroelectric *(63,76)*. Among these, guest–host LCs-based switchable windows technology is commercialized since 1990 *(63)*. Cholesterics liquid crystals are also employed for IR regulation for a variety of reflection-based technologies. The drawback with the use of regular cholesterics is their restricted d band width in the IR region (100 nm) but this problem can be solved by laminating a number of layers of this material during processing.

Khendelwal et al. have reported synthesis of organic-based IR reflectors for energy conservation. They have reported that energy consumption for temperature moderation in a building can be reduced by more than 12% by using a switchable IR reflector compared to a normal double glazing window and 9.3% compared to a static IR reflector *(78)*. The authors have fabricated an electrically switchable dye (coumarin)-impregnated cholesteric

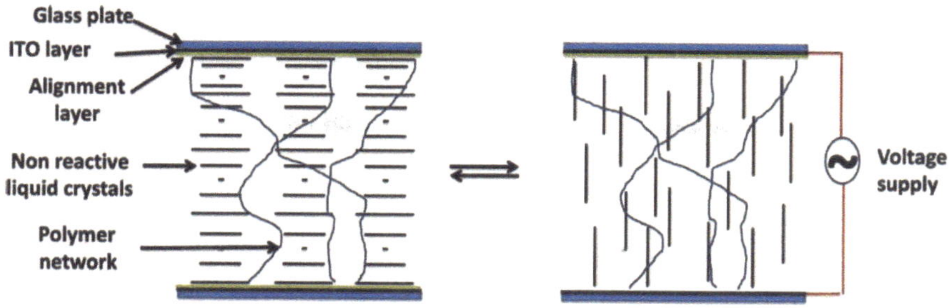

Figure 2.4 ***Schematic diagram shows the planar state (left) and homeotropic state on applying the electric voltage.***

polymer gel-based coating. The schematic of this reflector is depicted in Fig. 2.4. Upon removing the electric voltage, orientation of the molecules reverts back to the original state in the presence of an alignment layer and polymer network.

A broadband IR reflector is also reported by Chen et al. by polymerizing LC cross-linker in the presence of azobenzene derivative *(79)*. Nanoparticle-polymer composites are also used as IR regulating materials. The additional advantage of the polymer based reflective materials is ease of processing for commercial applications.

5 Active, passive, transparent, and opaque coating

In Sections 2.2 and 2.4, interaction of infrared radiation with metals and nanostructures, and its regulation using organic and inorganic materials were discussed in detail. This section endows details on the applicability of various IR coatings considering future energy supply and demands.

The wavelengths at which different optical coatings (thin films/nanostructures) perform varies over a large range. The selection of materials for coating depends on the application for which the particular system is being used under certain environmental conditions. The material coatings mainly are discussed in the context of coating for energy saving window applications. Cooling or heating strategy that keeps indoor comfort without electricity requirement could have a significant impact on global energy consumption. In order to cater to future energy supply and demands, and to slow down the insidious effects of global warming, energy-efficient buildings, vehicles, and windows are gaining importance *(80,81)*. In that regard, a smart window is an energy proficient device that regulates heat according to weather conditions (summer/winter) at minimum demand of paid energy (air conditioning in summer) *(82–84)*. Windows are most important part of the building as they can be active, passive, opaque, and transparent at the same time. Many studies have been carried out in the context of smart windows *(54,80,84–86)*.

Glass is used as a window on which no coating is performed. It allows the passage of visible light transmission as well as solar heat that increases the heating as well as cooling of the building structure. Materials coating is being performed on the glass to eliminate solar heat and reducing energy consumption. The materials are selected based upon their optical properties. They should be highly transparent in the visible region (380–760 nm) with high reflectance in the infrared spectral region (800–2500 nm) *(80)*. The coating should be such that the long-wave radiation should not enter in the room from outdoors during summers, and should not escape from inside during winters, thus leading toward lower energy consumption during both the seasons (Fig. 2.5) *(54,80,85)*. The spectral properties of such perfect windows are shown in Table 2.3. The energy serving approaches can be divided into two forms of active and passive coatings.

The active methods include improvement in heating and ventilation conditions. However, passive methods are related to the improvement in the properties and thermal performance of the building envelops *(84)*. The performance can be affected by adding thermal insulation to wall, using cool coatings on roofs and coated window glazing. These modifications lead to the change in the thermal properties and act as passive ways. These coatings can be applied by using various ways such as low-emission (low-e) coatings, thermos-chromic, aerogel, photochromic, electrochromic, gasochromic, and suspended particle and liquid crystal techniques.

Among various techniques mentioned, the low emissivity coatings are one of the potent techniques in regulating the long-wave thermal emission properties. Such low-e coatings aspire to allow the visible light pass through and block the IR and UV wavelengths and solve the purpose *(87)*. There are various material coatings that are carried out on the glass to form the smart windows. Few of them are discussed in the 2.5.1 and 2.5.2 sections.

5.1 Dielectric/metal/dielectric coatings

In dielectric/metal/dielectric-based structures, dielectrics are made from metal oxides such as TiO_2, SnO_2, ZnO, HfO_2, Cu_2O, and ZrO_2 whereas, the metal component typically comprises of silver (Ag), Au (gold), copper (Cu), and nitride-based materials *(87)*. The D-M-ubD-based mutilayered coatings on glass substrates serve that purpose for energy saving, as shown in Fig. 2.6. The thin metal layer is used for IR reflection and dielectric layers

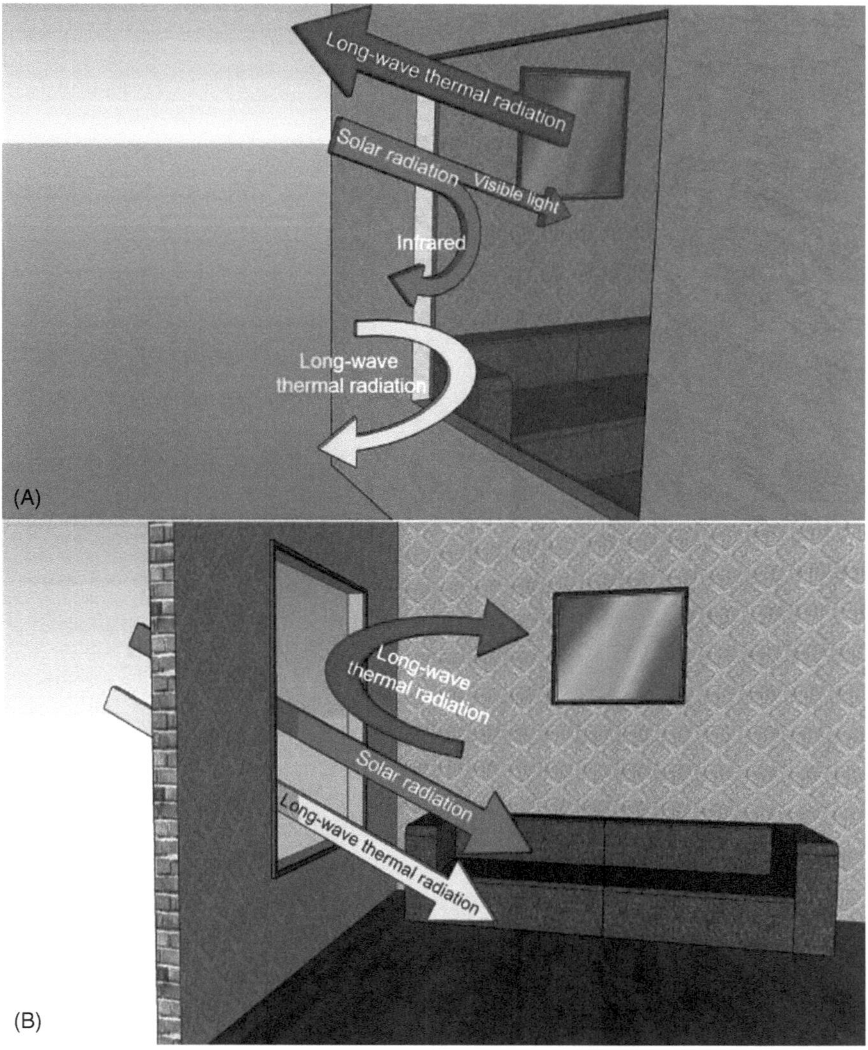

Figure 2.5 ***Diagrammatic sketch of perfect windows.*** The perfect windows for (A) summer and (B) winter are shown *(80)*.

for surface passivation (88–91). The thin metal layer must have low refractive index in the visible region. The D-M-D multilayered coatings regulate the heat transmission along with the transparency; hence, less energy is consumed in heating and cooling the building. The techniques of growth, structure, and morphology play an important role in regulating the IR radiation. Therefore, materials selectivity play an important role in regulating the heat with visible transparency.

Table 2.3 Spectral properties of perfect windows *(80)*.

		Solar spectrum		Long-wave thermal radiation ($\lambda > 2.5\ \mu m$)	
Season	**Spectral properties**	**Visible light (0.4–0.7 nm)**	**$\lambda < 0.4\ \mu m$ or $0.7 < \lambda < 2.5\ \mu m$**	**Outdoor**	**Indoor**
Summer	Spectral transmittance (τ_λ)	100%	0%	0%	100%
	Spectral reflectance (τ_λ)	0%	100%	100%	0%
	Spectral absorptivity (α_λ)	0%	0%	0%	0%
Winter	Spectral transmittance (τ_λ)	100%	100%	100%	0%
	Spectral reflectance (τ_λ)	0%	0%	0%	100%
	Spectral absorptivity (α_λ)	0%	0%	0%	0%

Apart from the optical transparency in the multilayered coatings, durability also plays an important role. The durability to resist from scratches and corrosion from long-term environmental exposure has been reported by Hu et al. *(92)* by providing the concept of "first durability then reflectivity." Various transition metals such as Ti, Zr, Hf, V, Nb, Ta, Cr, and Mo were suggested as the durable base, and reflectivity was improved by impinging silver and gold multilayers *(92)*. According to their findings, the traditional mirror coating such as Al/SiO_2 films is best suited for moderate environments; however, Ag-doped HfN film works for harsh environments (Fig. 2.7). In other work by Mishra et.al, it is found that Al mirror reflectivity can be enhanced by using multilayers of Al-SiO_2, Al-SiO_2-TiO_2 and Al-Ni-Ag-SiO_2-TiO_2. No change was found in the reflectivity behavior of these multilayers in harsh (accelerated weathering condition and under salt spray with different hours of exposure) conditions *(55)*. The D-M-D-based thin-film encapsulation was designed by Kwon et al. using hybrids of Al_2O_3/Ag/Al_2O_3/S-H (nanocomposite)/Al_2O *(93)*. Here, Al_2O_3 is used as a dielectric layer and Ag as an insertion metal layer. This novel design and technology

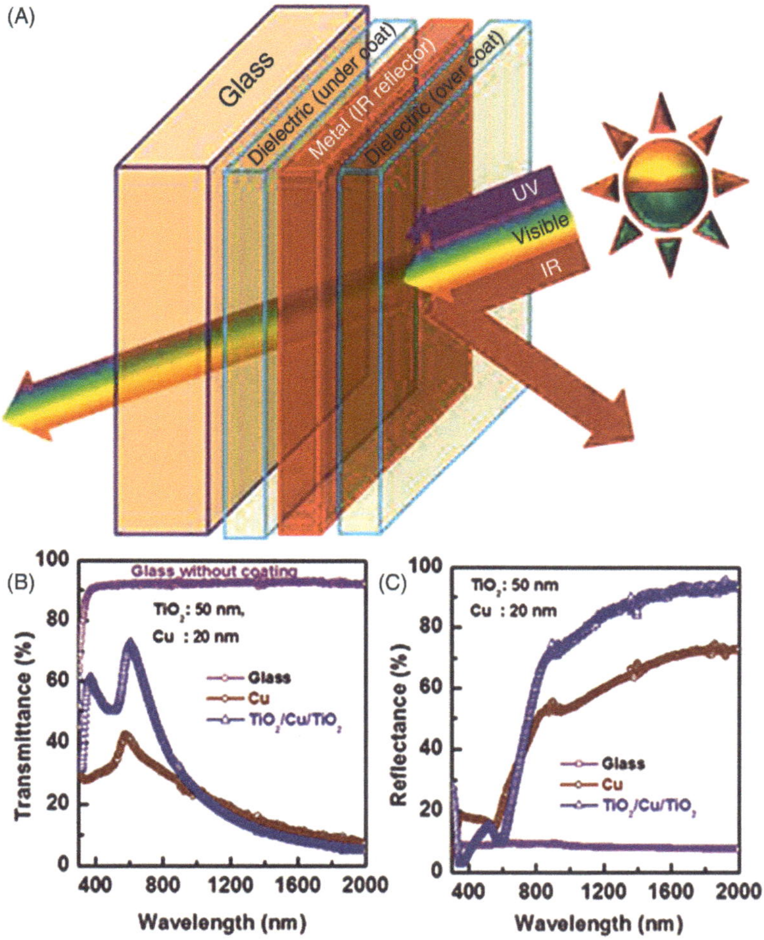

Figure 2.6 (A) Schematic diagram of DMD bases transparent heat mirror. Comparison of optical (B) transmittance and (C) reflectance of glass, thin film Cu (20 nm) and TiO_2/Cu/TiO_2 (50 nm/20 nm/50 nm) multilayered thin films without thermal treatment *(54,87)*.

provided the flexible and wearable structure in context of stress relaxation, and heat dissipation *(93)*.

From the above discussion, it is inferred that the growth and microstructure of the dielectric layer play an important role in the D-M-D-based coatings. Thus, there is need of designing and optimizing the growth parameters and techniques to fabricate the dielectric layers. Apart from the growth of dielectric layer, the selection and thickness of metal layer also play an important role. The thickness of the metal layer affects the optical properties and hence emissivity in I-R coatings. Furthermore, the metal layer should

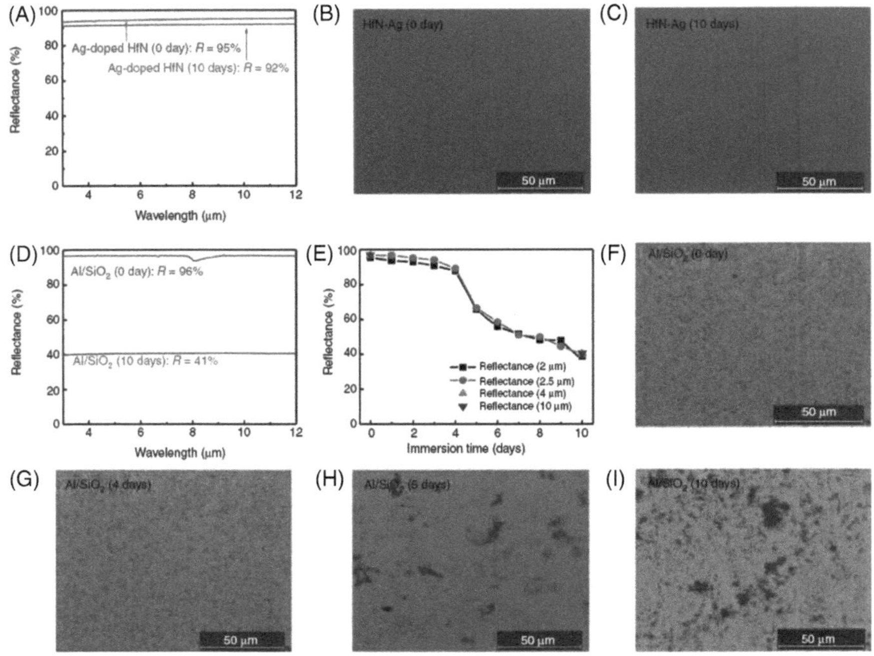

Figure 2.7 (A) The reflectance spectra of the Ag-doped HfN films before/after immersion in a NaCl solution at 35°C for 10 days. Surface of these Ag-doped HfN films (B) before immersion (0 day) and (C) after immersion for 10 days. (D) The reflectance spectra of Al/SiO_2 film before/after immersion in the same NaCl solution for 10 days. (E) Reflectance of the Al/SiO_2 films at various infrared wavelengths (2, 2.5, 4, and 10 μm) as a function of the immersion time. Typical surface of Al/SiO_2 films (F) before immersion (0 day), (G) after immersion for 4 days, (H) 5 days, and (I) 10 days *(92)*.

be chosen so that the interdiffusion of the metal layer is avoided during deposition and hence the stability gets enhanced *(94)*. Thus, the optimization of metal layer thickness is also an important factor in D-M-D-based I-R coatings.

5.2 Thermo-chromic materials and coatings

These are the materials whose optical properties are varied as a function of temperature. The properties of the materials are changed upon heating and while cooling, the material returns to its original state. The change in optical properties on heating and cooling is related to the change in the phase structure of the material. The transmittance and reflectance are modified due to phase transition. This change in phase, and hence optical properties of a material by heating and cooling, results in generation of the color, hence the name thermo-chromic (T-C) materials.

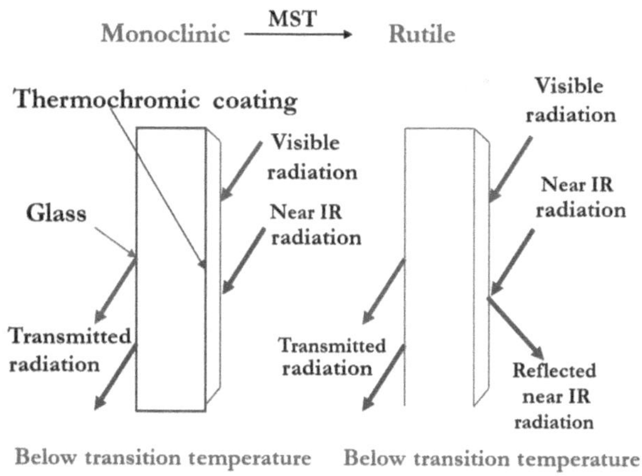

Figure 2.8 ***Schematic representation of thermos-chromic materials applied as an intelligent windows coating (84,99).***

The T-C materials play an important role in reducing the energy consumption and target toward lowering future energy demands. The T-C materials optimize the optical properties of a material and reduce the unwanted solar energy gain *(80,87,95,96)*. There are various materials that are used for T-C coating. However, vanadium dioxide (VO_2) proves to be an ideal material for such coatings. VO_2 is initially at monoclinic state. The monoclinic state changes to rutile at 68°C. This phase change with the change in temperature shows the reversible metal to semiconductor phase transition *(81,86,87)*, see Fig. 2.8. The change in phase with temperature of VO_2 leads to stupendous optical properties *(81)*. At low temperature (below transition temperature), VO_2 is in insulating state and allows transmission of the visible and IR. As the temperature rises above transition temperature, the VO_2 film is converted to metallic phase that allows visible light but blocks IR *(87,98)*. This transition of phase with temperature allows VO_2 to be a T-C material for smart windows application.

Yang et al., in their studies, show the integration of thermochromic VO_2 on flexible substrates forming hybridized VO_2/graphene, see Fig. 2.9. These T-C coating-based smart windows may respond to environmental temperatures to regulate light transmittance *(86)*.

In order to modify the optical properties of VO_2 for smart window applications, it is doped with various metals *(87,98)*. Doping with metals changes the transition temperature of VO_2 films *(87)*. Tungsten (W) has proven to be a wonderful metal for doping as it reduces the transition temperature VO_2 to 25°C by 2% doping *(98)*. Apart from W, gold (Au) and fluorine have also been reported to be used as dopant materials *(84)*.

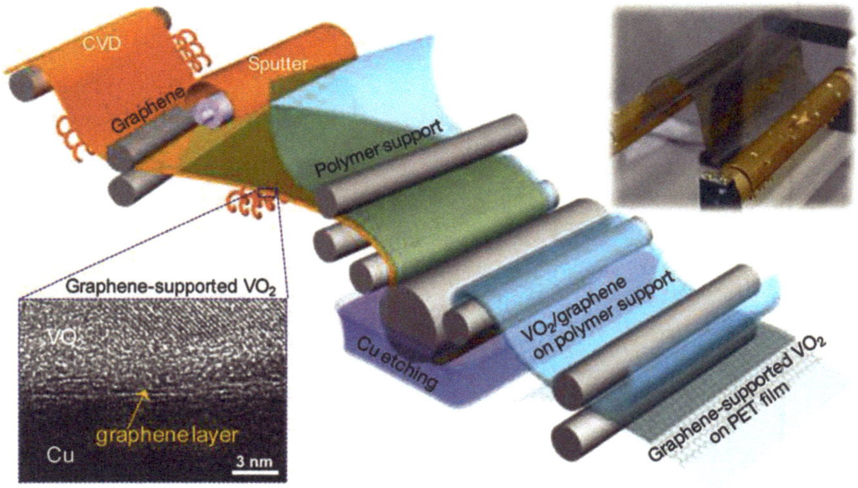

Figure 2.9 ***Schematic representation of thermos-chromic hybridized VO_2/graphene (86).***

6 Conclusions

The interaction of IR radiation with various materials for different coatings is discussed in this chapter. It is inferred that multilayer coatings play an important role in energy-efficient smart window applications by variation in their optical properties. The wavelengths at which different optical coatings (thin films/nanostructures) perform vary over a wide range of wavelengths. Different materials coatings, such as dielectric-metal-dielectric (D-M-D) and thermo-chromic (T-C) are discussed. The chapter discusses in brief about the selection of materials, synthesis techniques, for the different IR coatings.

References

1. Diffey, B. L. Sources and Measurement of Ultraviolet Radiation. *Methods* **2002**, *28* (1), 4–13.
2. Grenfell, T. C.; Warren, S. G.; Mullen, P. C. Reflection of Solar Radiation by the Antarctic Snow Surface at Ultraviolet, Visible, and Near-Infrared Wavelengths. *J. Geophys. Res. Atmos.* **1994**, *99* (D9), 18669–18684.
3. Bitran M.; Murray W. Infrared, Visible, and Ultraviolet Radiation. In *E. Bingham, B. Cohrssen and C. Powell, Eds., Patty's Toxicology; John Wiley & Sons, Inc.,* New York, **2001**, 249–314.
4. Wang, J., et al. Recent Progress on Localized Field Enhanced Two-Dimensional Material Photodetectors from Ultraviolet—Visible to Infrared. *Small* **2017**, *13* (35), 1700894.
5. Ring, E. F. J. The Discovery of Infrared Radiation in 1800. *Imaging Sci. J.* **2000**, *48* (1), 1–8.

6. Manley, M.; Baeten, V. Chapter 3—Spectroscopic Technique: Near Infrared (NIR) Spectroscopy. In Da-Wen Sun, 2nd ed.; *Modern Techniques for Food Authentication* pp. 51–102.
7. Dwivedi, B. James Clerk Maxwell and His Equations. *Resonance* **2003**, *8*, 4–16.
8. Tian, L., et al. Broad Range Energy Absorption Enabled by Hydrogenated TiO_2 Nanosheets: From Optical to Infrared and Microwave. *J. Mater. Chem. C* **2017**, *5* (19), 4645–4653.
9. Fang, V., et al., A Review of Near Infrared Reflectance Properties of Metal Oxide Nanostructures. GNS Science Report, 2013.
10. Caplan, J. D., et al. Near-Infrared Spectroscopy for the Detection of Vulnerable Coronary Artery Plaques. *J. Am. Coll. Cardiol.* **2006**, *47* (8 Suppl.), C92–C96.
11. Feng, R., et al. Visible- and NIR-Light Responsive Black-Phosphorus-Based Nanostructures in Solar Fuel Production and Environmental Remediation. *Adv. Mater.* **2018**, *30*, 1804770.
12. Alibabaei, L.; Brennaman, M. K.; Meyer, T. J. Light-Driven Water Splitting in the Dye-Sensitized Photoelectrosynthesis Cell. In *H. Tian, G. Boschloo and A. Hagfeldt, Molecular Devices for Solar Energy Conversion and Storage;* Springer, **2018**, 229–257.
13. Fujishima, A.; Honda, K. Electrochemical Photolysis of Water at a Semiconductor Electrode. *Nature* **1972**, *238* (5358), 37.
14. Fung, C. K. M., et al. Nanoresonant Signal Boosters for Carbon Nanotube Based Infrared Detectors. *Nanotechnology* **2009**, *20* (18), 185201.
15. Grätzel, M. Recent Advances in Sensitized Mesoscopic Solar Cells. *Acc. Chem. Res.* **2009**, *42* (11), 1788–1798.
16. Kastek, M., et al. Test Bed for Measurement of Angular Parameters of Passive Infrared Sensors. *Infrared Phys. Technol.* **2007**, *49* (3), 198–201.
17. Matsui, T., et al. A Novel Method to Prevent Secondary Exposure of Medical and Rescue Personnel to Toxic Materials Under Biochemical Hazard Conditions Using Microwave Radar and Infrared Thermography. *IEEE Trans. Biomed. Eng.* **2004**, *51* (12), 2184–2188.
18. Rogalski, A. *Infrared detectors* CRC Press: Boca Raton, FL, 2011.
19. Yu, P. Y.; Cardona, M. *Fundamentals of Semiconductors: Physics and Materials Properties.* Springer, 1996.
20. Chen, X., et al. Semiconductor-Based Photocatalytic Hydrogen Generation. *Chem. Rev.* **2010**, *110* (11), 6503–6570.
21. Chen, X., et al. Nanomaterials for Renewable Energy Production and Storage. *Chem. Soc. Rev.* **2012**, *41* (23), 7909–7937.
22. Vollmer, M.; Möllmann, K. -P. *Infrared Thermal Imaging: Fundamentals, Research and Applications Wiley-VCH; 2 edition.*
23. Gupta, R. P. *Remote sensing geology*. Springer-Verlag, 2017.
24. Kocer, H., et al. Reduced Near-Infrared Absorption Using Ultra-Thin Lossy Metals in Fabry-Perot Cavities. *Sci. Rep.* **2015**, *5*, 8157.
25. Bergström, D. *The Absorption of Laser Light by Rough Metal Surfaces.* Luleå tekniska universitet, 2008.
26. Subasri, A., et al. Economical Development of IR Coatings in Asbestos Sheet. *Int. J. Curr. Res. Life Sci.* **2018**, 7 (5), 2062–2066.
27. Nilsson, C. and T. Halberg, (2001) Multilayer structures for low emissive paint, Sensor Technology.
28. Hu, T., et al. Correlation Between Reflectivity and Resistivity in Multi-Component Metallic Systems. *Appl. Phys. Lett.* **2012**, *101* (1), 011902.
29. Faulkner, E. B.; Schwartz, R. J. *High Performance Pigments.* John Wiley & Sons, 2009.
30. Nayfeh, M. H. Optics in Nanotechnology. In *M. D. Al-Amri, Mohamed El-Gomati, M. S. Zubairy, Optics in Our Time;* Springer, **2016**, pp. 223–264.
31. Dong, L. *Optical Properties of Nanoparticles in Composite Materials.* KTH Royal Institute of Technology, 2012.

32. Rugh, J. P., et al. *Reduction in Vehicle Temperatures and Fuel Use From Cabin Ventilation, Solar-Reflective Paint, and a New Solar-Reflective Glazing*. SAE Technical Paper.
33. Brady, R. F., Jr.; Wake, L.V. Principles and Formulations for Organic Coatings with Tailored Infrared Properties. *Prog. Org. Coat.* **1992**, *20*, 1–25.
34. Kumar, S.; Verma, N. K.; Singla, M. L. Study on Reflectivity and Photostability of Al-doped TiO_2 Nanoparticles and Their Reflectors. *J. Mater. Res.* **2013**, *28* (3), 521–528.
35. Ranade, A. *Synthesis and Characterization of Transparent Oxide Coatings with Enhanced Near Infrared Reflectance*. Northwestern University, 2011.
36. Fangli, Y., et al. Preparation and Properties of Zinc Oxide Nanoparticles Coated with Zinc Aluminate. *J. Mater. Chem.* **2003**, *13* (3), 634–637.
37. Bendiganavale, A.; Malshe, V. . *Infrared Reflective Inorganic Pigments* **2010**, 1.
38. Haacke, G. Transparent Electrode Properties of Cadmium Stannate. *Appl. Phys. Lett.* **1976**, *28*, 622.
39. Zheng, B., et al. Identifying Key Factors Towards Highly Reflective Silver Coatings. *Adv. Mater. Sci. Eng.* **2017**, *2017*, 12.
40. Jung, K.T.; Shul, Y. G. Preparation of Transparent TS-1 Zeolite Film by Using Nanosized TS-1 Particles. *Chem. Mater.* **1997**, *9* (2), 420–422.
41. Lagorio, M. G. Why Do Marbles Become Paler on Grinding? Reflectance, Spectroscopy, Color, and Particle Size. *J. Chem. Educ.* **2004**, *81* (11), 1607.
42. Rocquefelte, X., et al. Investigation of the Origin of the Empirical Relationship between Refractive Index and Density on the Basis of First Principles Calculations for the Refractive Indices of Various TiO_2 Phases. *Inorg. Chem.* **2004**, *43* (7), 2246–2251.
43. Malshe, V. C.; Bendiganavale, A. K. Infrared Reflective Inorganic Pigments. *Recent Patents Chem. Eng.* **2008**, *1* (1), 67–79.
44. Wang, P., et al. Effect of Silicon Carbide Ceramic Coating Process on the Mirror Surface Quality. Eighth International Symposium on Advanced Optical Manufacturing and Testing Technology (AOMATT2016).
45. Jeevanandam, P., et al. Near Infrared Reflectance Properties of Metal Oxide Nanoparticles. *J. Phys. Chem. C* **2007**, *111* (5), 1912–1918.
46. Shrestha, K. M.; Sorensen, C. M.; Klabunde, K. J. Synthesis of CuO Nanorods, Reduction of CuO into Cu Nanorods, and Diffuse Reflectance Measurements of CuO and Cu Nanomaterials in the Near Infrared Region. *J. Phys. Chem. C* **2010**, *114* (34), 14368–14376.
47. Li, Y. -Q., et al. Highly Solar Radiation Reflective $Cr_2O_{3-}3TiO_2$ Orange Nanopigment Prepared by a Polymer-Pyrolysis Method. *ACS Sustain. Chem. Eng.* **2014**, *2* (2), 318–321.
48. Nikolic´, N. D.; Rakočević, Z.; Popov, K. I. Reflection and Structural Analyses of Mirror-Bright Metal Coatings. *J. Solid State Electrochem.* **2004**, *8* (8), 526–531.
49. Goossens, V., et al. Predicting Reflections of Thin Coatings. *Surf. Coat. Technol.* **2009**, *204* (5), 551–557.
50. Lee, C. -C.; Chen, S. -H.; Jaing, C. -C. Optical Monitoring of Silver-Based Transparent Heat Mirrors. *Appl. Opt.* **1996**, *35* (28), 5698–5703.
51. Yoldas, B. E.; O'Keefe, T. Deposition of Optically Transparent IR Reflective Coatings on Glass. *Appl. Opt.* **1984**, *23* (20), 3638–3643.
52. Woong-Sun, Y., et al. Ag/Ni/Ag Multilayer Reflector for GaN-Based Vertical Light-Emitting Diode. *Jpn J. Appl. Phys.* **2013**, *52* (5R), 052101.
53. Martín-Palma, R. J., et al. In-Depth Optical and Structural Study of Silver-Based Low-Emissivity Multilayer Coatings for Energy-Saving Applications. J. Phys. D 2004, 37 (11), 1554.
54. Dalapati, G. K., et al. Color Tunable Low Cost Transparent Heat Reflector Using Copper and Titanium Oxide for Energy Saving Application. *Sci. Rep.* **2016**, *6*, 20182.
55. Mishra, S. K., et al. Development and Degradation Behavior of Protective Multilayer Coatings for Aluminum Reflectors for Solar Thermal Applications. *Thin Solid Films* **2016**, *619*, 202–207.

56. Kure-Chu, S., et al. Multilayered Sn/Ag3Sn electroplating on Cu alloys for high reliable electronic/electric materials. 2014 International Conference on Electronics Packaging (ICEP).
57. Wang, X., et al. Thermal Annealing Effect on Optical Properties of Binary TiO_2-SiO_2 Sol-Gel Coatings. *Materials (Basel, Switzerland)* **2012**, *6* (1), 76–84.
58. Liu, Y. Y., et al. Natural Superhydrophilic TiO_2/SiO_2 Composite Thin Films Deposited by Radio Frequency Magnetron Sputtering. *J. Alloys Compounds* **2009**, *479* (1), 532–535.
59. Hodroj, A., et al. Thermal Annealing of Amorphous Ti–Si–O Thin Films. *J. Mater. Res.* **2011**, *23* (3), 755–759.
60. Nilchi, A., et al. New TiO_2/SiO_2 Nanocomposites—Phase Transformations and Photocatalytic Studies. *Colloids Surf. A* **2010**, *361* (1), 25–30.
61. Liu, X., et al. CaF_2:Ce^{3+}/Yb^{3+} Hollow Spheres Luminescence Downconversion Property Optimize Anti-Reflective Coatings for Solar Cells. *Solar Energy* **2016**, *134*, 45–51.
62. Pedroni, M., et al. Water (H_2O and D_2O) Dispersible NIR-to-NIR Upconverting Yb^{3+}/ Tm^{3+} Doped MF_2 (M = Ca, Sr) Colloids: Influence of the Host Crystal. *Cryst. Growth Des.* **2013**, *13* (11), 4906–4913.
63. Baetens, R.; Jelle, B. P.; Gustavsen, A. Properties, Requirements and Possibilities of Smart Windows for Dynamic Daylight and Solar Energy Control in Buildings: A State-of-the-art Review. *Sol. Energy Mater. Sol. Cells* **2010**, *94* (2), 87–105.
64. Shi, Y., et al. Crystal Structure and Near-Infrared Reflective Properties of Fe^{3+} Doped $AlPO_4$ Pigments. *Ceram. Int.* **2017**, *43* (8), 5979–5983.
65. Yang, R., et al. Synthesis, Characterization and Thermal Performance of Fe/N co-doped $MgTiO_3$ as a Novel High Near-Infrared Reflective Pigment. *Sol. Energy Mater. Sol. Cells* **2017**, *160*, 307–318.
66. George, G.; Vishnu, V. S.; Reddy, M. L. P. The Synthesis, Characterization and Optical Properties of Silicon and Praseodymium Doped Y_6MoO_{12} Compounds: Environmentally Benign Inorganic Pigments with High NIR Reflectance. *Dyes Pigments* **2011**, *88* (1), 109–115.
67. Sorokin, O., et al. Joining of Silicon Carbide Ceramic by Hybrid Spark Plasma Sintering. *Russ. J. Appl. Chem.* **2015**, *88*, 839–845.
68. Komarevskiy, N., et al. Potential of Glassy Carbon and Silicon Carbide Photonic Structures as Electromagnetic Radiation Shields for Atmospheric Re-entry. *Opt. Express* **2012**, *20*, 14189–14200.
69. Bilokur, M., et al. Optical Properties of Refractory TiN, AlN and (Ti,Al)N Coatings. SPIE Micro+Nano Materials, Devices, and Applications, SPIE.
70. Thirugnanasambantham, K. G.; Natarajan, S. Degradation Mechanism for High Temperature Erosion in Surface Modified IN718 Superalloy. *Surf. Eng.* **2015**, *31* (1), 24–28.
71. Chen, W., et al. Thiophene-Fused-Heteroaromatic Diones as Promising NIR Reflectors for Radiative Cooling. *Angew. Chem. Int. Ed.* **2018**, *57* (21), 6289–6293.
72. Lin, Y.; Li, Y.; Zhan, X. Small Molecule Semiconductors for High-Efficiency Organic Photovoltaics. *Chem. Soc. Rev.* **2012**, *41* (11), 4245–4272.
73. Kaur, B.; Bhattacharya, S. N.; Henry, D. J. Interpreting the Near-Infrared Reflectance of a Series of Perylene Pigments. *Dyes Pigm.* **2013**, *99* (2), 502–511.
74. Lampert, C. M. Chromogenic Smart Materials. *Mater. Today* **2004**, 7 (3), 28–35.
75. Lampert, C. M. Large-Area Smart Glass and Integrated Photovoltaics. *Sol. Energy Mater. Sol. Cells* **2003**, *76* (4), 489–499.
76. Lechner, R.; Thomas, L. K. All Solid State Electrochromic Devices on Glass and Polymeric Foils. *Sol. Energy Mater. Sol. Cells* **1998**, *54* (1), 139–146.
77. Hu, X., et al. Effective Electrically Tunable Infrared Reflectors Based on Polymer Stabilised Cholesteric Liquid Crystals. *Liquid Cryst.* **2018**, *46*, 1–8.

78. Khandelwal, H., et al. Electrically Switchable Polymer Stabilised Broadband Infrared Reflectors and Their Potential as Smart Windows for Energy Saving in Buildings. *Sci. Rep.* **2015**, *5*, 11773.
79. Chen, X., et al. Broadband Reflection of Polymer-Stabilized Chiral Nematic Liquid Crystals Induced by a Chiral Azobenzene Compound. *Chem. Commun.* **2014**, *50* (6), 691–694.
80. Long, L.; Ye, H. How to be Smart and Energy Efficient: A General Discussion on Thermochromic Windows. *Sci. Rep.* **2014**, *4*, 6427.
81. Gao, Y., et al. Nanoceramic VO_2 Thermochromic Smart Glass: A Review on Progress in Solution Processing. *Nano Energy* **2012**, *1* (2), 221–246.
82. Al-Shukri, A. M. Thin Film Coated Energy-Efficient Glass Windows for Warm Climates. *Desalination* **2007**, *209* (1), 290–297.
83. Sadineni, S. B.; Madala, S.; Boehm, R. F. Passive Building Energy Savings: A Review of Building Envelope Components. *Renew. Sust. Energy Rev.* **2011**, *15* (8), 3617–3631.
84. Kamalisarvestani, M., et al. Performance, Materials and Coating Technologies of Thermochromic Thin Films on Smart Windows. *Renew. Sust. Energy Rev.* **2013**, *26*, 353–364.
85. Fan, J. C. C., et al. Transparent Heat-Mirror Films of $TiO_2/Ag/TiO_2$ for Solar Energy Collection and Radiation Insulation. *Appl. Phys. Lett.* **1974**, *25* (12), 693–695.
86. Kim, H., et al. Flexible Thermochromic Window Based on Hybridized VO_2/Graphene. *ACS Nano* **2013**, *7* (7), 5769–5776.
87. Dalapati, G. K., et al. Transparent Heat Regulating (THR) Materials and Coatings for Energy Saving Window Applications: Impact of Materials Design, Micro-Structural, and Interface Quality on the THR Performance. *Prog. Mater. Sci.* **2018**, *95*, 42–131.
88. Dan, A., et al. Solar Energy Absorption Mediated by Surface Plasma Polaritons in Spectrally Selective Dielectric-Metal-Dielectric Coatings: A Critical Review. *Renew. Sust. Energy Rev.* **2017**, *79*, 1050–1077.
89. Al-Kuhaili, M. F., et al. Energy-Saving Transparent Heat Mirrors Based on Tungsten Oxide–Gold $WO_3/Au/WO_3$ Multilayer Structures. *Sol. Energy* **2012**, *86* (11), 3183–3189.
90. Kulczyk-Malecka, J., et al. Investigation of Silver Diffusion in $TiO_2/Ag/TiO_2$ Coatings. *Acta Mater.* **2013**, *66*, 396.
91. Peres, L. ZnS | Ag | TiO2 multilayer electrodes with broadband transparency for thin film solar cells. *RSC Adv.* **2016**, *6* (66), 61057–61063.
92. Hu, C., et al. New Design for Highly Durable Infrared-Reflective Coatings. *Light Sci. Appl.* **2018**, *7*, 17175.
93. Kwon, J. H., et al. Functional Design of Dielectric-Metal-Dielectric-Based Thin-Film Encapsulation with Heat Transfer and Flexibility for Flexible Displays. *ACS Appl. Mater. Interfaces* **2017**, *9*, 27062–27072.
94. Sharma, H.; Shukla, A. K.; Vankar, V. D. Effect of Titanium Interlayer on the Microstructure and Electron Emission Characteristics of Multiwalled Carbon Nanotubes. *J. Appl. Phys.* **2011**, *110* (3), 033726.
95. Greenberg, C. B. Undoped and Doped VO_2 Films Grown From $VO(OC_3H_7)_3$. *Thin Solid Films* **1983**, *110* (1), 73–82.
96. Parkin, I. P.; Manning, T. D. Intelligent Thermochromic Windows. *J. Chem. Educ.* **2006**, *83* (3), 393.
97. Blackman, C. S., et al. Atmospheric Pressure Chemical Vapour Deposition of Thermochromic Tungsten Doped Vanadium Dioxide Thin Films for Use in Architectural Glazing. *Thin Solid Films* **2009**, *517* (16), 4565–4570.
98. Wu, C.; Feng, F.; Xie, Y. Design of Vanadium Oxide Structures with Controllable Electrical Properties for Energy Applications. *Chem. Soc. Rev.* **2013**, *42* (12), 5157–5183.
99. Kiri, P.; Hyett, G.; Binions, R. Solid State Thermochromic Materials. *Adv. Mater. Lett.* **2010**, *1*, 86–105.

CHAPTER 3

Transparent heat regulation materials and coatings: present status, challenges, and opportunity

Mikhail Vasiliev, Mohammad Nur-E-Alam, Kamal Alameh
Electron Science Research Institute, School of Science, Edith Cowan University, Joondalup, WA, Australia

1 Introduction

Worldwide annual energy consumption is projected to exceed 700 quadrillion British thermal units (Btu) by 2040, with the energy generation contributions from fuels other than coal (mainly renewables) being on the increase currently *(1)*. Around 22.7 billion tons of anthracite coal fuel is needed to be burnt to release this annual energy consumption thermally. At the same time, the combustion of fossil fuels remains among the main concerns identified in relation to the past and current global warming and environmental pollution trends *(2–4)*. In light of this, the development of various types of energy-saving approaches and technologies is of increasing importance today, especially in the building and construction sectors, where a substantial fraction of the total energy generated worldwide is being used. In the United States and also the EU, buildings now account for over 40% of the total energy consumption *(5)*. Energy-efficient buildings, construction materials, windows, and vehicles are gaining significant importance today *(6–12)*. The concept of a zero-energy building (ZEB) was first mentioned in 2000 and became a mainstream idea already in 2006 *(13)*. Technologies for enabling widespread heating and also cooling-related energy savings in buildings through reducing the thermal emittance of glass surfaces have a much longer history, dating back to at least the early 1970s *(14,15)*. Since then, a large number of research works have been dedicated to achieving continually improved control over the various performance aspects of modern energy-efficient coatings and window glazings, such as their visible light transmission, solar heat gain coefficient (SHGC), thermal insulation U-value, the ability to control window tint actively or passively,

Energy Saving Coating Materials
http://dx.doi.org/10.1016/B978-0-12-822103-7.00003-0

and even the suitability for use in building-integrated photovoltaic (BIPV) solar windows. Excellent reviews of the current state of research fields relevant to the materials science of thin films as well as advanced architecture and the cutting-edge novel trends in transparent heat regulating (THR) coatings, and also in window-integrated PV systems, have recently been reported *(16–20)*. Fig. 3.1 illustrates the extremely wide (and growing) range of existing and future application areas for advanced solar-control glass and energy-generating solar window systems, which are beginning to appear on the market *(21–26)*.

The scale of the worldwide annual flat-glass installations is huge (est. 9.2 billion square meters for 2016 *(27)*), of which a significant fraction is based on advanced glazings, giving rise to a combined market volume of multiple billions of dollars. Increasingly advanced coating and glazing system designs featuring multiple combinations of properties optimized simultaneously, from reflected color to the required SHGCs and thermal insulation U-values desired in a large range of climate-specific, or industry-specific application types, continue to appear on the market. This is reflected in the ever-growing number of entries in the international glazing database

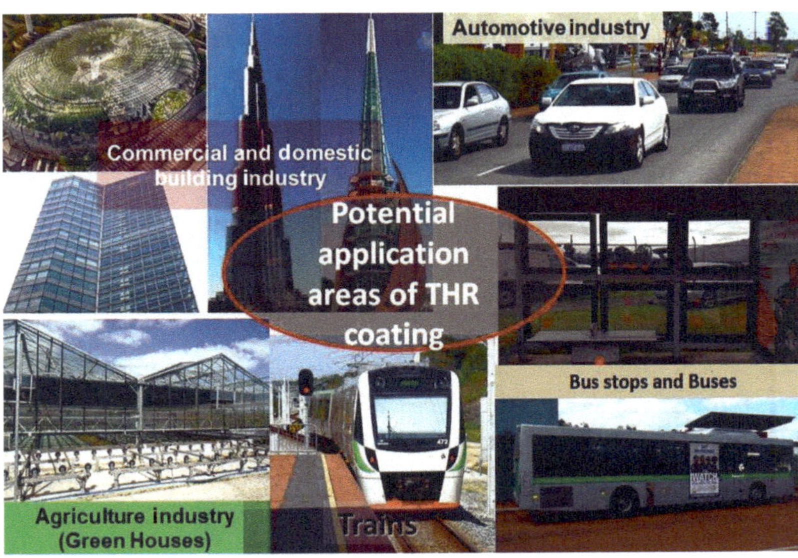

Figure 3.1 Modern architectural glass and coating products and their key applications, ranging from airport domes to advanced greenhouses and window-powered public infrastructure installations. A combination of energy savings, solar radiation control, and on-site distributed generation of electricity can be enabled (or assisted) by spectrally selective coatings of heat-mirror design type.

(IGDB *(28)*). *Optics* software package *(29)* distributed by Berkley Lab at Lawrence Berkley National Laboratory (LBNL) can be downloaded and used in conjunction with IGDB datasets to retrieve and analyze the datasets on the spectral transmission and related properties of all coated glazing types listed within IGDB.

Of particular importance is the manufacturability of environmentally stable and physically durable advanced low-emissivity (low-e) coatings for glass industry, which rely on nanoscale layer's quality and long-term stability in ultrathin metal (usually Ag) layers. Despite their long history of use in industrial low-e coatings, several questions remain unanswered in the areas related to the layer growth morphology, and especially the compatibility of Ag layers with surrounding materials, which limits the freedom of design in functional coatings as well as affects their thermal and environmental stability. This chapter describes the present status of advanced glazing systems development and outlines the associated challenges, both already addressed, and the ones still outstanding from the present point of view. Examples of existing technologies and their energy-saving potential are discussed, as well as new product development opportunities in areas related to energy-saving glazing systems and solar window technologies. Additional application areas of advanced coatings include new transparent BIPV products, which rely on the tailored spectral reflectivity profiles with high spectral selectivity, and also require improved thermal and environmental stability.

Leading industrial manufacturers of advanced low-e-coated glass (e.g., Viracon, Inc., Guardian Industries, Asahi Glass, and others) have developed and marketed a growing range of high spectral selectivity low-e products, which often feature high visible transmission (approaching 75%–80%) across a broad spectral band covering most of the visible range. Practically, all (or at least most) of these commercial high-transparency THR coatings rely on using sputtered multilayer metal-dielectric (MD) material systems containing two thin layers of silver. This is because achieving the neutrality of apparent color and simultaneously very high visible light transmission across the visible range is more difficult with gold layers, or other possible metallic constituents. Surrounding dielectric materials vary within some range (around a dozen) of dielectrics, which have so far been accepted widely in production environments. Yet, the overall range of the available optical thin-film coating materials (and their combinations) so far unexplored in industrial glass manufacturer settings is still very broad. This provides the motivation (and also an opportunity) for the lab-based academic experimentalists in research environments to design and develop new and unconventional types of advanced THR coatings, which can rely on a much broader range of the available material combinations than are in common use today. The generic challenges now faced by the manufacturers of advanced low-e coatings still remain largely the same as in the past—new coating products possessing very high chemical, environmental, and mechanical stability simultaneously with application-specific and highly customized optical properties are highly desirable in many applications.

It is interesting to observe this still young, yet already mature, research field of THR materials and coatings evolve continuously, while always expanding the range of their possible applications, from broadband mirrors and customized heat shields toward the most advanced of construction materials, which are now enabling global-scale energy savings in built environments. The same THR coatings, which used to provide only a heat-reflecting surface, can also be used as core components in the emergent categories of transparent and semitransparent solar windows, constructed either as concentrators or patterned cell modules. The following sections within this chapter (as well as the next chapter) deal with the materials science and technology-related aspects of THR materials and coatings in more detail; some examples of the specific design approaches and the emerging applications of these material systems are also provided.

2 THR materials and coatings—The current status and challenges

In order to design and manufacture the high-performance solar and thermal control coatings, the MD thin-film material system used as an optical interference filter needs to be optimized carefully in terms of the following main performance indicators:

- The intercompatibility of all material layers used in terms of their mutual adhesion behavior, mechanical stress robustness, achievable interface quality, and chemical stability in each other's proximity.
- The environmental stability and robustness of all individual layers used as well as that of the entire coating system.
- The quality of first coating layer adhesion to the substrate, especially considering that low substrate temperatures are generally required for the deposition of MD coatings.
- The mechanical robustness, required minimum thickness, and scratch resistance of the last coating layer (outer protection layer).

- The stability of designed coating's spectral properties with respect to small thickness control errors as well as refractive index errors, which are inevitable in real deposition processes.
- The ability of the manufactured coating system to withstand ambient temperature variations within the approximate range of −50 °C to 200 °C, as well as realistic ambient humidity variations and solvent contact. These properties are important even if the coating is to be protected by its embedded placement.
- The chemical stability of ultrathin metal (typically silver) layers during and after the deposition process, which depends on the chemical nature of the surrounding layers, and the possibility of protecting metallic layers from rapid oxidation and longer term out-diffusion or reactivity.
- The morphology and dielectric function character of ultrathin (~5–15 nm in thickness) metallic layers deposited, and its dependency on the deposition process parameters as well as on the chemical nature of materials that surround the metallic layers.

2.1 MD material system development for THR coatings—The present trends and past developments

The "low-e" coatings represent the most common type of THR coatings industrialized so far, and these are typically structured as MD thin films deposited by various methods onto flat glass *(30)*, which are designed to promote the strong and very wide-band reflection of thermal infrared wavelengths, and this prevents heat leakage from inside of buildings into the environment. A comprehensive review of the technologies, processes, and the various experimental methodologies and techniques relevant to the manufacture of various optical coatings on glass substrates can be found in ref. *(31)*.

In order to ensure an extended high-reflection region from the near-IR well into the thermal infrared range, the presence of metal-based reflective layers is necessary within the material system of coatings. Therefore, MD coatings have been studied extensively over the last several decades, mainly for the applications in buildings and construction industry. Current industry trends tend to strongly favor THR coatings, since there is significant regulatory momentum dedicated globally to ensuring that new buildings to be constructed after 2020 are designed with the goal of providing net zero energy consumption *(32)*. Advanced THR coatings of high spectral selectivity and very low thermal emissivity are, in conjunction with the modern designs of glazing and window framing systems, at the core of the diverse group of technologies necessary to realize this NZEB goal. This is because modern construction practices rely increasingly on using flat glass products throughout the outer surface areas of building envelopes. The core advantage of this all-important high reflectivity throughout the IR range (often coupled with the required high degree of VIS transparency and/or precise control over the apparent color) is due to the enabling thermal and also electric energy savings in buildings by providing low thermal emissivity (low-e) properties. The ultra-broadband high reflectivity of ultrathin nanoscale plasmonic metal (such as Au or Ag) layers in the near-IR and also the far-IR (thermal) regions relies, in part, on their very high extinction coefficients (the imaginary part of complex refractive index), which also increase with wavelength, throughout the entire IR range. This is due to the excitation of surface plasmons (collective oscillation modes of the electron gas excited by the absorbed photons) in the surface regions of plasmonic metal layers. The high reflectivity at long wavelengths is caused by the high electrical conductivity of the metal and this, in turn, by the number and mobility of its free s-electrons *(14)*. A necessary condition is that the film is not made up of isolated "islands" of plasmonic metal, but instead, it is homogeneous and continuous *(14)*. Silver behaves as a free electron medium in the far infrared (thermal radiation) range and its reflectivity is in excess of 96%–98% for wavelengths above several microns. In THR coatings, both the plasmonic-metal properties intrinsic to the metallic material type used and the optical interference effects occurring within nanoscale multilayers are harnessed to provide a combination of high visible-range transparency and ultra-broadband effective reflection of heat. Strong reflection is not possible from non-contiguous metallic layers, especially if metal islands are separated widely across the layers, which can happen due to a variety of factors, such as growth morphology problems, thermal exposure, or chemical reactivity.

Most of the above-listed coating performance criteria are generally more difficult to fulfill when depositing MD coatings, when compared to dielectric-only multilayer films. The ultrathin metal layer morphology quality criterion is known academically to be a rather difficult one, especially when silver layers are deposited onto the preceding layer of an oxide nature. This is due to the thermodynamic surface energy-related intermaterial compatibility issues affecting the substrate-surface "wettability" properties. It has been reported by multiple literature sources and has also been confirmed by our team (at ESRI, ECU) that ultrathin layers of silver tend to agglomerate on top of many oxide-material surfaces, forming nanoparticles-type coverage rather than uniform, continuous, and roughness-free layers, until a minimum thickness of about 20 nm is reached (Fig. 3.2 and refs. *(33–35)*).

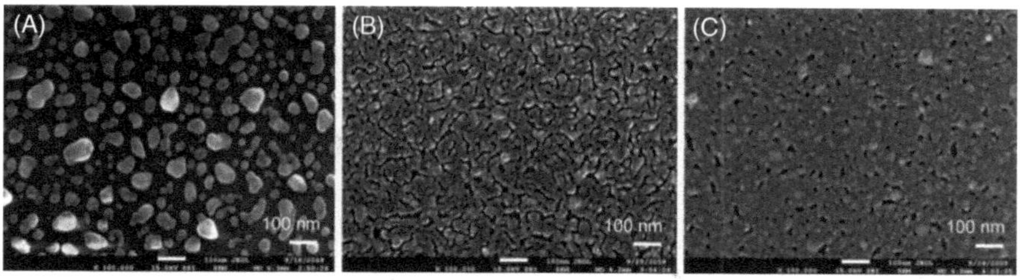

Figure 3.2 Morphologies of ultrathin silver layers at different growth stages and thicknesses deposited onto a fused silica substrate. (A) After 11.2 s of growth; (B) after 12.8 s; (C) after 16.8 s. *The images and corresponding descriptive data were taken from* (35), *and are reproduced with permission from the copyright owner and the authors.*

More information on the various ultrathin layer growth modes and their features is available from ref. *(36)*.

This special behavior of ultrathin silver layers (which are superior in terms of the dielectric function properties to all other metals when used in heat-reflecting MD systems) is notably problematic, and occurs regardless of the choice of deposition technique, if the metal layer thickness is less than about 20 nm. Ultrathin metal thicknesses are, however, required, if the visible transparency of coating is crucial.

There are solutions to improve or control the ultrathin silver morphology, typically involving either the use of additional ultrathin "seeder," or "wetting layers" *(37–39)*, or the avoidance of silver-oxide interfaces by design. Both approaches have been explored by many researcher groups worldwide. Oxide materials remain among the most popular dielectric constituents in dielectric-metal-dielectric (DMD) films; titanium dioxide (TiO_2), silicon dioxide (SiO_2), zinc oxide (ZnO), tin oxide (SnO), and aluminum oxide (Al_2O_3) are the dielectrics most often used in glass coatings *(40,41)*. The use of nickel oxide (NiO) and also the indium tin oxide (ITO) has also been reported *(42,43)*. The important advantages of using oxide materials include a wide range of available refractive indices as well as very high transparency across the visible and near-infrared spectral ranges. One disadvantage, which affects most oxide material types RF deposited from ceramic targets, or thermally (or e-beam) evaporated in the absence of oxygen, is related to the oxygen loss occurring during the deposition process, which leads to reduced transparency *(44)*.

Diffusion processes can affect the quality and especially the long-term stability of MD interfaces, since the metal concentration gradient drives the diffusion of metal atoms into the dielectric layers at elevated temperatures *(45)*. In order to protect MD interfaces from the diffusion-driven degradation, additional ultrathin "barrier layers" can be deposited to surround the ultrathin Ag, composed either of chemically stable metals such as Ti or

NiCr, or in some cases, oxide materials selected to provide a barrier-type effect through minimized diffusion coefficient.

An excellent overview of the large range of optical materials and deposition techniques used in the fabrication of the MD films of solar control type is presented in ref. *(39)*, where a discussion on the use of the so-called chromatic coated-glass materials is also included. These fall into several categories, such as photochromic glasses and polymers, thermochromic metal oxides, and electrochromic materials. Achieving passive (temperature-driven), or active (voltage regulated) control over the color and transparency of THR coatings on glass remains among the most progressive directions of industrial research in the glass industry today.

The nature of the multiple and often conflicting materials-related requirements faced by the designers of MD coatings led to the fact that still only a relatively limited number of low-e glass products possessing a high spectral selectivity properties are available on the market today. Among these, very few are environmentally stable in the long term, and therefore require protected storage conditions, especially with respect to the ambient humidity. Fig. 3.3

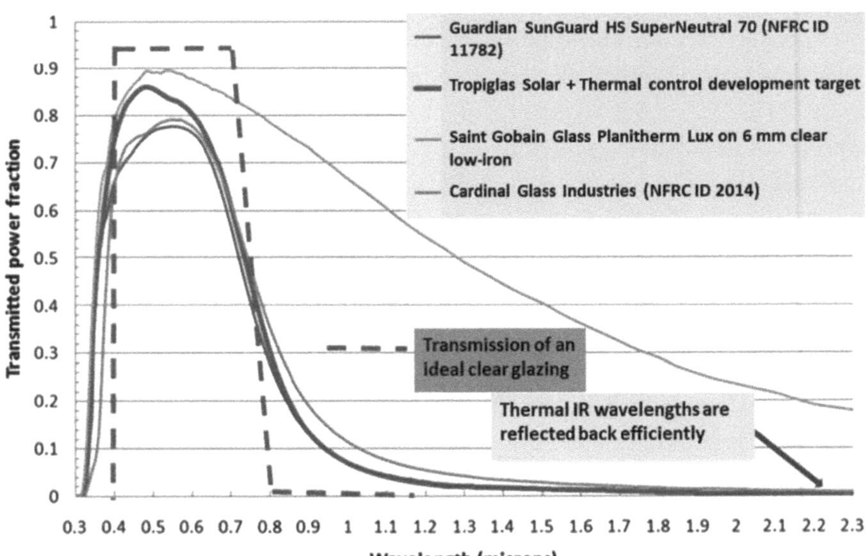

Figure 3.3 Transmission spectrum of an idealized energy-efficient clear glazing and also that of a "Tropiglas Advanced Glazing System" developed by Edith Cowan University (ECU) for Clearvue Technologies Ltd. in 2013. The transmission spectra of two similar commercial coated-glass low-e systems of similar type are also shown. *The spectral transmission data of several industry-leading coatings available today were obtained from IGDB and processed using* Optics *software* (29,30).

illustrates the "idealized" spectral response of high-performance THR coatings in comparison with several available THR products, and also the current developmental version of a future THR product manufactured by our team at ECU since 2013 (but only in small-area samples so far), for use in future advanced glazings being developed by Clearvue Technologies Ltd. (Perth, Australia).

Double-Ag DMD multilayers are probably the best coated-glass systems of today for providing the maximized "LSG" (light to solar gain, the ratio of visible light transmission to the SHGC) parameter, which is of importance for designing energy-efficient buildings in hot and tropical climates. Due to their strong reflectivity across the nonvisible solar wavelengths, these THR coating types are also attractive for use as spectrally selective back reflectors, which stimulate the internal redistribution of nonvisible solar energy in transparent concentrator-type solar window systems.

Advanced low-e coatings, coupled with other glazing system design features (e.g., embedded air gaps, vacuum spacers, specialized glass types, etc.), enable significant savings of electric energy expenditures for the heating, ventilation, and air conditioning (HVAC) in most climate types, from tropical to northern *(5,17,46)*. This is, in part, because the high reflectivity of these coatings across the entire IR wavelength range (solar IR and also thermal IR) leads to both keeping heat inside buildings in winter and preventing a large fraction of the incoming solar heat from entering buildings. Additionally, when used in conjunction with other window features, for example, air-filled spaces, a wide range of glazing U-factor values can be designed, if using quality THR coating products. Thus, advanced low-e THR coatings give the glazing system designer a significant degree of control over the required optical and thermal insulation properties of windows. The ways in which these properties can be adjusted through the design of glazing system construction are considered in more detail in Section 3.

2.2 All-dielectric, highly stable heat-mirror coatings for specialized applications and harsh environments

Transparent heat mirrors can also be designed using common dielectric thin-film materials only. A number of applications (from laser mirrors to solar windows) exist, in which very high thermal and chemical stability is required from a THR coating, and thus the all-dielectric system designs are favored, especially if the filtering of only near-infrared radiation wavelengths needs to be controlled. It is interesting to consider both a theoretical design example of a special (all-dielectric) class of THR coatings and the practical advantages/performance of thin films of this type.

We designed and modeled several 3-material multilayer systems demonstrating the potential for achieving very high visible-range transparency concurrently with very high reflectivity in the near-infrared optical range, with a very steep spectral transition step between the visible (VIS) and near-IR regions. Such designs are of interest for spectrally selective mirrors usually employed in transparent solar windows *(23,47,48)*, which collect the radiation energy incident onto glazing systems and reroute a part of this energy toward the near-edge mounted solar cells, using mechanisms, such as luminescence, scattering, diffraction, total internal reflection (TIR), or a combination thereof.

When modeling the transmission and reflection spectra of light entering the multilayer from air and exiting into a glass substrate, the following simplified description of heat mirror (edge filter) design of interest can be used. We neglected the wavelength dispersion of refractive indices in this calculated design example, presuming that ZnS, MgF_2, and Y_2O_3 are used as optical materials of constant refractive index 2.34, 1.38, and 1.9 accordingly. The absorption effects are also neglected across the wavelength range of interest, so we can simplify the design description (using standard terminology and notation from the field of thin-film design) by using the following parameter set and design structure:

n(Substrate) = 1.0; n(Exit Medium) = 1.51; n(H) = 2.34; n(M) = 1.9; n(L) = 1.38;

Base design wavelength = 860 nm.

The following software-optimized design structure inspired by the multilayer coating design approaches described in ref. *(49)* results in an impressive broadband edge-type filter of the heat mirror type:

S{1.1}(L/2HL/2)1{1}(L/2HL/2)5{1.25}(L/2HL/2)1{0.57}(LMHML)8{0.642}(LMHHML)8{1}(L/2)1

In this formula describing a 79-layer film of total physical thickness 8.48 μm, the design formula notation reported in ref. *(49)* is also used. In the above notation, **S** stands for the substrate thus defining the order of the layer deposition sequence, whereas the indices within curled brackets define the design-wavelength modification (scaling) relative to the base design wavelength applicable to within the layers in the immediately following group of round brackets. The number following the group of layers within round brackets is called the repetition index, or the number of times the deposition sequence from within the preceding round brackets gets repeated *(47)*.

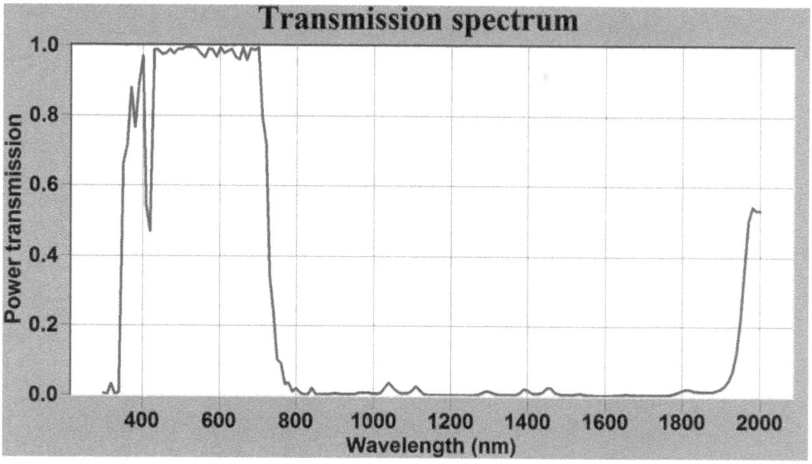

Figure 3.4 Software-modeled optical transmission (UV-NIR) spectrum of a triple-material edge-filter coating of a 79-layer structure described by formula S{1.1}(L/2HL/2)1{1}(L/2HL/2)5{1.25}(L/2HL/2)1{0.57}(LMHML)8{0.642}(LMHHML)8{1}(L/2)1, based on non-absorbing materials of constant refractive indices 1.38, 1.9, and 2.34.

The modeled transmission spectrum of this theoretical edge-filter design (which can be obtained by entering the film design data into any thin-film modelling software of choice) is shown in Fig. 3.4.

This remarkable 3-material thin-film filter demonstrates a unique combination of high VIS transparency combined with very high spectral selectivity, leading to the efficient reflection of the incident solar radiation heat (near-IR energy) in the spectral range between ~800 nm and 2000 nm. Practical all-dielectric heat-mirror designs, however, cannot extend their spectral reflectivity range much further into the far-IR thermal range. At the same time, high reflectivity in the spectral range up to about 40–50 μm is required for low-e building glass applications, which is achieved by using MD filter design approaches.

We have developed, at ESRI-ECU (and prototyped at an external manufacturer back in 2011), a specialized 96-layer all-dielectric, chemically stable, >95% transparent (VIS) THR coating on a fused silica substrate capable of withstanding in excess of 1 h at 600 °C in a conventional annealing box furnace. The measured transmission spectra of the as-deposited and also the thermally processed film samples are shown in Fig. 3.5. The optical materials used in this heat-mirror coating design were amorphous SiO_2, Ta_2O_5, and Al_2O_3 layers deposited by thermal evaporation.

This remarkably stable thin film of total thickness near 12 μm was designed to efficiently reflect the near-IR solar radiation within the spectral

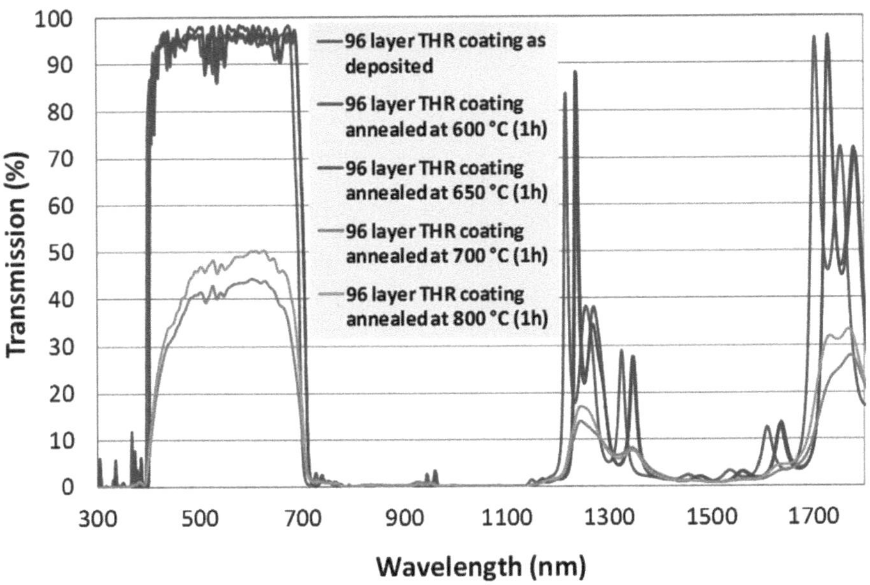

Figure 3.5 Spectral transmission and thermal exposure stability properties of a 96-layer all-dielectric heat-mirror coating developed by ECU.

responsivity band of $CuInSe_2$ (CIS) solar cells, for use in a luminescent solar concentrator (LSC)-type solar window application featuring edge-mounted CIS cell modules collecting the near-IR light propagated from a fraction of clear transparent glass area adjacent to cells. Multiple internal reflections of near-IR solar radiation light of wavelengths between about 700 and 1200 nm occurring within this window sample of size 200 mm × 200 mm (Fig. 3.6) were promoted by way of applying the coating of Fig. 3.5 to the back surface of a glass-based luminescent concentrator *(47)*.

In this early (2011) prototype of a highly transparent energy-generating solar window, the all-dielectric THR coating contributed to the near-IR energy collection by solar cells, by working in conjunction with a large-area diffraction grating and several inorganic luminescent pigments incorporated into internal structure of glazing (epoxy-based interlayer). The presence of this THR coating has contributed in excess of 30% of the total electric power output at peak, which reached about 615 mW (V_{oc} = 18.5 V, I_{sc} = 64 mA, FF = 0.52 using a system of four parallel-connected CIS cell modules of nominal efficiency 9.19%).

Additional benefit of this all-dielectric coating design type for window-type applications is related to practically zero absorption of light within the coating's own materials, whereas in most metal-dielectric THR designs of high spectral selectivity, significant absorption occurs in the spectral region

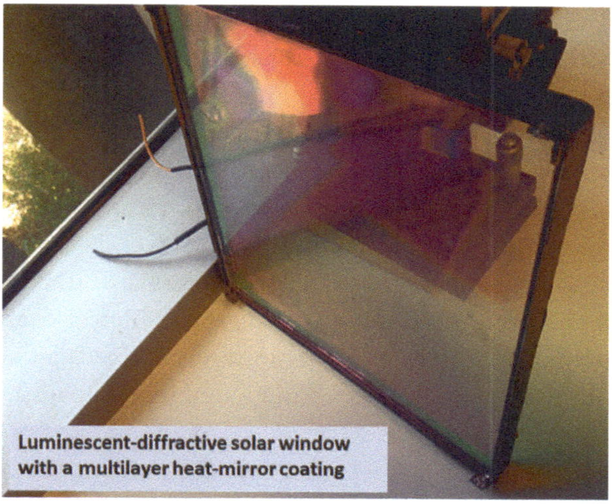

Figure 3.6 A 200 mm × 200 mm semitransparent solar concentrator window sample, in which a microstructured diffractive element is used in conjunction with inorganic luminescent materials and a THR coating of Fig. 3.5, to redirect a part of the incident solar radiation toward the edge-mounted strip-shaped $CuInSe_2$ solar cell modules.

near 750–800 nm, where the transition from the high transmission (HT) region to high reflection (HR) takes place.

3 Control over the solar heat gain, visible transmittance, and thermal insulation quality in advanced glazing systems for modern buildings reliant on THR coatings

The prospects of achieving, through the glazing system design, the best possible year-round energy-saving performance in buildings are governed by factors such as the geography (type of climate), THR coating type, and the engineering of glazing structure. Crucial for maximizing the energy efficiency is the performance of THR low-e coating(s), in terms of their optical transmission and reflection properties, regardless of climate type. Usually, the goals of glazing system design are related to providing, to the best extent possible economically, the required combination of visible light transmission and external reflected color, in conjunction with the best achievable thermal insulation (low U-factor). The main parameters of importance for the characterization of any glazing system are SHGC, visible light transmission, light-to-solar gain ratio (LSG = VLT/SHGC), and the rate of thermal energy exchange per unit time, glass area, and temperature difference (U-factor). Solar heat gain properties are also very important, but these can vary in a rather broad range, adjustable dependent on the climate type and building wall orientation. The best results are typically achieved, in simple glazing systems not requiring vacuum spacers or multiple coatings, by incorporating high spectral selectivity low-e coatings into a double-pane window system. The glazing design procedure usually starts from selecting a suitable product from the IGDB *(28)*, followed by a close inspection of its optical properties using *Optics* software *(29)*. The optical properties of interest are the transmission, front reflection, and back-reflection spectra. For custom newly developed coatings, an additional IGDB entry can be generated for modeling purposes, following their optical characterization by spectrophotometry, by way of preparing a formatted data file containing the optical data of any given substrate-film combination. Fig. 3.7 shows the

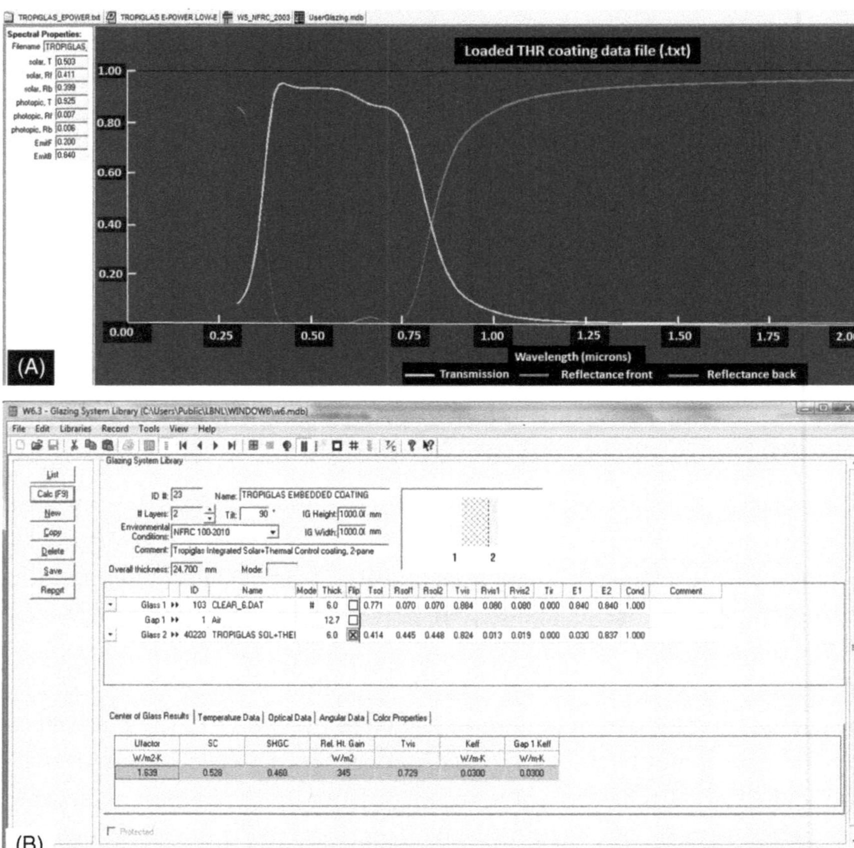

Figure 3.7 (A) Transmittance, reflectance, and back-reflectance data of an ECU-THR coating on a clear low-iron glass substrate across the visible and broader infrared range (up to 40 μm) imported into *Optics6* software for modeling of the glazing system properties. (B) *WINDOW 6.3* Software model and its main calculated output parameters for a double-pane air-filled window system using an ECU-THR coating.

graphical result from *Optics6* software obtained after importing the optical properties data file of a "ECU-THR"-coated glass. The spectral transmission of this coated glass type developed at Edith Cowan University was also shown among the data traces presented in Fig. 3.3 in Section 2.1.

After running the model calculations, which account for both the substrate and the coating properties, *Optics* software allows exporting a specially formatted "glazing system file" into a database, for use with other related software packages, for example, *WINDOW (50)*. This software allows standardized modeling of the optical properties, as well as the solar heat gain-related and thermal insulation parameters for arbitrary glazing system configurations composed of a sequence of coated or uncoated glass substrates separated by air, gas-filled, or vacuum spaces.

Apart from the visible light transmission and the glazing's apparent color, which are of practical and also aesthetic importance, the most important energy efficiency-related parameters of any window type are the U-factor, SHGC, and relative heat gain (RHG). U-factor allows direct energy-efficient comparisons between the various insulating surface types (e.g., brick walls, roofing materials, simple glass surfaces, and custom-designed coated glazing systems of varying complexity). RHG is the amount of heat gain through a glass product taking into consideration the effects of both the solar heat gain (sometimes expressed alternatively as shading coefficient, linked to optical transmission) and the conductive heat gain (linked to the U-value). The lower the RHG, the greater the glass restricts heat gain. Table 3.1 shows a comparison dataset for the thermal insulation performance and also the heat transmission properties of several example glazing system design types, calculated using *WINDOW* software.

It is interesting to note that these important thermal insulation-related and also the solar radiation-shielding properties of glazing systems can be varied within a rather broad range by way of selecting different THR products and system configurations. At the same time, achieving the U-factor values near 1 $W/(m^2 \star K)$ and below (which is necessary for ZEB windows in most climates) requires specialized design approaches, such as building triple-pane, inert gas-filled glazings, or using vacuum spaces, also in conjunction with using industry-leading THR low-E coatings.

4 Energy generating THR window glazing systems

Energy generation (or energy harvesting) has not traditionally been associated with windows, or any glazing products, until recently. Various approaches to the incorporation of photovoltaic (PV) systems into building

Table 3.1 Thermal performance data comparisons made for several glazing system configuration types using WINDOW software package *(50)* and the spectral data of example THR coatings.

Window system types	U-factor (W/(m²*K))	SHGC	RHG (W/m²)
Single-pane 6 mm thick low-e THR coated glass ("ECU-THR"; this coated glass had LSG = T_{vis}/SHGC = 1.898)	3.202	0.434	332
Saint-Gobain Glass (SGG) Planitherm Lux on 6 mm clear glass (IGDB ID 11015)	3.386	0.738	554
Plain uncoated low-iron clear glass, 6 mm (IGDB ID 103)	5.808	0.818	634
Double-pane ECU Solar Control Low-E ("ECU THR") coated 6 mm clear low-iron glass	1.818	0.421	317
Double-pane plain-glass (IGDB ID 103) with a 1/2" air gap	2.684	0.704	533
SGG Planitherm Lux (placed onto internal side of ½"air gap (surface 3) of a double-pane system)	1.739	0.659	491
ECU Solar Control Low-E ("ECU THR") coating on low-iron 6 mm, placed onto surface 3 inside a double-pane system with a ½" air gap	1.639	0.46	345

SHGC, solar heat gain coefficient; *RHG*, relative heat gain in W/m^2. All air gap widths used in the calculations were standardized to ½" (12.7 mm) for all double-pane systems.

envelopes have started being explored actively in the last decade, leading to a new field of BIPV. Historically, BIPV installations started as façade- and wall-integrated conventional (Si or $CuInSe_2$) PV modules occupying the building envelope areas other than roof surfaces, and continued toward the development of semitransparent, glass-integrated PV systems using patterned amorphous-silicon modules, or dye-sensitized solar cells, for example *(19,20,25,26)*.

More recently, the fields of LSCs, and also the development of semitransparent organic solar cells and perovskite-based solar cell materials have started showing the signs of a "renaissance" *(51,52)*, driven by the opportunities to capture the growing markets in both the distributed electricity generation and also in advanced construction *(18,19,23,26,48)*. Windows and glazing systems, despite being a practically ancient technology at its core, are now starting to be viewed and recognized as being the renewable energy platform and resource of the near future *(26)*. This potential for enabling extremely widespread energy harvesting at the point of use is practically unparalleled, considering the entire history of industrial energy generation, and is expected to develop hand-in-hand with the growing use of THR coatings for energy-saving applications. Expanding the worldwide energy generation facilities into the almost-uncharted, yet vast territory of glass and windows will require significant time, research efforts, broad-based investments, and long-term strategic thinking on behalf of the governments and private companies. This is due to the fundamental and crucial ways, in which the energy industry differs from all other industries, which has been underscored in a recent publication *(53)* by the Breakthrough Energy Coalition, authored by B. Gates. Advanced spectrally selective THR coatings are well-positioned to play a crucial role in enabling a synergistic and fortunate combination of the two core window functionalities required to transform the glass industry—energy savings simultaneous with energy harvesting. This is because the spectral components of solar energy reflected off these coatings (UV and near-IR) can either be further routed toward the near-edge-mounted PV modules (in LSC-type concentrator windows) or propagated backward through the semitransparent glass area-based PV, promoting improved absorption within the semiconductor materials.

Fig. 3.8 details the potential for harvesting of the nonvisible solar radiation components using THR coatings. The transmission spectrum of an ECU-THR coating on a low-iron clear glass substrate is shown as an example of enabling both spectral filtering of the transmitted solar energy density spectrum and efficient reflection of up to 89% of the incident solar IR radiation. A part of this energy reflected off the coating can then propagate toward the solar modules installed near glass panel edges, if other mechanisms, for example, luminescence, Mie scattering, or deflection due to diffraction, are also present, stimulating partial light trapping through a series of TIR events at the front-side glass–air interfaces.

While fundamental factors limiting the efficiency of light trapping in refraction-based concentrators (such as the Yablonovitch limit and etendue conservation theorem *(54)*) do invariably affect the process of internal light redistribution in planar window-type concentrators, the addition of a heat-mirror coating alone helps "lock" the IR energy escape pathways from at least one surface. Dependent on the concentrator area and its construction type, the addition of ECU-THR, or similar (e.g., VNE 2463 by Viracon, Inc.) advanced THR coating alone to an LSC-type solar window system

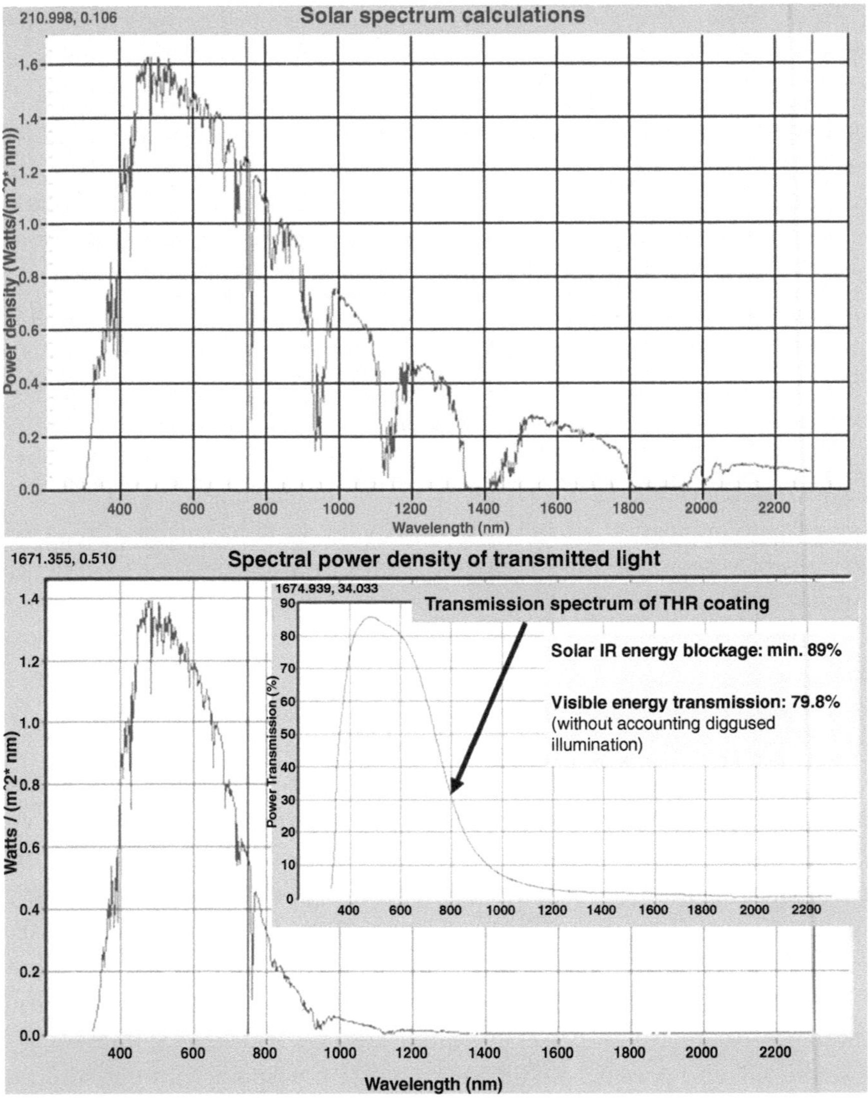

Figure 3.8 Spectral filtering of the normally incident standard (AM1.5G) transmitted solar irradiation enabled by the ECU-THR coating. The results of integrating the spectral power density numerically show that transmission of up to ~89% of the incoming solar IR energy is blocked, thus preventing the overheating of internal building spaces.

using inorganic luminescent materials *(23,48)* leads to generating up to a third more electric power. Fig. 3.9 shows examples of transparent solar windows prototypes constructed by the ESRI (ECU) team, which used either the ECU-THR or the VNE 2463 coatings placed at the last surface of each glazing system.

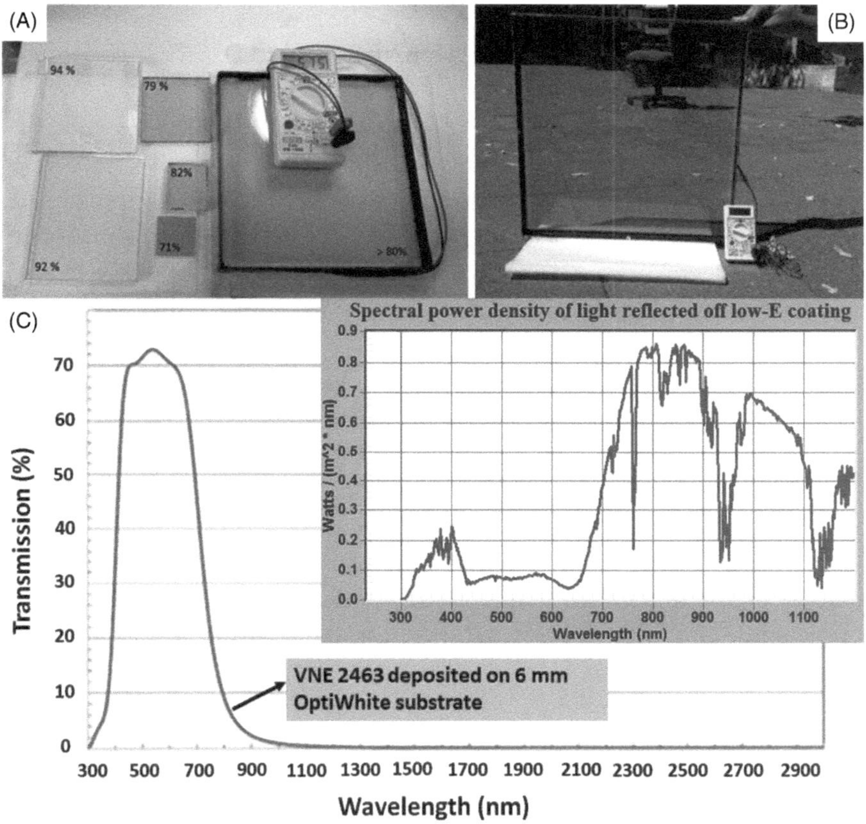

Figure 3.9 (A) Visual appearance of the various THR-coated, or uncoated (e.g., acrylic sheet, 94% transmittance vs low-iron ultraclear glass, 92%) glazing systems, and a 200 mm × 200 mm ECU-THR-coated transparent solar window prototype of visible light transmission > 80%; (B) A 500 mm × 500 mm THR-coated (VNE 2463, Viracon Inc.) solar window prototype of visible light transmission near 65% with edge-mounted stripe-shaped Avancis, Inc. $CuInSe_2$ solar cell modules; (C) Measured transmission spectrum of Viracon, Inc. VNE 2463 THR low-E coating on a low-iron clear 6 mm substrate; subset shows the spectral power density distribution of standard (AM1.5G), normally incident solar radiation reflected off this coated glass substrate *(48)*.

$CuInSe_2$ (CIS) solar cell modules we selected for use in this application for multiple reasons are described in ref. *(21)*, one of the main reasons being their spectral response extending well beyond 1100 nm, which is particularly attractive, considering the availability of significant solar energy incoming between 1100 nm and 1300 nm, which is also reflected efficiently without strong absorption by THR coatings. The CIS modules used in the experiments were produced, using a proprietary connectorization and re-encapsulation processes, from Avancis, Inc. modules of ~1 m^2 area.

We have sliced the active areas of large-area CIS panels to reshape these into strip-shaped narrow modules, each containing multiple internally series-connected CIS cells, to enable attaching these custom CIS modules to the glazing system edges.

A large range of large-area, highly transparent solar window prototypes of different internal structure design types have been designed, constructed, and characterized by our team in the recent few years. Several important examples of these are illustrated in Fig. 3.10, which have been reproduced from our recent publication *(23)*. Notably, incorporating either the THR-coated near-edge macro-deflectors, or THR-type "tilted embedded coatings" covering a macro-patterned, cm-scale periodic internal deflectors, or the wide-area microstructured elements, have all led to improvements in the efficiency of solar energy collection.

THR coatings of high spectral selectivity have been, in our team's experience, central to enabling both the energy-saving and the energy collection performance functionalities in transparent solar windows. THR coatings of multiple design types, from omnidirectional color-neutral heat reflectors, to simple edge-type heat-mirrors, have been found to assist in the spectrally selective management of light in solar windows, whether deployed across the entire glass area, or if only covering discrete light-deflecting structures embedded into glazing systems.

A number of characterization experiments were conducted by our research group, aimed at quantifying the transparent solar window performance in various operation conditions, and also aimed at identifying the most appropriate end-use applications, for either single framed windows or their combinations *(21,22)*. Fig. 3.11 illustrates the outdoor performance of a framed highly transparent solar window of glass panel size 500 mm × 500 mm employing the edge-mounted CIS solar modules only. Two different functional types of THR coatings were used in this solar window prototype: a Viracon VNE 2463 high-selectivity large-area heat mirror on an OptiWhite 6 mm ultra-clear substrate placed across the entire back-side window surface, and an all-dielectric omnidirectional near-IR reflector-type coating deposited by e-beam evaporation at ESRI (ECU) cleanroom. The purpose of that coating placed onto special wedge-type angled internal deflectors was to enhance the photon collection probability of the near-IR solar radiation at the edge-mounted solar modules.

The window sample shown in Fig. 3.11 generated about 3 W of electric output power when positioned vertically into a sun-facing configuration, at near-peak sunny weather. We measured about 43 V (Voc), up to ~115 mA

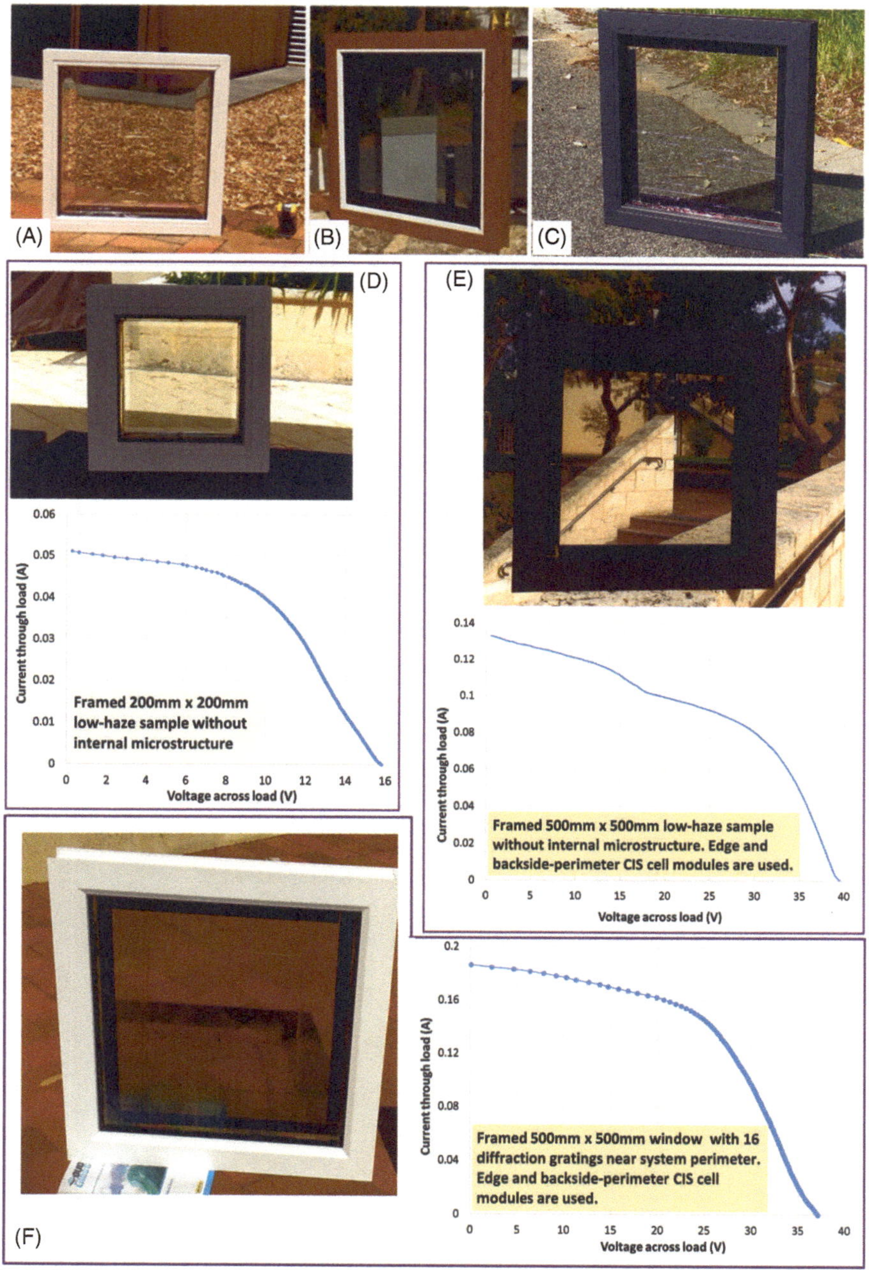

Figure 3.10 Several types of framed transparent solar window prototypes using glass panels of size 200 mm × 200 mm and 500 mm × 500 mm designed and fabricated by ESRI (ECU) researchers between 2014 and 2016 *(23)*. Different internal macro- and microstructures embedded into glass panel area (as well as solar cell modules) resulted in different power collection efficiencies. Sample PV *I*–*V* curve characterization results are also shown. (A–C) Various solar window samples of glass panel dimensions 500 mm x 500 mm containing either the internal-coated structures, or flat-glass panes only; (D–F) measured PV I-V curve datasets obtained from other solar window samples containing either flat glass, or added microstructured diffractive elements.

Figure 3.11 Solar window demonstration experiments aimed at visualization of the energy generation and use conducted at Edith Cowan University (ECU), Joondalup Campus, in 2016.

(Isc), at a module fill factor (FF) near 0.6, with the solar cell temperatures stabilized at above 45 °C.

Up to five series-connected DC (nom. 12 V) battery fans (without any batteries installed) were able to run when powered by a 500 mm × 500 mm transparent solar window prototype positioned vertically. The same window was also used to change the transparency of a glass-integrated polymer-dispersed liquid crystal (PDLC) panel of 1 m^2 area.

We demonstrated that multiple series-connected inductive loads such as DC motor-driven ventilation fans can be powered by this moderate-area solar window device, at a sufficiently fast blade rotation rate to generate audible noise heard from some distance away. Another interesting application opportunity of solar windows of similar types and size is related to actively controlling the window transparency in buildings, by way of applying either the on-site generated power or the energy stored in battery systems charged by windows, to the integrated liquid crystal-based transparencies.

5 SUMMARY

Transparent optical coatings designed to regulate the propagation of heat form a special class of functional materials, which continue to present unique opportunities and challenges in multiple industries, from modern construction to the energy generation in novel renewables and related technologies. Despite the relatively long history of research, development, and industrialization of THR materials, active research efforts are ongoing. New component substances, their combinations, and the optimized deposition techniques are continually being studied to further expand the suitability of the various THR coatings for the ever-growing range of applications. While the fundamental properties of known THR system components and their limitations remain the same, new and novel ways of overcoming the various practical technology-related difficulties in the use of these materials are being found continually. Due to this, we are at present witnessing a remarkable "quantum leap" in the widespread utilization of THR materials and coatings, by which they enter the previously uncharted territory of transparent energy-generating window development.

The following chapter delves into the treatment of multiple technical details related to the materials science aspects of DMD building blocks of THR multilayers, as well as the properties of advanced, recently developed THR coatings.

Acknowledgments

The authors are grateful to the Australian Research Council for their continued support of our research efforts (linkage grants LP130100130 and LP160101589). We also acknowledge the financial support from Clearvue Technologies Ltd. (Perth, Australia).

References

1. Capuano, L. International Energy Outlook 2018 (IEO2018), US Energy Information Administration. https://www.eia.gov/pressroom/presentations/capuano_07242018.pdf Accessed 26.09.2019.
2. Chu, S.; Majumdar, A. Opportunities and Challenges for a Sustainable Energy Future. *Nature* **2012**, *488*, 294–303.
3. Hoffert, M. I.; Caldeira, K.; Benford, G., et al. Advanced Technology Paths to Global Climate Stability: Energy for a Greenhouse Planet. *Science* **2002**, *298*, 981.
4. Poizot, P.; Dolhem, F. Clean Energy New Deal for a Sustainable World: From Non-CO2 Generating Energy Sources to Greener Electrochemical Storage Devices. *Energy Environ. Sci.* **2011**, *4*, 2003–2019.
5. Cao, X.; Xilei, D.; Liu, J. Building Energy-Consumption Status Worldwide and the State-of-the-Art Technologies for Zero-Energy Buildings During the Past Decade. *Energy Build.* **2016**, *128*, 198–213.
6. Long, L.; Ye, H. How to be Smart and Energy Efficient: A General Discussion on Thermochromic Windows. *Sci. Rep.* **2014**, *4*, 6427.
7. Gao, Y.; Luo, H.; Zhang, Z.; Kang, L.; Chen, Z.; Du, J., et al. Nanoceramic VO_2 Thermochromic Smart Glass: A Review on Progress in Solution Processing. *Nano Energy* **2012**, *1*, 221–246.
8. Llordes, A.; Garcia, G.; Gazquez, J.; Milliron, D. J. Tunable Near-Infrared and Visible-Light Transmittance in Nanocrystal-in-Glass Composites. *Nature* **2013**, *500*, 323–326.
9. Raman, A. P.; Anoma, M. A.; Zhu, L.; Rephaeli, E.; Fan, S. Passive Radiative Cooling Below Ambient Air Temperature Under Direct Sunlight. *Nature* **2014**, *515*, 540–544.
10. Wang, J.; Zhang, L.; Yu, L.; Jiao, Z.; Xie, H.; Lou, X. W., et al. A Bi-Functional Device for Self-Powered Electrochromic Window and Self-Rechargeable Transparent Battery Applications. *Nat. Commun.* **2014**, *5*, 4921.
11. Wang, S.; Liu, M.; Kong, L.; Long, Y.; Jiang, X.; Yu, A. Recent Progress in VO_2 Smart Coatings: Strategies to Improve the Thermochromic Properties. *Prog Mater Sci.* **2016**, *81*, 1–54.
12. Dalapati, G. K.; Masudy-Panah, S.; Chua, S. T.; Sharma, M.; Wong, T. I.; Tan, H. R., et al. Color Tunable Low Cost Transparent Heat Reflector Using Copper and Titanium Oxide for Energy Saving Application. *Sci. Rep.* **2016**, *6*, 20182.
13. Torcellini, P.; Pless, S.; Deru, M.; Crawley, D. *Zero Energy Buildings: A Critical Look at the Definition* National Renewable Energy Laboratory and Department of Energy: US, 2006.
14. Groth, R.; Reichelt, W. Gold Coated Glass in the Building Industry. *Gold Bull.* **1974**, 7 (3), 62–68.
15. Lampert, C. M. Heat Mirror Coatings for Energy Conserving Windows. *Solar Energy Mater.* **1981**, *6* (1), 1–41.
16. Dalapati, G. K.; Kushwaha, A. K.; Sharma, M.; Suresh, V.; Shannigrahi, S.; Zhuk, S., et al. Transparent Heat Regulating (THR) Materials and Coatings for Energy Saving Window Applications: Impact of Materials Design, Micro-Structural, and Interface Quality on the THR Performance. *Prog. Mater. Sci.* **2018**, *95*, 42–131.

17. Rezaei, S. D.; Shannigrahi, S.; Ramakrishna, S. A Review of Conventional, Advanced, and Smart Glazing Technologies and Materials for Improving Indoor Environment. *Sol. Energy Mater. Sol. Cells* **2017**, *159*, 26–51.
18. Traverse, C. J.; Pandey, R.; Barr, M. C.; Lunt, R. R. Emergence of Highly Transparent Photovoltaics for Distributed Applications. *Nat. Energy* **2017**, *2*, 849–860.
19. Shin, D. H.; Choi, S. -H. Recent Studies of Semitransparent Solar Cells. *Coatings* **2018**, *8*, 329.
20. Cornaro, C.; Basciano, G.; Puggioni, V. A.; Pierro, M. Energy Saving Assessment of Semi-Transparent Photovoltaic Modules Integrated into NZEB. *Buildings* **2017**, 7, 9.
21. Vasiliev, M.; Alameh, K.; Nur-E-Alam, M. Spectrally-Selective Energy-Harvesting Solar Windows for Public Infrastructure Applications. *Appl. Sci.* **2018**, *8*, 849.
22. Vasiliev, M.; Alameh, K.; Badshah, M. A.; Kim, S. -M.; Nur-E-Alam, M. Semi-Transparent Energy-Harvesting Solar Concentrator Windows Employing Infrared Transmission-Enhanced Glass and Large-Area Microstructured Diffractive Elements. *Photonics* **2018**, *2018* (5), 25.
23. Vasiliev, M.; Alghamedi, R.; Nur-E-Alam, M.; Alameh, K. Photonic Microstructures for Energy-Generating Clear Glass and Net-Zero Energy Buildings. *Sci. Rep.* **2016**, *6*, 31831.
24. Rosenberg, V.; Vasiliev, M.; Alameh, K. A Spectrally Selective Luminescence Concentrator Panel with a Photovoltaic cell. Patent EP 2 726 920 B1, **2017**.
25. Available from: https://www.forumforthefuture.org/greenfutures/articles/massive-solar-façade-swiss-convention-centre Accessed 02.05.2018.
26. Extance, A. The Dawn of Solar Windows, IEEE Spectrum. **2018**. Available from: https://spectrum.ieee.org/energy/renewables/the-dawn-of-solar-windows. Accessed 10.02.2018.
27. Schreuder, N. Glass Windows for Energy Efficient Buildings. AGC Glass Europe, 2014. Available from: www.iea.org. Accessed 05.10.2018.
28. International Glazing Database, Berkeley Lab, USA. Available from: https://windows.lbl.gov/software/igdb. Accessed 23.10.2018.
29. *Optics* Software Package, Berkeley Lab, USA. Available from: https://windows.lbl.gov/software/optics Accessed 23.10.2018.
30. Gläser, H. J. *Large Area Glass Coating* Von Ardenne Anlagentechnik GmbH: Dresden, Germany, 2000.
31. Pulker, H. K. *Coatings on Glass*, 2nd edition; Elsevier Science B.V: The Netherlands, 1999.
32. Schimschar, S.; Surmeli, N.; Hermelink, A. Guidance Document for National Plans for Increasing the Number of Nearly Zero-Energy Buildings. The European Commission. **2013**. Available from: https://ec.europa.eu/energy/en/topics/energy-efficiency/buildings/nearly-zero-energy-buildings Accessed 11.07.2018.
33. Campbell, C. T. Ultrathin Metal Films and Particles on Oxide Surfaces: Structural, Electronic and Chemisorptive Properties. *Surf. Sci. Rep.* **1997**, *27*, 1–111.
34. Doraiswamy, N.; Jayaram, G.; Marks, L. D. Unusual Island Structures in Ag Growth On Si(100)-(2×1). *Phys. Rev. B* **1995**, *51*, 10167–10170.
35. Bulir, J.; Novotny, M.; Lynnykova, A.; Lancok, J. Preparation of Nanostructured Ultrathin Silver Layer. *J. Nanophoton.* **2011**, *5*, 051511–51521.
36. Oates, T. W. H. *Metal Plasma Immersion Ion Implantation and Deposition Using Polymer Substrates, Chapter 7, PhD Thesis* University of Sydney: Sydney, NSW, Australia, 2003.
37. Chen, W.; Thoreson, M. D.; Ishii, S.; Kildishev, A. V.; Shalaev, V. M. Ultra-Thin Ultra-Smooth and Low-Loss Silver Films on a Germanium Wetting Layer. *Opt. Exp* **2010**, *18* (5), 5124–5134.
38. Ni, C. *Effects of Different Wetting Layers on the Growth of Smooth Ultra-Thin siLver Thin Films. Ph. D. Thesis* University of Dayton: Dayton, Ohio, USA, 2014.

39. Kanu, S. S.; Binions, R. Thin Films for Solar Control Applications. *Proc. R. Soc. A* **2010**, *466*, 19–44.
40. West, G. T.; Kelly, P. J. Improved Durability of Dielectric Coatings for Large-Area Applications on Glass via Ion Beam Pretreatment of the Substrate. *J. Vac. Sci. Technol. A* **2006**, *24*, 334–339.
41. Zhang, L.; Li, J.; Zhang, X. W.; Jiang, X. Y.; Zhang, Z. L. High-Performance ZnO Thin Film Transistors with Sputtering $SiO_2/Ta_2O_5/SiO_2$ Multilayer Gate Dielectric. *Thin Solid Films* **2010**, *518*, 6130–6133.
42. Al-Kuhaili, M. F.; Ahmad, S. H. A.; Durrani, S. M. A.; Faiz, M. M.; Ul-Hamid, A. Application of Nickel Oxide Thin Films in NiO/Ag Multilayer Energy-Efficient Coatings. *Mater. Sci. Semicond. Process.* **2015**, *39*, 84–89.
43. Lee, J. -H.; Lee, S. -H.; Hwangbo, C. K. Optical and Structural Properties of TiO_2/Ti/Ag/TiO_2 and TiO_2/ITO/Ag/ITO/TiO_2 Metal-Dielectric Multilayers by RF Magnetron Sputtering for Display Application. *J. Korean Phys. Soc.* **2004**, *44* (3), 750–756.
44. El Mouedden, Y.; Alghamedi, R.; Vasiliev, M.; Alameh, K. Thin Film Coatings for Solar and Thermal Radiation Control Prepared by Physical Vapour Deposition. In *In Proceedings of the 9th International Conference on High Capacity Optical Networks and Enabling Technologies (HONET)*; 2012Istanbul, Turkey. 12-14 December 2012.
45. He, M.; Lu, T. -M. *Metal-Dielectric Interfaces in Gigascale Electronics*, 157. Springer Series in Materials Science: New York, 2012doi: 10.1007/978-1-4614-1812-2_2.
46. Qu, J.; Song, J.; Qin, J.; Song, Z.; Zhang, W.; Shi, Y., et al. Transparent Thermal Insulation Coatings for Energy Efficient Glass Windows and Curtain Walls. *Energy Build.* **2014**, 77, 1–10.
47. Rosenberg, V.; Vasiliev, M.; Alameh, K. A Spectrally Selective Panel. Patent WO2013003894A1, **2013**.
48. Alghamedi, R.; Vasiliev, M.; Nur-E-Alam, M.; Alameh, K. Spectrally-Selective All-Inorganic Scattering Luminophores for Solar Energy-Harvesting Clear Glass Windows. *Sci. Rep.* **2014**, *4*, 6632.
49. Thelen, A. *Design of Optical Interference Coatings* McGraw-Hill Book Company: New York, 1989.
50. *WINDOW* Software Package, Berkeley Lab, USA. Available online: https://windows.lbl.gov/software/window Accessed 25.10.2018.
51. Mazzaro, R.; Vomiero, A. The Renaissance of Luminescent Solar Concentrators: The Role of Inorganic Nanomaterials. *Adv. Energy Mater.* **2018**, 1801903.
52. Reinders, A.; Kishore, R.; Slooff, L.; Eggink, W. Luminescent Solar Concentrator Photovoltaic Designs. *Jpn. J. Appl Phys* **2018**, *57*, 08RD10.
53. Gates, B. *Energy Innovation – Why We Need It and How to Get It.* Breakthrough Energy Coalition, 2015 Available from: https://www.google.com/url?sa(t&rct(j&q(&esrc(s&source(web&cd(1&cad(rja&uact(8&ved(2ahUKEwjQjvmfiaPeAhVWdCsKHVsMCbQQFjAAegQICRAC&url(https%3A%2F%2Fwww.gatesnotes.com%2F~%2Fmedia%2FFiles%2FEnergy%2FEnergy_Innovation_Nov_30_2015.pdf%3Fla%3Den&usg(AOvVaw1VZwJaefqQT1p4Qrm-UzYm.
54. Rau, U.; Paetzold, U. W.; Kirchartz, T. Thermodynamics of Light Management in Photovoltaic Devices. *Phys. Rev. B* **2014**, *90*, 035211.

CHAPTER 4

Dielectric/metal/dielectric (DMD) multilayers: growth and stability of ultra-thin metal layers for transparent heat regulation (THR)

Mohammad Nur-E-Alam, Mikhail Vasiliev, Kamal Alameh
Electron Science Research Institute, School of Science, Edith Cowan University, Joondalup, WA, Australia

1 Introduction

Dielectric/metal/dielectric (DMD) thin-film multilayers have been studied widely in the recent decades due to their extensive application potential in the fields of optics, photonics, plasmonics, and opto-electronics. Despite the relatively extensive history of these studies *(1–25)*, active research in this area is still ongoing *(21–34)*. In the modern-glass industry alone, the potential applications of these multilayer structures are numerous. Also, the sputtering of thin-film coatings on large-size glass substrates is still relying on materials systems researched decades ago. This is due to production-related economic factors, such as the high-cost of large-scale sputtering sources of unconventional materials, and also due to the ongoing need for the translation of recent academic research results (nano-engineered thin-film coating structures) into industry practices. Our recent review of the international glazing data-base (IGDB, *(35)*) as well as multiple energy-saving (Low-E) coated glass products available both locally in Australia and overseas (e.g., China) indicated that only a small number of these products possesses high spectral selectivity simultaneously with high visible light transmission (VLT). Moreover, the coated glass products possessing these desirable properties are typically very unstable environmentally and chemically, thus requiring installation into the humidity-controlled sections of advanced glazing products to be performed shortly after the coating deposition. Therefore, research efforts aimed at the development of materials systems addressing these issues in practical applications are essential.

DMD multilayer thin-film structures also can provide significantly lower electrical resistivity compared to the conventional conductive highly

Energy Saving Coating Materials
http://dx.doi.org/10.1016/B978-0-12-822103-7.00004-2

transparent oxide materials that are suitable for opto-electronic devices. In these types of multilayer structures, both the metallic and dielectric materials play important roles in achieving the combination of optical and electrical properties required for application-specific devices, for example, flat panel displays, liquid crystal displays, and organic light-emitting devices (OLEDs). For example, RF magnetron sputtered TiO_2/Ag/TiO_2 multilayer structures have been investigated by many groups, and it has been reported that both the optical and electrical properties were layer-thickness dependent *(36)*. The metal layer in the multilayer structure mainly plays a significant role in terms of not only controlling the optical transparency but also the electrical properties. The thicker the metal layer is, the lower the sheet resistance and transparency are For a range of opto-electronics applications, multi-property optimization of DMD structures is an important active area of research for further development of modern devices in the future. Jeong et al., have reported on the optimized 70 nm Al_2O_3/10 nm Ag/70 nm Al_2O_3 multilayer passivation prepared by plasma damage-free linear facing target sputtering process to improve the life-time of OLEDS, while the Ag layer was deposited using thermal evaporation technique *(37)*. Like all or most DMD developers and industry practitioners, the aluminum oxide layers were deposited using metallic sources in conjunction with oxygen gas, rather than using a single ceramic (Al_2O_3) source.

Many modern glass and window products are critically dependent on the performance of their optical coatings, and the glazing systems of today are capable of controlling properties such as emissivity, solar heat gain, apparent color, and transparency. Of particular importance is the manufacturability of stable and durable advanced low-emissivity dielectric/metal/dielectric multilayer coatings for glass industry, which rely on nanoscale layer thickness control and accurate stoichiometry as well as long-term stability, especially for the ultrathin plasmonic metals (typically Au or Ag) layers.

Silver is a noble metal that possesses the lowest optical losses across a very broad spectral range spanning from the blue to the near-infrared wavelengths. This is why Ag is the metal of choice in modern Low-E coatings, which feature clear, almost color-neutral visual appearance *(38)*.

Another fast-paced research and development area of today is related to plasmonics nano-sensor development, in which multiple significant advances have been reported recently *(39–46)*. These surface plasmon resonance-based optical sensors have numerous commercial applications in oil and gas, biochemical, medical and chemical sensing, due to their high sensitivity to the refractive index change in dielectric media placed on top of ultra-thin plasmonics metal films (Ag). Therefore, the development of DMD structures containing Ag layers is also of importance for these applications. Table 4.1 lists several typical DMD structure designs reported in the literature.

The deposition of ultra-thin Ag layers with low surface roughness, uniform layer morphology and precise thickness control is the key to achieving the required application-specific optical properties. However, the deposition of ultra-thin Ag layers (of less than 20 nm thickness) tends to lead to the formation of nano-islands of silver, even when using the most advanced

Table 4.1 Examples of DMD structure designs and their deposition techniques.

DMD multilayer structure type	Deposition process	Metal layer thickness (nm)	Substrate type	Reference
$InZnSnO_x$/Ag/ $InZnSnO_x$	RF magnetron Sputtering, and Thermal evaporation	8–20	Polyethylene terephthalate (PET)	*(24)*
Al-doped ZnO/ Ag/Al-doped ZnO	RF magnetron Sputtering	19	Polyethylene terephthalate (PET)	*(32)*
ITO/Ag/ITO	Roll-to-Roll (RTR) Sputtering	8	Polyethylene terephthalate (PET)	*(33)*
TiO_2/Ag/TiO_2	DC/RF magnetron controlled Sputtering	8	K9 glass	*(36)*
Al_2O_3/Ag/ Al_2O_3	Linear facing target sputtering (LFTS), and thermal evaporation	10	Glass, and polyethylene terephthalate (PET)	*(37)*
ITO/Ag/ITO/ Ag/ITO	RF magnetron sputtering	13, 18	Glass	*(82)*
TiO_2/Ag/Ti/ TiO_2/Ag/Ti/ TiO_2	RF magnetron sputtering	13–18	Glass	*(82)*
TiO_2/ITO/Ag/ ITO/TiO_2/ ITO/Ag/ ITO/TiO_2	RF magnetron sputtering	13–18	Glass	*(82)*

deposition process, such as magnetron sputtering *(47–50)*. The formation of Ag nano-islands instead of contiguous ultrathin layers significantly affects the optical performance of multilayer thin film structures. Additionally, the material layers deposited on top of ultrathin Ag layers tend to sink into the spaces in-between the Ag nano-islands, thus affecting both the morphology and the effective optical thickness of the adjacent layers. It has also been reported by several groups of researchers that the thin silver layers growth according to the Volmer–Weber mode leads to forming three-dimensional islands *(51–53)* rather than contiguous layers. This is a common problem, which is related to the surface energy-dependent inter-material compatibility between Ag and its underlying layers *(23–27,31–36,46,53–55)*. A vast body of literature published so far is dedicated to finding the optimally suitable "wetting layers" or "seeder layers" designed to engineer the contiguous growth of ultrathin Ag films on a range of substrate materials for use in different applications *(56–60)*.

Another issue associated with Ag is that it tarnishes and/or oxidizes very quickly; therefore the thin Ag layers have to be protected from oxidation or other chemical reactions during the deposition of multilayers and also during the end-use of coated products. Additionally, in many applications, it is of interest to design metal-dielectric multilayers possessing good mechanical properties, such as scratch resistance and adhesion strength.

Despite their long history of use in industrial low-emissivity coatings, questions remain in the areas related to the layer growth morphology, and especially the compatibility of Ag layers with surrounding materials, which limits the freedom of design in functional coatings as well as affects their thermal and environmental stability. In this Chapter, we report on the additional degrees of freedom available for the design of material systems for a range of advanced coatings based on the DMD approach.

We focus on the properties of co-sputtered metal-dielectric nanocomposite material systems, which can serve as building blocks for the design of highly transparent low-emissivity coatings that demonstrate excellent environmental stability. These are also compatible with a much greater range of surrounding layer materials compared to ultrathin Ag layers grown conventionally. In particular, these nanocomposites also offer improved morphological properties (due to both avoiding the island-type growth issues and larger typical thicknesses required to be deposited). Other important benefits of using metal-dielectric nanocomposites are related to the ability to accurately control the apparent color properties of low-emissivity coatings, which is important for the designers of tomorrow's energy-efficient building products. Several reports have recently been published, which investigated the properties of various metal–dielectric composite (MDC) thin film layers including Ag + SiO_2, Ag + TiO_2, and Ag + MgF_2 for realizing energy-efficient optical coatings, due to their unique combinations of optical, thermal and mechanical properties. Most of these focused on studying the metal-dielectric nanoparticle array properties of interest for various applications *(61–69)*. Magnesium fluoride (MgF_2) has been widely used for various optical applications due to its high hardness, good chemical stability and high optical transmittance, which makes it a promising material to design new composite type materials, such as Ag + MgF_2 (in optimized volumetric fractions) *(67,70)*. To the best of our knowledge, the majority of published literature describing Ag–MgF_2 and similar nano-composites research relates to applications other than the development of advanced THR coatings for use in modern construction.

We have optimized RF sputtering deposition process parameters for the growth of MDC (Ag + 2–10 vol.% MgF_2) single-layer thin films of different dielectric dilution levels. In addition we have prepared several metal-dielectric multilayer coating structures containing at least one MDC layer as a building block, which promotes high stability and durability for the entire coating systems. The transmission and reflection spectra of MDC structures have been characterized after the deposition and also after running various environmental exposure experiments, such as oven annealing and environmental chamber-ageing. Experimental results show that the ultra-thin Ag-containing layers protected by dielectric dilution and by the selection of surrounding layer types (sputtered from ceramic metal-oxide targets) inside multilayer structures can survive up to 400°C in air without any signs of optical or physical performance degradation observed. Additionally some of these multilayer coatings were found to survive prolonged exposure to moisture, water, and some organic solvents.

2 DMD coatings: materials, designs, and architectures

The selection of thin film materials for use in any type of multilayer coating design is of primary importance, since it normally determines the main functionality of the coating-substrate system and also imposes the limitations on the required number of layers and their inter-compatibility. The selection of dielectric materials most suitable for any particular coating application and coating functionality type can be assisted by multiple excellent available literature sources, such as *(4,5,51,71,72)*. Multilayer thin film design practices are well covered in *(72)* for most types of coating designs, and many excellent software packages (e.g., OptiLayer, TF Calc and others) can assist in fine tuning of the spectral properties required for most applications. The selection of metallic materials for thin-film applications is mainly governed by the specific application type (e.g., Low-E coatings, plasmonic or magneto-plasmonic sensors or even simple mirrors). This is due to both the limitations of metallic materials (e.g., high extinction coefficients), and their intrinsic functionalities, such as the potential to support the surface plasmon propagation (Au, Ag), or magnetic properties (Co, Ni).

Of particular importance is the availability and quality of data on the complex refractive index for each material required, tabulated across the entire spectral range of relevance for any selected application. For example, these data are required for the wavelength range from about 300 nm up to ≥ 10 μm when designing transparent THR coatings on glass substrates. One of the principal data sources on the optical constants of numerous materials has been the book by Palik *(73)*. It is important to note that the available online and also published literature sources are still limited, mainly due to the original data sources being either limited spectrally or linked to a particular deposition technique or material phase. This is why it is crucially important to analyze critically the external data sources on the optical properties data, and the experimental capabilities of measuring, fitting or otherwise reproducing the complex refractive index data are necessary for all designers. Even for common metals, such as Ag, a number of recent literature sources appear to use either the theoretically modeled index dispersion functions, or rely on joining the datasets published by different original sources. The latter leads to data inconsistencies in the optical constants of Ag between 1 and 2 μm, which in turn sometimes results in modeling errors seen in the transmission and reflection spectra. In the subsequent parts of this Chapter, we attempt to fill this gap by presenting our own dataset for Ag. The data sources for compiling this dataset were *(73, 74)*, and also our group's data fitting experiments followed by experimental confirmation procedures. These procedures involved the deposition and characterization of multiple Ag-containing films and multilayers. The availability of reliable optical constants data is also of prime importance for the designers of metal-dielectric nanocomposites of all types, as well as the multilayers reliant on DMD building blocks.

Transparent heat-regulating coatings that enable energy savings in modern construction rely on using various metals, such as Ag, Au, Al, Ti, Ni, and Cr, to form both the reflecting layers and also barrier layers. Numerous

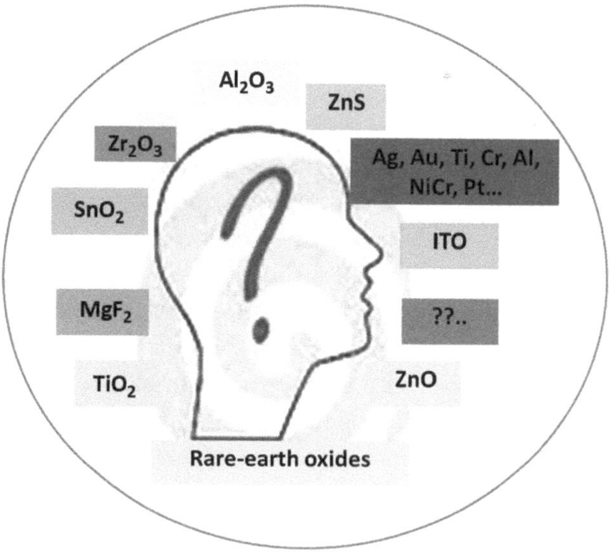

Figure 4.1 ***A schematic diagram showing various potential metal and dielectric materials considerable to design, optimize, and fabricate DMD coatings for various applications including energy generating transparent heat regulating window systems.***

dielectrics, such as SnO_2, Bi_2O_3, TiO_2, Zr_2O_3, ZnS, Ta_2O_5, MgF_2, MgO, SiO_2, ZnO, Al_2O_3, and Dy_2O_3, have been trialed in THR applications by multiple groups, and many new dielectric materials are yet to be trialed *(1–5)*. Our group started designing and testing THR multilayers since 2012 *(75)* and developed several types and compositions of DMD materials systems.

The schematic diagram shown in Fig. 4.1 provides an outlook on the large range of possible metals and dielectric materials suitable for design and development of high quality multilayer thin-film coatings.

Despite the huge range of material choices provided by nature, there still exists a need for designing new ones. DMD materials can be classified into two different structural groups: (1) triple layer blocks, composed of discrete-material sequences, and (2) co-deposited nanocomposites of solid solution type, where the metal component is diluted by dielectric filler, surrounded by dielectric layers. In this Chapter we describe the synthesis techniques and material properties of both DMD material classes, focusing on Ag–MgF_2 chemistry.

The successful manufacturing of DMD multilayers containing Ag layers relies not only on having a right choice of dielectric materials (as most of them are oxides and lose a part of their oxygen content during the deposition process, if no extra oxygen input is used inside the chamber to compensate the oxygen loss) but also on the management of the disadvantages of using oxide materials. The prevention of oxidation degradation of the ultrathin metal (Ag) layers, due to the adjacent dielectric layers being present, or due to oxygen process gas inside deposition chambers, is of importance. Another important obstacle for ultrathin Ag growth is related to the surface energy properties of many oxide materials causing imperfect growth morphology by affecting the wetting properties. These issues can be addressed effectively through the design of co-deposited silver-dielectric nanocomposites.

The growth of ultra-thin (less than 20 nm) and smooth Ag layers using any kind of vacuum deposition technology is still challenging, unless the layer(s) serving as a substrate for Ag deposition are selected carefully, considering their chemical properties. Additionally, the layers following ultrathin Ag in the deposition sequence also need to be selected carefully to ensure both the Ag layer stability and the cohesion of the entire coating. Several research works have been conducted and reported in which the various seeder layers (such as Ge or others *(57–60)*) and barrier layers (Ti, Cr, NiCr, or others *(76–78)*) have been trialed. However, using these special types of seeder layers is obviously challenging in terms of maintaining the high visible transmission in THR multilayers of high spectral selectivity. The design and growth of discrete DMD blocks with ultra-thin Ag layers are detailed in next section.

2.1 Dielectric/metal/dielectric multilayer structure designs

Many studies have been focused on the optimization and development of thin-film dielectric/metal/dielectric coatings suitable for use as heat mirrors useful for heat-flow regulation applications, which work by reflecting the UV, near-infrared, and thermal radiations while maintaining a maximized transmission in the visible range or a controlled color. There are multiple software packages available commercially, which allow design optimization of application specific multilayer coatings. The design procedure itself, however, begins from the selection of several mutually compatible metallic and dielectric material constituents of the coating. Successful selection of the most suitable optical materials governs the degrees of freedom available in design variations, and ultimately all of the coating's final properties, including the durability. The structure of THR coating designs can vary greatly, depending on the required spectral properties in the UV, VIS, Solar IR, and Thermal IR regions as well as the degree of spectral selectivity. The latter is characterized by the spectral sharpness of the transition between the regions of high transmission and high reflection, since most THR designs are essentially edge-type filters.

Fig. 4.2 shows a schematic diagram of several different approaches to the design of practical DMD multilayer coatings.

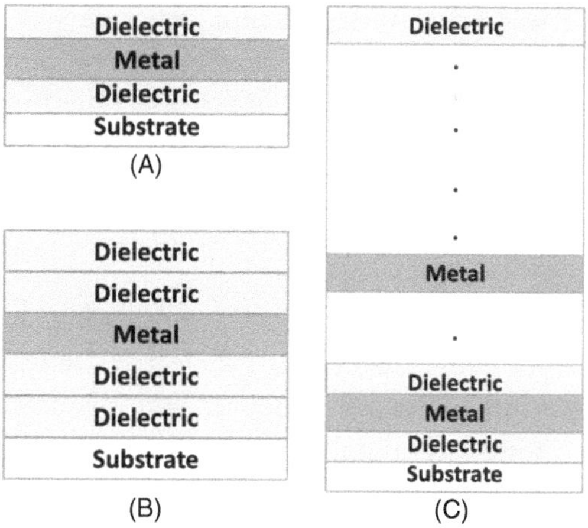

Figure 4.2 ***Schematic diagram of popular design structure types of DMD multilayer coatings.*** (A) A simple tri-layer structure containing a single metal layer sandwiched in-between two identical dielectric materials; (B) a single metal layer embedded in-between two dissimilar dielectric material types; (C) generalized design approach, in which multiple metal layers can be surrounded by different dielectric layers.

Tri-layer DMD coatings, where generally a single metal (often Ag) thin layer is sandwiched in between two (identical in material type) dielectric layers, are common, yet are very limited in the design flexibility in terms of achieving the required optical properties. Several tri-layer DMD coating designs *(17,79,80)* have been reported for use as metal-dielectric multilayer coatings for future applications as heat mirrors, suitable for the fast fabrication and providing cost effectiveness. However, with these types of coating structure (containing a single metal layer), it is very hard to achieve finely controlled and spectrally selective optical features (in terms of high peak transmission in the visible spectral range coincident with strong IR reflectivity). Also, it is rather difficult to protect the Ag layer from oxidation and tarnishing occurring due to the diffusion of substances such as H_2S and O_2 from air through the top layer. In addition, generally there is a risk of moisture exposure causing silver to degrade due to the diffusion of water vapor, which requires quickly protecting these coatings after the manufacture by way of encapsulation or installation into humidity-controlled spaces. In some cases the adjacent oxide layer, which is used to protect the thin metal layer, can also affect the silver layer chemically, leading to a drop in the performance efficiency of these coatings. To address these issues, we have developed several material systems of this simple tri-layer type modified to include ultrathin (0.5–2 nm) barrier layers to protect the metal layer.

The design flexibility in DMD multilayers is expanded greatly by introducing multiple metal layers surrounded by customized dielectric material combinations and sequences into the film structure (see Fig. 4.2(B, C)). The additional degrees of THR coating design freedom can complicate the manufacturing process somewhat, yet provide a significant advantage in controlling the coatings properties. Most important of these are the adhesion, inter-layer cohesion, chemical and thermal stability, together with high spectral selectivity which is still not a common feature in most commercially available advanced glazing systems.

We have prepared various types of DMD multilayer coatings using nanoscale Ag layer surrounded by a range of dielectrics (oxides, fluorides and sulphides) arranged into different sequences and investigated their optical properties. We also tested the durability of different coating types designed for use in harsh environments.

3 Examples of ultra-thin metal layer growth and DMD multilayer fabrication

Single ultrathin metal layers and DMD structures are usually prepared in labs by using either an E-Beam or Thermal evaporation, or a RF Magnetron sputtering system. In industrial settings, however, sputtering is much more common. All of these are vacuum-based physical vapor deposition (PVD) technologies. These are the best suited for the production of high quality, high spectral selectivity modern advanced THR coatings on glass. Alternative ways of THR and generic Low-E coating manufacture involves technologies other than PVD, for example, pyrolytic coating. To the best of our knowledge, achieving significant energy savings in near-zero-energy buildings relies today and will continue to rely tomorrow, on the multi-metal layer THR coatings fabricated by PVD processes.

The literature history about the thin Ag layers reveals that the physical properties including micro-structure, uniformity, residual stress, and electrical conductivity of sputter-deposited Ag thin films and their growth process parameter dependencies have been studied extensively world-wide. However, most of the older studies were based on characterizing the bulk Ag layers (above 20 nm), rather than ultra-thin ones (less than 15 nm thick). The reason behind this is the hurdle of reaching a growth mode (known as Stranski–Krastanov growth mode *(51,81)*) to obtain a smooth and continuous silver thin film morphology. Also, silver is very well-known to grow as separated adatoms and islands at very early stages of deposition process, which spread and percolate with increasing film thickness *(47)*. By implementing a "wetting" or "seeder" layer onto the substrate, or by making MD nano-composites, researchers have conducted various experimental studies to overcome these issues. Multiple reports are available from the literature *(56–60)*. The data sources on the optical constants (refractive index and absorption coefficient) dispersion function of Ag layers are found to be patchy, due partially to the film thickness dependency, and also due to the studies often being limited to a short wavelength range. During our coatings development work (conducted between 2011 and 2015, largely unpublished so far) we accumulated the available information on the optical constants of Ag layers from all available sources, and found that there were data discrepancies in the region 1000–2000 nm. We modeled and fitted multiple transmission spectra of deposited Ag layers of various thicknesses (12.5–40 nm) experimentally to obtain an acceptable quality of the optical constant data between 300 and (~7000) nm. Fig. 4.3 shows the refined dataset, in which our fitted data is used in conjunction with the data from

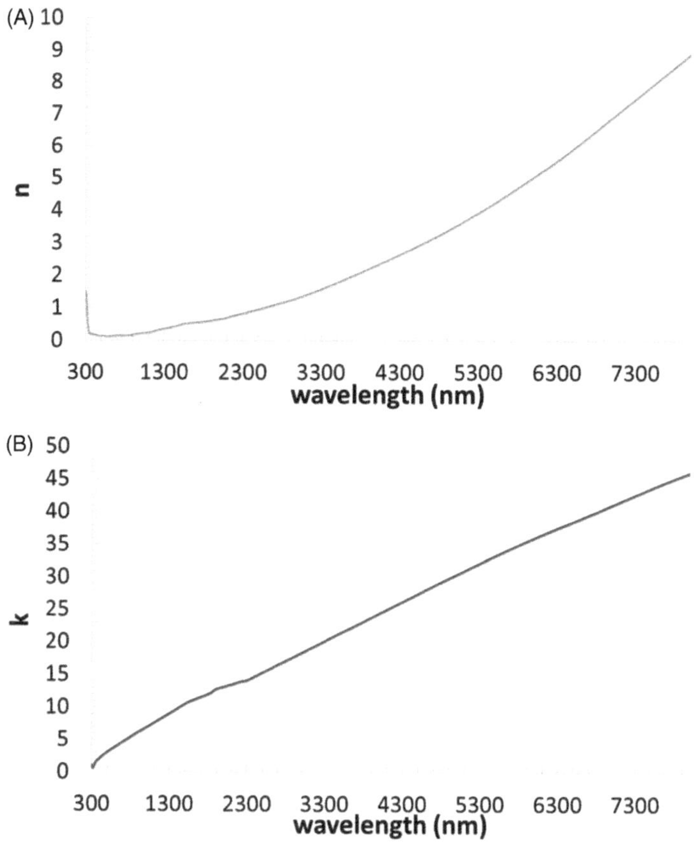

Figure 4.3 ***Extrapolated digitized optical constants dispersion data for thin sputtered Ag layers.***

(73,74), which we found to result in close matching between the modeled and the measured transmission spectra for all film thicknesses. Using this data set, we have experimentally fitted the ultra-thin Ag film transmission between 300 and 2500 nm for different deposited onto two different types of optically clear glass substrates, as shown in Fig. 4.4. The actual physical layer thickness of the growing films was monitored in-situ during the deposition, using a finely calibrated quartz crystal microbalance sensor. The precise calibration of sensor (tooling factor, TF) was obtained after running multiple deposition and characterization experiments followed by precise transmission spectrum fitting in the spectral region, for which reliable (n, k) data are widely available.

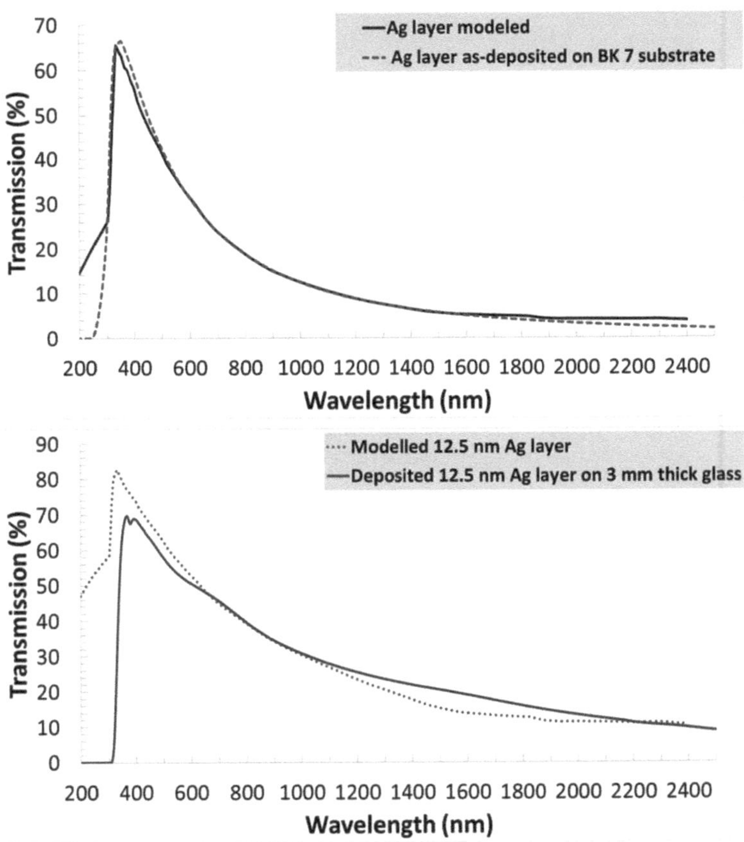

Figure 4.4 ***Experimentally measured transmission spectra fitted with the modeled transmission spectra.*** (A) ~19 nm Ag layer deposited onto 2 mm thick BK7 glass substrate; (B) 12.5 nm ultra-thin Ag layer deposited onto a 3 mm thick low-iron clear glass substrate.

Fig. 4.4A shows the fitted transmission spectrum of an as-deposited Ag layer of about 20 nm thickness obtained from the OptiLayer modeling and spectrophotometry (Agilent Technologies, Cary 5000) measurements. From this fitting procedure, we confirmed that our originally deposited film thickness was about 19 nm, which is accurate to within ±5%. Even though the TF calibration accuracy was well within ±2%, there were still some thickness dependent morphology non-uniformities in this Ag layer. This is because the Ag layer was deposited onto the glass substrate directly, rather onto a purpose-optimized wetting layer or any suitable dielectric underlayer. Fig. 4.4B shows the measured transmission spectrum of a 12.5 nm thick Ag layer deposited onto a 3 mm thick low-iron clear glass substrate plotted together with its modeled transmission spectrum. From the plots, it is very clear that our extrapolated (n, k) data are suitable for designing THR multilayers, at least for Ag layer thicker than about 12 nm.

Forming high quality tri-layer building blocks of DMD type requires not only using reliable optical constants data for all materials used in the sequence, but also careful attention dedicated to the inter-compatibility of all materials and their interfaces. The prevention of metal layer oxidation during growth or thereafter is very crucial for developing thermally and environmentally stable DMD coatings. Throughout our experimental experience with RF magnetron sputtered ultra-thin (≤15 nm) Ag layers, we have noticed that the underlying dielectric layer adjacent to silver can play a significant role in improving the metal layer morphology and the overall dielectric/metal interface quality by preventing the island-type growth and excessive interface roughness. In one of our previous published works, we were able to successfully grow a 12.5 nm Ag layer onto a ZnS sublayer and also protected the Ag layer with a ZnS overlayer without having any adverse morphological changes within the structure that could affect the long range surface plasmon-polariton performance of a custom-designed plasmonics structure. In addition, we also applied an amorphous alumina (Al_2O_3) sputtered thin layer to improve the quality and adhesion of the first sub-layer of DMD sequence to the adhesor layer *(46)*. Also, in one of our magneto-plasmonic sensor development works, we tested Al_2O_3 layers as barrier-type layers for isolating the metal (Ag) layers chemically from the effects of diffusing oxygen and avoiding the oxidation-related compositional changes, while the overall film structures were treated using high temperature oven-annealing process to crystallize the magnetic garnet layers of the multilayer structure *(45)*.

In accordance with our group's experience with the sputter-deposition of dielectric and metal layers, and also based on the easily commercially available sputtering target materials, we designed, and experimentally developed multiple DMD multilayer structure-types, starting from simple tri-layer (single silver layer) structures to more complex double-metal multi-layer coatings (Fig. 4.3B and C) and characterized and tested their physical and optical properties. Fig. 4.5 shows the transmission spectrum of a typical simple tri-layer DMD structure. Fig. 4.5A illustrates the optical performance

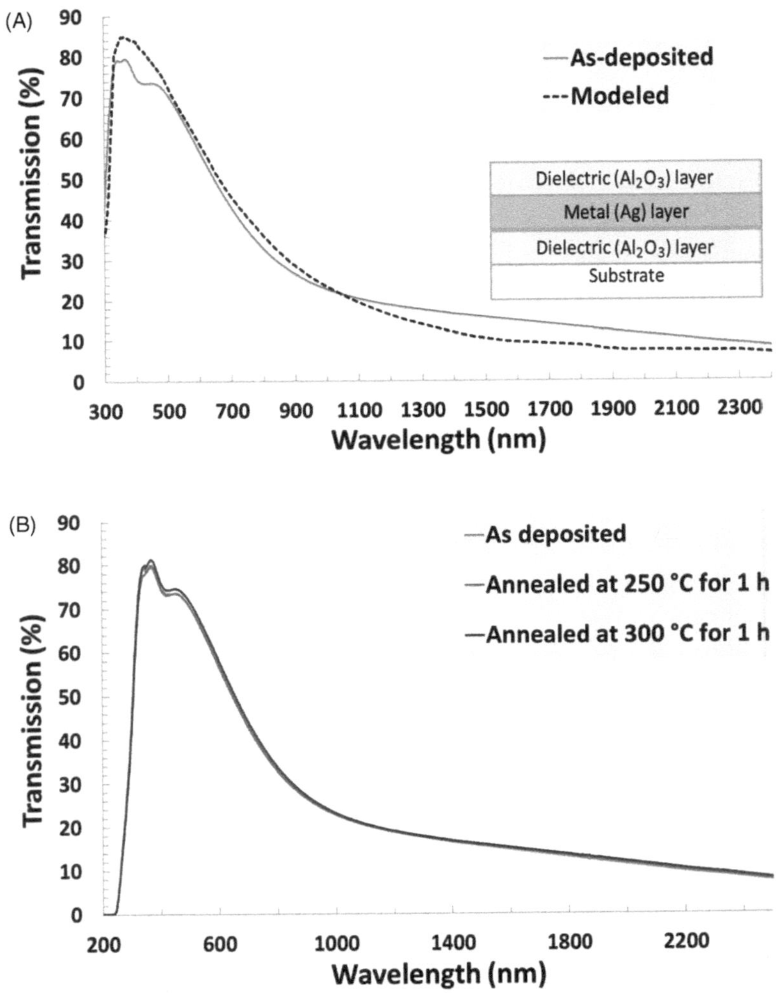

Figure 4.5 ***Measured transmission spectrum of a typical DMD multilayer structure.*** (A) A single metal (Ag) thin layer sandwiched in-between two dielectric layers; (B) physical and thermal stability performance test results; the sample showed almost identical transmission spectra after being subjected to an air-atmosphere high temperature oven annealing treatment.

of an as-deposited 30 nm Al_2O_3/15 nm Ag/30 nm Al_2O_3 multilayer structure (where these thicknesses were chosen arbitrarily during deposition) compared with its OptiLayer-modeled spectrum. The experimental results show that the sample had a slightly higher transmittance, mostly in the IR spectral region (above 950 nm–2500 nm) compared to the simulated result. However, the spectral response of the same sample in between 500 and 900 nm was found to fit very closely to its modeled spectrum, which confirms the high quality growth of the ultra-thin metal layer and also the overall DMD multilayer structure. The data discrepancies in the near-IR region may have originated from the insufficiently precise optical constants dataset for Al_2O_3.

The deposition of tri-layer structure of Fig. 4.5, took place in a sequence in the same RF sputtering chamber at high vacuum near 1×10^{-5} Torr. Only Ar (1–2 mTorr) was used as process gas, since the alumina sputtering target was based on sintered oxide ceramics. Al_2O_3 is in our experience one of only few oxide materials which practically do not suffer oxygen loss during sputtering deposition. This is why no in-situ oxidation of the thin silver layer was expected during the deposition. Also, the relatively thick (30 nm) amorphous alumina layers were expected to provide sufficient diffusion barrier to protect silver from oxidation in air. In order to test this, as well as assess the overall thermal exposure stability of this tri-layer, we conducted air-atmosphere annealing tests at elevated temperatures. Fig. 4.5B shows that practically no detectable changes have occurred after being exposed to high temperatures up to 300°C for 1 h. These tests have confirmed that it was possible to demonstrate a useful, thermally stable building block of advanced DMD multilayers.

We also tested the long term durability and humidity-exposure stability of a more advanced DMD structure type (structure of type shown in Fig. 4.2B), which are parameters of interest for applications in modern glazing systems. Fig. 4.6 shows the test results obtained. Interestingly, after 1-year long storage in ambient (lab conditions), without any humidity control, no degradation in the transmission spectrum features. Transmission increased slightly in the visible range, possibly due to the changes in the refractive indices in some layers due to strain relaxation.

Following this one year long storage, the same sample was tested inside an environmental exposure chamber, where the air was kept at 70°C and at 65% relative humidity for 1 hour. Again, no signs of coating degradation were detected either on visual inspection or by measuring the transmission spectrum. Therefore, it has been possible to sufficiently insulate even

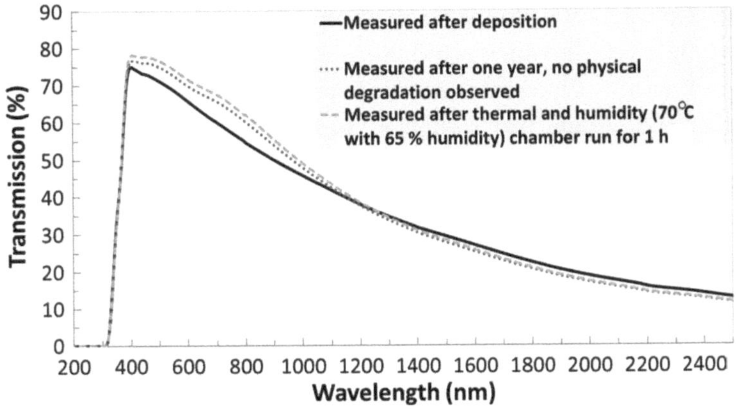

Figure 4.6 Transmission spectrum of a DMD multilayer structure of type shown in Fig. 4.2B where a single metal (Ag) ultrathin (12.5 nm) layer is used together with a specialized protective oxide over a layer of ≤3 nm thickness, where the protected silver layer is sandwiched in-between double-dielectric layers of dissimilar materials. The transmission spectra measured after the deposition, and also after a 1-year lab storage followed by an elevated-temperature, high humidity chamber exposure.

an ultrathin sputtered silver layer from a range of adverse environmental factors during long term storage by means of carefully engineering the all dielectric encaptulating structure.

Currently, it is of substantial scientific and also commercial interest to prove that practical multilayer DMD structures can be designed and fabricated, which demonstrate very high color unbiased visible range transmission simultaneously with low solar heat gain and also low thermal emissivity. Only some of the products currently listed in IGDB *(35)* approach this combination of requirements, at VLT near 80%. To the best of our knowledge, practically all double-silver, high spectral selectivity Low-E coated glass products commercially available today can neither be stored at ambient conditions for longer than 1–3 weeks, nor can they survive oven exposure at temperatures higher than about 180°C.

We designed and fabricated (in 2013) an optimized DMD multilayer coating, the structure of which followed the generic design type of Fig. 4.2C, using two different dielectric material types and two ultrathin silver layers, which demonstrated an exceptional combination of properties. Fig. 4.7 summarizes these properties as well as the high temperature exposure test data.

The exceptionally high visible transmission of this optimized DMD structure deposited onto a 1 mm thick Corning Eagle XG substrate was

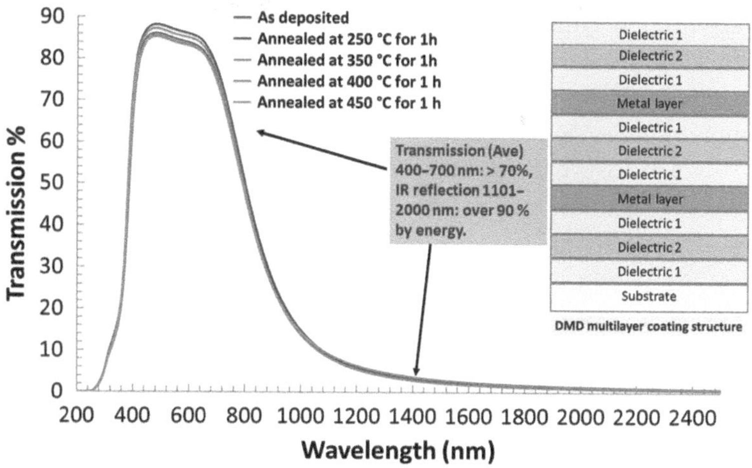

Figure 4.7 Transmission spectrum of an optimized DMD multilayer coating structures (containing two Ag layers) deposited on a glass substrate. The spectral characteristics, which show quality growth of ultra-thin metal layer as well as material interfaces with the dielectric materials into the multilayer structure, are confirmed by the displayed high temperature stability performance.

retained at above 80% across most of the visible range, even after running a sequence of high temperature (up to 450°C for 1 h) annealing processes. Translating this specialized Low-E coating design into the commercial industry practices will however take some time, due to the design being reliant on the optical material combination, which is uncommon in the coating industries today.

4 Metal-dielectric nanocomposite materials for the design of advanced, high-stability dielectric/metal/dielectric multilayer coatings

A class of nanocomposite materials termed "cermets" is represented by a range of co-deposited metallic and ceramic phase constituents. Typical examples of these are co-sputtered or co-evaporated layers of stoichiometries such as Ag–TiO_2, Ag–SiO_2, Ag–MgF_2, and others *(61,62,64,65,67–69)*. When designing cermet type nanocomposites for THR coating applications, the plasmonic metal constituents are not typically diluted with a dielectric significantly, since it would otherwise be difficult to preserve the high infrared reflectivity from thin layers. Cermet type nanocomposites, in which the volumetric content of the dielectric phase stays low (within 10–15 vol.%), are essentially semi-metallic substances (metal phase is weakly diluted with the dielectric) of the solid solution type. Hence, we term these special co-deposited thin film substances "metal-dielectric nanocomposites", or MDC. In THR application area, we focused on the development of Ag–MgF_2 MDCs possessing metal dilution levels between 2 and 10 vol.%. Two principal factors governed the choice of this stoichiometry type: (1) the low refractive index of MgF_2 being compatible with the low (real part) refractive index of Ag, and (2) MgF_2 forming dense, mechanically strong layers during the deposition, without fluorine loss. Therefore, it is reasonable to expect that the addition of MgF_2 to a soft ultrathin metal (Ag) layers could accomplish all of the following: (1) promote adhesion of metallic layer to its substrate, (2) promote mechanical hardness of soft plasmonic metal layer, (3) improve metal layer morphology by way of "filling the pores" between metal nano-islands, and (4) potentially smoothen the interface roughness in the metallic layers. Very broad optical transparency range of magnesium fluoride is also favorably suitable for preserving the transparency of thin plasmonic metal layers. For these reasons, we report in detail on the development and properties of several Ag–MgF_2 nanocomposite material types and their performance in practical DMD (D/MDC/D) THR multilayer applications within the next sections. We have also studied the MDC material systems of type Ag–ZnS; however, our principal finding was that these MDC layers were much more difficult to incorporate into high performance Low-E THR designs.

4.1 Growth of composition-optimized single layer MDC thin films

Several batches of MDC (Ag + MgF_2) films were deposited onto glass substrates by using co-sputtering process in a RF magnetron sputtering system (Korea Vacuum Technology, KVS T-4065) which had 3 sputtering gun sources. The volumetric fraction of dielectric (MgF_2) phase was controlled by adjusting the RF power densities applied to the MgF_2 sputtering target during the co-sputtering process, after carefully calibrating the partial deposition rates for both MgF_2 and Ag. A standard quartz crystal microbalance deposition monitor was used to monitor the combined thin film deposition rate in real time. The thicknesses of ultrathin composite films trialed were in the range between 12 and 25 nm. Table 4.2 summarizes the deposition process parameters used to deposit the metal-dielectric nanocomposite layers. All MDC layers were deposited at room temperature, that is, without heating the substrates during the sputtering process.

Table 4.2 Sputtering process parameters used to deposit metal-dielectric composites.

Process parameters	Single layer MDC
Sputtering targets	Ag, MgF_2 (3-inch diameter)
RF Power	Ag (55–65 W), MgF_2 (83–95 W)
Base pressure (Torr)	5×10^{-6}
Process gas	Argon (Ar)
Process pressure	2–3 mTorr
Substrate temperature (°C)	Room temperature
Post deposition heat treatment	N/A

The transmission and reflection spectra of all MDC layers were measured after the deposition using an Agilent Cary 5000 UV-VIS-NIR spectrophotometer. The reflection measurements were carried out using a custom-built reflection-mode optical measurement module attached inside the spectrophotometer, which contained four mirrors adjusted at 45° tilts to form a periscope-type device. In this measurement arrangement, the four mirrors were used to reflect the spectrophotometer's incident light beam off a calibrated reference mirror before reaching the detector. For reflection measurement using a nominally transmission-mode spectrophotometer, first the baseline was measured using the four-mirror setup. In the next step, one of the mirrors was replaced by the thin-film sample of interest (reflecting silver layer or MDC layer on a glass substrate of size 40×40 mm^2 to remeasure the transmission spectrum. We then apply our custom Excel-based computational algorithm to both transmission data sets (one generated with reference mirror, and another with the reflecting sample) to derive the sample's reflection spectrum.

Fig. 4.8A shows the measured transmission spectrum of a Ag + MgF_2 (10 vol.%) composite (~20 nm) thin-film compared to that of pure Ag (20 nm) thin-film layer on glass. The transmission peak in this Ag + MgF_2 (10 vol.%) composite layer exhibited short-wavelength features similar to the 20 nm Ag layer, except a higher transmission in the IR spectral region. This experimental investigation revealed that the co-sputtered Ag layer with an optimized volumetric fraction of MgF_2 content can provide an avenue for using comparatively thicker MDC layers with high metal volume content (which act as a quasi-metallic layer) into any type of DMD structures for possibly improving THR film properties and maintaining the high structural stability of the coating while allowing the design of high spectral selectivity coatings with adequate or high optical transparency in the visible region. Fig. 4.8B and C show the broad-range (200–2500 nm wavelength)

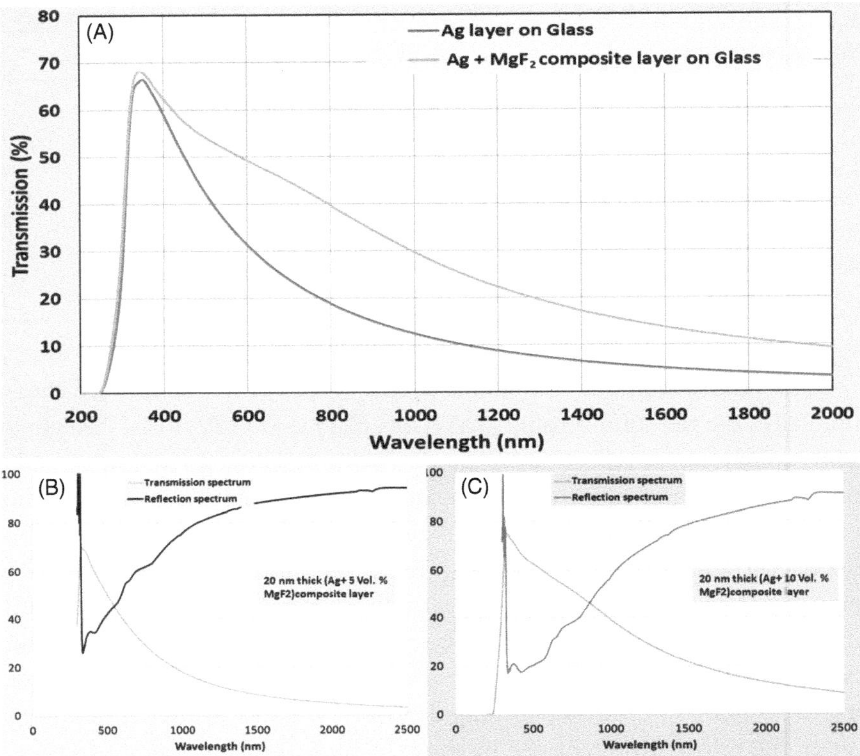

Figure 4.8 Measured (A) transmission spectra of ~20 nm thick Ag and also the Ag + MgF_2 (10 vol.%) composite layers; (B) transmission and reflection spectra of Ag+ 5 vol.% MgF_2 layer, and (C) Ag+ 10 vol.% MgF_2 layer, deposited (with the same thickness) onto glass substrates.

transmission and reflection spectra of Ag + 5 vol.% MgF_2 and also Ag + 10 vol.% MgF_2, ~20 nm thick MDC layers, respectively. The optical absorption was calculated by subtracting the transmittance and reflectance of the composite layers using the formula A = 1−T−R (%), where A is the absorbed power fraction, T is the power transmission coefficient, and R is the power reflectivity. It was found that the Ag + 5 vol.% MgF_2 sample had a comparatively lower absorption (less than 9% by power) in the UV-Visible spectral range, which is close to that of the pure Ag layer. This indicates that the addition of 5 vol.% MgF_2 into the Ag layer during the co-sputtering process can potentially improve the coating adhesion and stability aspects without compromising the plasmonics metal properties. Note that the visual inspection confirmed that the composite layer showed no signs of surface degradation even after 1 month of post-deposition storage in ambient air, while pure unprotected Ag layers normally start to degrade (oxidize and tarnish) on the surface shortly after the deposition.

We also found that metal layer dilutions near 2 vol.% of added MgF_2 also led to obtaining MDC type nanocomposites of potential interest for use in THR multilayers. We report on using these materials in the subsequent section.

4.2 DMD multilayer coating structures containing MDC layers

Promising material system development results achieved with single layer MDC films indicated that another degree of freedom was now available in the design of advanced THR coatings containing multiple optimized sequences of dielectric, metallic, and also MDC layers. Several DMD multilayer structures containing either one or more than one MDC layers were developed, and their performance was characterized. Fig. 4.9A shows the transmission spectrum of an as-deposited DMD multilayer coating, where the layer structure consists of 30 nm Al_2O_3/25 nm MDC (Ag + 5 vol.% MgF_2)/30 nm Al_2O_3 in comparison to that of a typical tri-layer silver based DMD coating (Fig. 4.5(A), described in Section 3). It can be noted that the use of 25 nm MDC layer instead of a pure 15 nm Ag layer resulted in better thermal radiation filtering performance without compromising the visible spectral transmission. Notably also, the MDC based structure demonstrated much lower thermal emissivity while maintaining the same superior visible transmission compared to the silver based coating.

In order to further investigate the thermal stability properties of the MDC-based coatings subjected to high temperature oven annealing processes, a test multilayer structure containing 30 nm Al_2O_3/[6 nm MDC + 15 nm Ag + 5 nm MDC]/30 nm Al_2O_3 was developed and subjected to high temperature oven annealing process. This structure was designed as a more complex generalized MDC type multilayer building block, to study whether the advantages of pure silver and MDC approach can be combined. Another reason for testing this structure was also to see whether a high quality ultrathin silver layer grows well on top of an MDC sublayer. The performance of the developed structure shows that the optical properties are comparable to some of the conventional multilayer structures containing pure Ag layers *(6,82)*. The transmission spectra of these thin film structures (as deposited and also oven annealed at 400°C for 1 h) are shown in Fig. 4.9C. It can be seen clearly that no optical performance degradation was observed in the oven annealed samples, which shows that the very thin MDC barrier layers can play a vital role in protecting the ultra-thin core

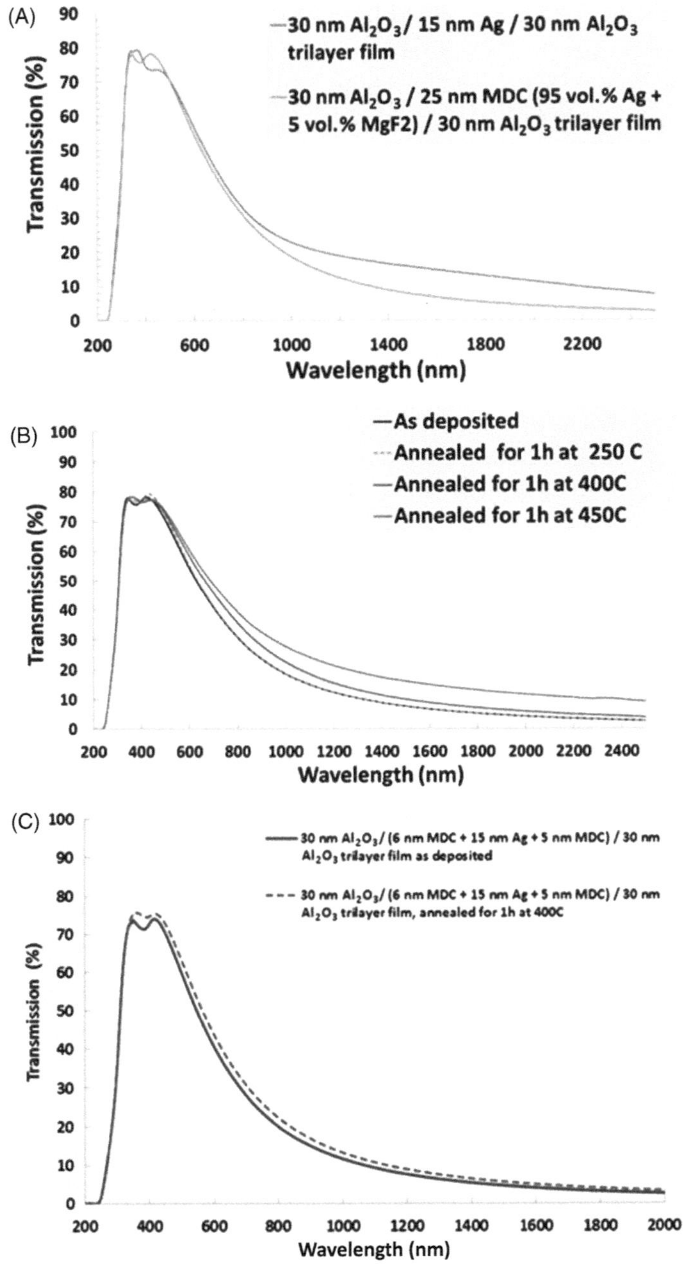

Figure 4.9 Measured transmission spectra of multilayer thin film structures containing a single metallic or MDC layer. (A) Transmission of a tri-layer MDC structure compared to that of a typical Ag layer within a DMD structure; (B) thermal performance stability of the tri-layer MDC structure; and (C) thermal performance stability of a complex DMD multilayer structure containing both the MDC outer sublayers and pure-Ag thin core layer.

Ag layer within the multilayer structures. This confirms that using MDC layers represents a very promising and fruitful approach to the design and development of highly durable optical coatings and thin film structures for various applications, including energy-efficient glass window products and environmentally stable optical sensors.

The energy-saving potential of advanced coated glass products featuring simple MDC based metal-dielectric structures can be illustrated by considering the solar heat filtering performance of a simple generic coating described in Fig. 4.10. Without even constructing any type of purpose designed double-glazing system, it can be shown that a simple single-pane window using only one MDC based coating can demonstrate significant reduction of radiant heat transmission into the internal spaces of buildings.

As can be seen from the data of Fig. 4.10, it is possible to obtain a highly transparent non-color-biased coated glass of very high spectral selectivity using a rather simple coating design based on a single MDC layer. Moreover,

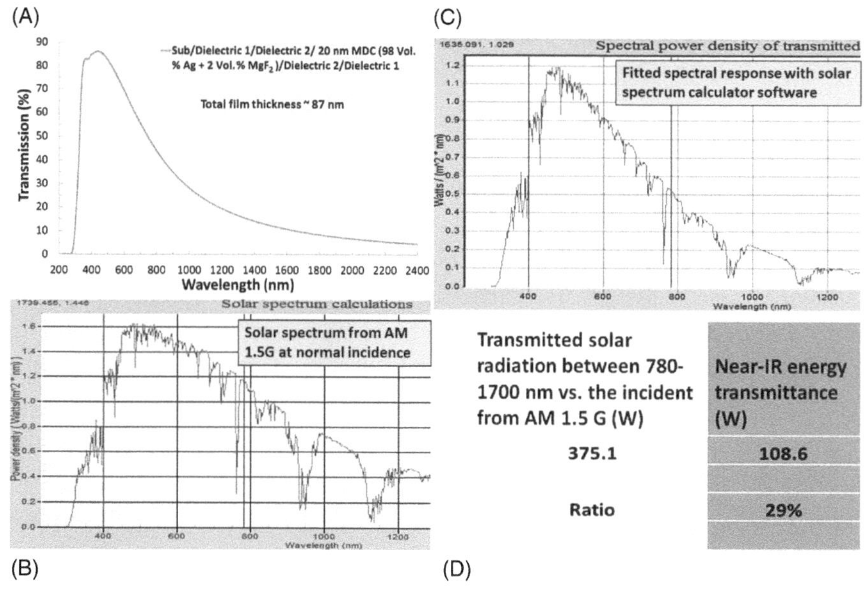

Figure 4.10 Spectral response and solar radiation filtering performance of a sample ultra-thin D/MDC/D multilayer structure. (A) Transmission spectrum of this structure (total film thickness is around 87 nm) deposited onto a low-iron glass substrate; (B) standardized spectral power density distribution of solar irradiation (AM 1.5 G); (C) power density spectrum of solar radiation modified on transmission through the coated glass of part (A); (D) power transmittance of near-IR solar radiation between 780 nm and 1700 nm (per 1 m^2 of glass area) calculated by integration of the spectral power density distribution.

the results of integrating the standardized solar radiation spectrum presumed normally incident onto this coated glass and applying the same procedure to the transmitted spectrum show that only about 29% of solar IR heat can enter the internal building spaces through glass. Dependent on more specific, climate-type related energy saving requirements, it can be concluded easily that an appropriate glazing system can be designed straightforwardly for most climate types.

In practical energy-efficient building applications, another significant issue of importance exists, which is often addressed through the design of Low-E coatings: the apparent color of coated glass, especially in reflection. Even though complete color neutrality is highly desirable in many coated glass products, it is the degree of control over the precise color appearance, which is of utmost importance for both the coating manufacturers and architects. Most advanced software packages allow modeling of the apparent color of coatings, using different colorimetric representations (e.g., CIE chromaticity coordinates, in which the neutral white color corresponds to a point with x, y, z coordinates (1/3, 1/3, 1/3)). Fig. 4.11 illustrates the CIE coordinates and shows the visual representation of the color appearance of a tri-layer DMD coating of Fig. 4.5(A) based on a pure Ag layer. The red triangle indicates the apparent color of this coating structure on glass. By way of optimizing layer thicknesses in DMD multilayer of arbitrary types, it is possible to approach the neutral color point much closer, as illustrated in Fig. 4.11B.

It is often very important to be able to reach through the coating design, and with high accuracy, any particular chromaticity coordinates point in many architectural applications. The degrees of design freedom available to the coating engineers in solving this problem are often limited to only the layer thickness variations, while the common coating materials rarely vary in production. In our THR coating design experience, regardless of the number of metallic or MDC layers within the structure, using the co-deposition (MDC) approach instead of pure Ag can offer a valuable additional degree of flexibility in designing the coatings of predefined required color.

5 Summary

Dielectric/metal/dielectric (DMD) thin film coating sequences continue to represent a cornerstone building block of advanced transparent heat regulation systems since several decades ago. It can be foreseen that these material systems and their nanocomposite-type derivatives will continue to underpin

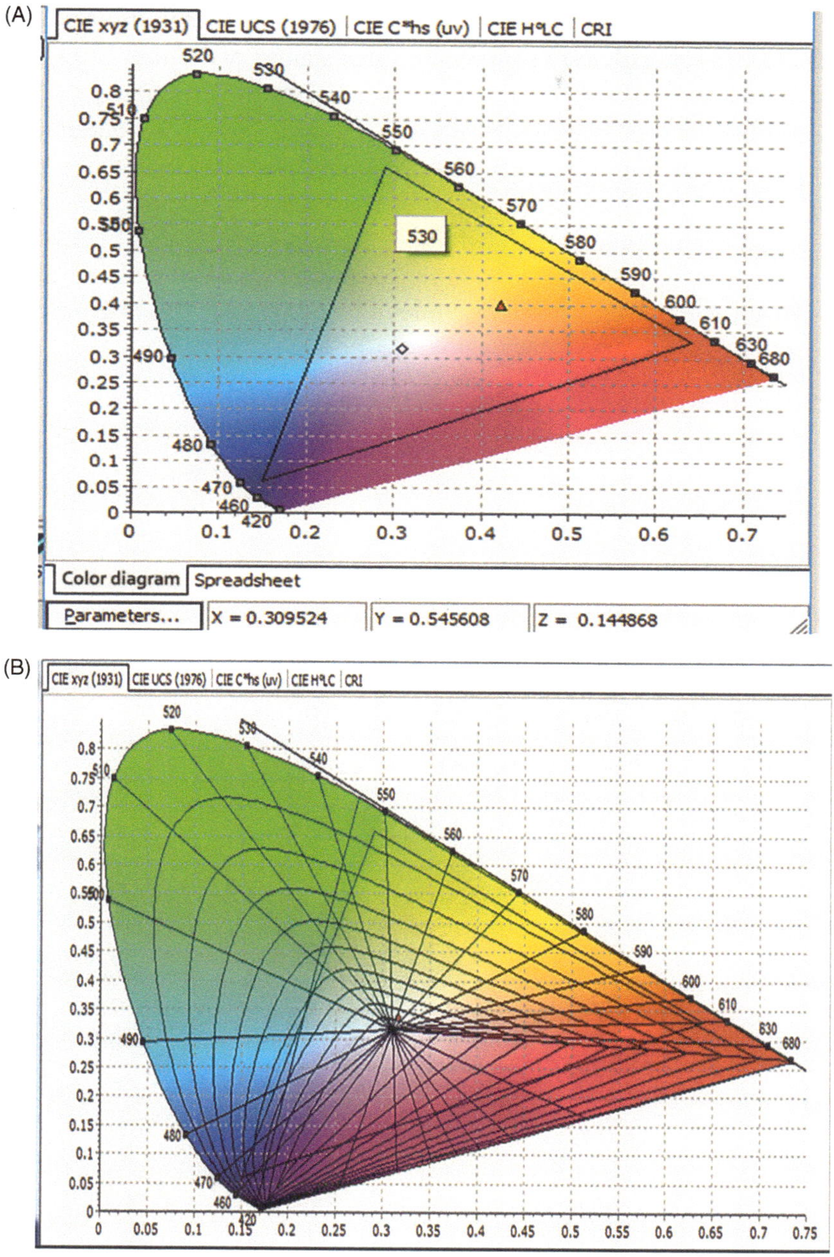

Figure 4.11 Optimized DMD structures employing either one or multiple different dielectric materials in sequence surrounding (A) a single 15 nm thick sputtered Ag layer surrounded by one dielectric material type, or (B) a single 12.5 nm Ag layer surrounded by two different underlying dielectric layers and covered by three different dielectric layers. This THR design of total thickness below 200 nm achieved an almost completely color-neutral appearance with Tvis (max) ≥85%. These graphical datasets were generated using OptiLayer software.

the ongoing development of new energy saving and heat regulation components. DMD materials can be fabricated using a range of modern PVD techniques, and are of interest for use in advanced glazings. While many research questions still remain in areas related to the materials science of ultrathin films and interfaces, significant progress has been demonstrated recently by multiple groups world-wide. The issues addressed so far were related primarily to the nano-scale morphology, growth modes, and process quality. These practical issues have in the past impeded the widespread utilization of advanced THR technologies, which are currently undergoing a rapid evolution.

Acknowledgements

The authors would like to acknowledge the contributions of our fellow researchers (Dr. Yamna El-Mouedden and Dr. Ramzy Alghamedi, formerly also at ESRI, ECU) for their committed efforts dedicated to the deposition process development, and also material characterization experiments, the results of which have greatly assisted us in the preparation of this chapter.

References

1. Macleod, H. A. A New Approach to the Design of Metal-Dielectric Thin-Film Optical Coatings. *Optica Acta: Int. J. Opt.* **1978**, *25* (2), 93–106.
2. Karlsson, B.; Valkonen, E.; Karlsson, T., et al. Materials for Solar-Transmitting Heat-Reflecting Coatings. *Thin Solid Films* **1981**, *86* (1), 91–98.
3. Guenther, K. H. Physical and Chemical Aspects in the Application of Thin Films on Optical Elements. *Appl. Opt.* **1984**, *23* (20), 3612–3632.
4. Bass, M. 2nd ed.; *Handbook of Optics, 1st Volume*McGraw-Hill: America, 1995.
5. Pulker, H. K. *Coatings on Glass*, 2nd ed.; Elsevier: The Netherlands, 1999.
6. Liu, X.; Cai, X.; Mao, J., et al. ZnS/Ag/ZnS Nano-Multilayer Films for Transparent Electrodes in Flat Display Application. *Appl. Surf. Sci.* **2001**, *183* (1–2), 103–110.
7. Liu, X.; Cai, X.; Qiao, J., et al. The Design of ZnS/Ag/ZnS Transparent Conductive Multilayer Films. *Thin Solid Films* **2003**, *441* (1–2), 200–2006.
8. Sivaramakrishnan, K.; Alford, T. L. Metallic Conductivity and the Role of Copper in ZnO/Cu/ZnO Thin Films for Flexible Electronics. *Appl. Phys. Lett.* **2009**, *94* (052104)doi: 10.1063/1.3077184.
9. Lewis, J.; Grego, S.; Chalamala, B., et al. Highly Flexible Transparent Electrodes for Organic Light-Emitting Diode-Based Displays. *Appl. Phys. Lett.* **2004**, *85* (3450)doi: 10.1063/1.1806559.
10. Cho, H.; Yun, C.; Park, J. W., et al. Highly Flexible Organic Light-Emitting Diodes Based on ZnS/Ag/WO_3 Multilayer Transparent Electrodes. *Org. Electron.* **2009**, *10* (6), 1163–1169.
11. Sahu, D. R.; Lin, S. Y.; Huang, J. L. Investigation of Conductive and Transparent Al-Doped ZnO/Ag/Al-Doped ZnO Multilayer Coatings by Electron Beam Evaporation. *Thin Solid Films* **2008**, *516* (15), 4728–4732.
12. Li, X. F.; Chen, Y. R.; Miao, J., et al. High Solar Absorption of a Multilayered Thin Film Structure. *Opt. Exp.* **2007**, *15* (4), 1907–1912.
13. Thelen, A. Multilayer Filters with Wide Transmittance Bands. *J. Opt. Soc. Am.* **1963**, *63* (11), 1266–1270.

14. Lampert, C. M. Advanced Optical Materials for Energy Efficiency and Solar Conversion. *Solid Wind Technol.* **1987**, *4* (3), 347–379.
15. Yaremchuk, I. Y.; Fitio, V.M.; Bobitski, Y.B. Optical properties of multilayer thin-film interference filter. In *8th International Conference on Laser and Fiber-Optical Networks Modeling*, Kharkiv, Ukraine, 2006; pp. 117–120.
16. Leftheriotis, G.; Yianoulis, P.; Patrikios, D. Deposition and Optical Properties of Optimised ZnS/Ag/ZnS Thin Films for Energy Saving Applications. *Thin Solid Films* **1997**, *306* (1), 92–99.
17. Durrani, S. M. A.; Khawaja, E. E.; Al-Shukri, A. M., et al. Dielectric/Ag/Dielectric Coated Energy-Efficient Glass Windows for Warm Climates. *Energy Build.* **2004**, *36* (9), 891–898.
18. Kanu, S. S.; Binions, R. Thin Films for Solar Control Applications. *Proc. R. Soc. A* **2010**, *466*, 19–44.
19. Lampert, C. M. Heat Mirror Coatings for Energy Conserving Windows. *Solid Wind Technol.* **1981**, *4* (3), 347–379.
20. Jaksic, Z.; Maksimovic, M.; Sarajlic, M. Silver-Silica Transparent Metal Structures as Bandpass Filters for the Ultraviolet Range. *J. Opt. A: Pure Appl. Opt.* **2005**, *7*, 51–55.
21. Berini, P. Plasmon-Polariton Modes Guided by a Metal Film of Finite Width Bounded by Different Dielectrics. *Opt. Exp.* **2000**, 7 (10), 329–335.
22. Feng, S.; Elson, J.; Overfelt, P. Optical Properties of Multilayer Metal-Dielectric Nanofilms with All-Evanescent Modes. *Opt. Exp.* **2005**, *13* (11), 4113–4124.
23. Maier, S. A.; Atwater, H. A. Plasmonics: Localization and Guiding of Electromagnetic Energy in Metal/Dielectric Structures. *J. Appl. Phys.* **2005**, *98* (011101)doi: 10.1063/1.1951057.
24. Choi, K. C.; Nam, H. J.; Jeong, J. A., et al. Highly Flexible and Transparent $InZnSnO_x$/Ag/$InZnSnO_x$ Multilayer Electrode for Flexible Organic Light Emitting Diodes. *Appl. Phys. Lett.* **2008**, *92* (223302).
25. Jeong, J. A.; Kim, H. K. Low resistance and Highly Transparent ITO-Ag-ITO Multilayer Electrode Using Surface Plasmon Resonance of Ag Layer of Bulk-Heterojunction Organic Solar Cells. *Solar Energy Mater. Solar Cells* **2009**, *93* (10), 1801–1809.
26. Kim, S.; Lee, J. L. Design of Dielectric/Metal/Dielectric Transparent Electrodes for Flexible Electronics. *J. Photon. Energy* **2012**, *2* (1), 021215.
27. Cattin, L.; Bernede, J. C.; Morsli, M. Toward Indium-Free Optoelectronic Devices: Dielectric/Metal/Dielectric Alternative Transparent Conductive Electrode in Organic Photovoltaic Cells. *Phys. Status Solidi A* **2013**, *210* (6), 1047–1061.
28. Schubert, S.; Meiss, J.; Meskamp, L. M., et al. Improvement of Transparent Metal Top Electrodes for Organic Solar Cells by Introducing a High Surface Energy Seed Layer. *Adv. Energy Mater.* **2013**, *3*, 438–443.
29. Leftheriotis, G.; Koubli, E.; Yianoulis, P. Combined Electromic-Transparent Conducting Coatings Consisting of Noble Metal, Dielectric and WO_3 Multilayers. *Solar Energy Mater. Solar Cells* **2013**, *116*, 110–119.
30. Gaspera, E. D.; Peng, Y.; Hou, Q., et al. Ultra-Thin High Efficiency Semitransparent Perovskite Solar Cells. *Nano Energy* **2015**, *13*, 249–257.
31. Seo, K. W.; Kim, H. K. Plasma Damage Free Sputtering of Ti-Doped In_2O_3 Film on Ag Nanowire Network for Transparent and Flexible Electrodes. *Thin Solid Films* **2015**, *591*, 301–304.
32. Kim, J. H.; Kim, D. S.; Kim, S. K., et al. Highly Flexible Al-Doped ZnO/Ag/Al-Doped ZnO Multilayer Films Deposited on PET Substrates at Room Temperature. *Ceram. Int.* **2016**, *42*, 3473–3478.
33. Kim, T. H.; Park, S. H.; Kim, D. H., et al. Roll-to-roll Sputtered ITO/Ag/ITO Multilayers for Highly Transparent and Flexible Electrochromic Applications. *Solar Energy Mater. Solar Cells* **2017**, *160*, 203–210.

34. Dalapati, G. K.; Kushwaha, A. K.; Sharma, M., et al. Transparent Heat Regulating (THR) Materials and Coatings for Energy Saving Window Applications: Impact of Materials Design, Micro-Structural, and Interface Quality on the THR Performance. *Prog. Mater. Sci.* **2018**, *95*, 42–131.
35. International Glazing Database, Berkeley Lab, USA. Available online at: <https://windows.lbl.gov/software/igdb>, (accessed 23 October 2018).
36. Jia, J. H.; Zhou, P.; Xie, H., et al. Study of Optical and Electrical Properties of TiO_2/Ag/TiO_2 Multilayers. *J. Korean Phys. Soc.* **2004**, *44* (3), 717–721.
37. Jeong, J. A.; Kim, H. K. Al_2O_3/Ag/Al_2O_3 Multilayer Thin Film Passivation Prepared by Plasma Damage-Free Linear Facing Target Sputtering for Organic Light Editing Diodes. *Thin Solid Films* **2013**, *547*, 63–67.
38. Lampton, M. *Trade Study: Why Silver*, <https://www.ssl.berkeley.edu/~mlampton/WhySilver.pdf>, 2007 (accessed 06 July 2018).
39. Nikolajsen, T.; Leosson, K.; Salakhutdinov, I., et al. Polymer-Based Surface-Plasmon-Polariton Stripe Waveguides at Telecommunication Wavelengths. *Appl. Phys. Lett.* **2003**, *82* (5), 668.
40. Chen, J.; Yue, S.; Gong, Q. Hybrid Long-Range Surface Plasmon-Polariton Modes with Tight Field Confinement Guided by Asymmetrical Waveguides. *Opt. Exp.* **2009**, *17* (26), 23603–23609.
41. Nagaraj; Krokhin, A. A. Long-Range Surface Plasmons in Dielectric-Metal-Dielectric Structure with Highly Anisotropic Substrates. *Phys. Rev. B* **2010**, *81*, 085426.
42. Marini, A.; Skryabin, D.V.; Malomed, B. Stable Spatial Plasmon Solitons in a Dielectric-Metal-Dielectric Geometry with Gain And Loss. *Opt. Exp.* **2011**, *19* (7), 6616–6622.
43. Yan, J.Y.; Li, L.; Xiao, J. Ring-Like Solitons in Plasmonic Fiber Waveguides Composed of Metal-Dielectric Multilayers. *Opt. Exp.* **2012**, *20* (3), 1945–1952.
44. Michel, D.; Xiao, F.; Skillman, L., et al. Surface Plasmon Resonance sensor for in situ Detection of Xanthan Gum. *IEEE J. Sel. Top. Quantum Electron.* **2016**, *22* (3), 379–382.
45. Kuzmichev, A. N.; Kerilkamp, L. E.; Nur-E-Alam, M., et al. Tunable Optical Nanocavity of Iron-Garnet with a Buried Metal Layer. *Materials* **2015**, *8*, 3012–3023.
46. Sekatskii, S. K.; Smirnov, A. A.; Dietler, G., et al. Photonic Crystal-Supported Long-Range Surface Plasmon-Polaritons Propagation Along High-Quality Silver Nanofilms. *Appl. Sci.* **2018**, *8*, 248.
47. Carton, O.; Ghaymouni, J.; Lejeune, M., et al. Optical Characterization of Porous Sputtered Silver Thin Films. *J. Spectrosc.* **2013**, *1*, 307824.
48. Chen, W.; Thoreson, M. D.; Kildishev, A. V. Ultra-Thin Ultra-Smooth and Low Loss Silver and Silver-Silica Composite Films for Superlensing Applications. CLEO: QELS_ Fundamental Science 2010, San Jose, California, United States. <https://engineering.purdue.edu/~shalaev/Publication_list_files/QThE4.pdf>, 2010.
49. Anders, A.; Byon, E.; Kim, D., et al. Smoothing of Ultrathin Silver Films by Transition Metal Seeding. *Solid State Commun.* **2006**, *140* (5), 225–229.
50. Singer, R. R.; Leitner, A.; Aussenegg, F. R. Structure Analysis and Models for Optical Constants of Discontinuous Metallic Silver Films. *J. Opt. Soc. Am. B* **1995**, *12* (2), 2220–2228.
51. Ohring, M. *Materials Science of Thin Films; Deposition and Structure*, 2nd ed.; Academic Press: San Diego, 2002.
52. Fedorovich, R. D.; Naumovets, A. G.; Tomchuk, P. M. Electron and Light Emission from Island Metal Films and Generation of Hot Electrons in Nanoparticles. *Phys. Rep.* **2000**, *328*, 73–179.
53. Lazzari, R.; Juppile, J. Silver Layers on Oxide Surfaces: Morphology and Optical Properties. *Surf. Sci.* **2001**, *482*, 823–828.
54. Didier, F.; Jupille, J. Layer-by-Layer Growth Mode of Silver on Magnesium Oxide (100). *Surf. Sci.* **1994**, *307–309*, 587–590.

55. Roy, A.; Bhattacharjee, K.; Ghatak, J. Growth of Epitaxially Oriented Ag Nanoislands on Air-Oxidized Si (111)-(7 × 7) Surfaces: Influence of Short Range Order on the Substrate, *PACS* 68.55.-a; 81.07.Bc; 61.05.jh., 2010.
56. Mehrer, H. *Diffusion in Solids: Fundamentals, Methods, Materials, Diffusion-Controlled Processes* Springer Science and Business Media: Berlin, Germany, 2007.
57. Chen, W.; Thoreson, M. D.; Ishii, S., et al. Ultra-Thin Ultra-Smooth and Low-Loss Silver Films on a Germanium Wetting Layer. *Opt. Exp.* **2010**, *18* (5), 5124–5134.
58. Bulir, J.; Novotny, M.; Lynnykova, A., et al. Preparation of Nanostructured Ultrathin Silver Layer. *J. Nanophotonics* **2011**, *5*, 1–10 051511.
59. Chuan, N. Effects of Different Wetting Layers on the Growth of Smooth Ultra-Thin Silver Thin Films (Ph.D. thesis); Dayton, OH, USA: University of Dayton, 2014.
60. Ciesielski, A.; Skowronski, L.; Gorecka, E., et al. Growth Model and Structure Evolution of Ag Layers Deposited on Ge Films. *Beilstein J. Nanotechnol.* **2018**, *9*, 66–76.
61. Jiang, X.; Chen, S.; Mao, C. Synthesis of Ag/SiO_2 Nanocomposite Material by Adsorption Phase Nanoreactor Technique. *Colloids Surf. A* **2008**, *320*, 104–110.
62. Yu, B.; Leung, K. M.; Guo, Q., et al. Synthesis of Ag–TiO_2 Composite Nano Thin Film for Antimicrobial Application. *Nanotechnology* **2011**, *22*, 115603.
63. Cummings, K. D.; Garland, J. C.; Tanner, D. B. Optical Properties of a Small-Particle Composite. *Phys. Rev. B* **1984**, *30* (8), 4170–4182.
64. Nielsen, R. B.; Thoreson, M. D.; Chen, W., et al. Toward Superlensing with Metal–Dielectric Composites and Multilayers. *Appl. Phys. B* **2010**, *100*, 93–100.
65. Necula, B. S.; Apachitei, L. E. F.; Zaat, S. A. J., et al. In Vitro Antibacterial Activity of Porous TiO_2-Ag Composite Layers Against Methicillin-Resistant *Staphylococcus aureus*. *Acta Biomater.* **2009**, doi: 10.1016/j.actabio.2009.05.010.
66. Amin, K. A. M.; Panhuis, M. Reinforced Materials Based on Chitosan, TiO_2 and Ag Composites. *Polymers* **2012**, *4*, 590–599.
67. Chiang, H. P.; Leung, P. T.; Tse, W. S. Optical Properties of Composite Materials at High Temperatures. *Solid State Commun.* **1997**, *101* (1), 45–50.
68. Zhao, Y. L.; Xu, X.; Ming, H. Ag-MgF_2 Composite Films Deposited by RF Magnetron Sputtering. *J. Funct. Mater.* **2007**, *38* (3), 386–388.
69. Gu, D.; Zhang, C.; Wu, Y. K., et al. Ultrasmooth and Thermally Stable Silver-Based Thin Films with Subnanometer Roughness by Aluminum Doping. *ACS Nano* **2014**, *8* (10), 10343–10351.
70. Wakaki, M.; Soujima, N.; Shibuya, T. Optical Properties of MgF2 Nano-Composite Films Dispersed with Noble Metal Nanoparticles Synthesized by Sol-Gel Method. Proc. SPIE 9359, Optical Components and Materials XII, 2015, 93590A; DOI: 10.1117/12.2077045.
71. Willey, R. R. *Practical Design and Production of Optical Thin Films*. 2nd ed.; CRC Press, 2002.
72. Thelen, A. *Design of Optical Interference Coatings*. McGraw-Hill Book Company: New York, 1989.
73. Palik, E. D. *Handbook of Optical Constants of Solids*. Academic Press, 1991.
74. Available on <http://www.luxpop.com/> (accessed 12 June 2011).
75. El-Mouedden, Y.; Alghamedi, R.; Nur-E-Alam, M. Thin Film Coatings for Solar and Thermal Radiation Control Prepared by Physical Vapour Deposition. In *9th International Conference on High Capacity Optical Networks and Enabling Technologies (HONET)*, Istanbul, Turkey, 2012.
76. Nicolet, M. A. Diffusion Barriers in Thin Films. *Thin Solid Films* **1978**, *52*, 415–443.
77. Dalapati, G. K. A Multilayer Coating. US Patent Application, 15/568,456, 2018.
78. Marechal, N.; Quesnel, E. Silver Thin Films Deposited by Magnetron Sputtering. *Thin Solid Films* **1994**, *241*, 34–38.

79. Kuhaili, M. F. A.; Aswad, A. H. A.; Durrani, S. M. A., et al. Energy-Saving Transparent Heat Mirrors Based on Tungsten Oxide-gold WO_3/Au/WO_3 Multilayer Structures. *Solar Energy* **2012**, *86*, 3183–3189.
80. Mohelnikova, J. Materials for Reflective Coatings of Window Glass Applications. *Constr. Build. Mater.* **2009**, *23*, 1993–1998.
81. Prieto, J. E.; Markov, I. Stranski–Krastanov Mechanism of Growth and the Effect of Misfit Sign on Quantum Dots Nucleation. *Surf. Sci.* **2017**, *664*, 172.
82. Lee, J. H.; Lee, S. H.; Hwangbo, C. K. Optical and Structural Properties of TiO_2/Ti/Ag/TiO_2 and TiO_2/ITO/Ag/ITO/TiO_2 Metal-Dielectric Multilayers by RF Magnetron Sputtering for Display Application. *J. Korean Phys. Soc.* **2004**, *44* (3), 750–756.

CHAPTER 5

Novel materials and concepts for regulating infra-red radiation: radiative cooling and cool paint

Yee-Fun Lim
Institute of Materials Research and Engineering, A*STAR (Agency for Science, Technology and Research), Singapore

1 Introduction

Air conditioning and refrigeration are estimated to consume 17% of all electricity usage worldwide *(1)*. This figure becomes much higher in hot tropical countries; in Mumbai, India, air conditioning alone accounts for 40% of the city's total power consumption *(2)*, while the average household in Singapore spends 30% of its energy bills on air-conditioning *(3)*. Furthermore, a combination of income growth and climate change is expected to drive up the energy demand for air-conditioning by 72% by the year 2100 *(4)*. Clearly, efforts to mitigate this demand will have great impact in terms of energy and cost savings as well as reduction in carbon emissions.

There is thus an urgent need to invest in energy saving window technologies, and also other technologies that mitigate the building heat gain envelope by regulating infra-red (IR) radiation *(5)*. Indeed, such technologies are the main focus of this book. Fig. 5.1 provides a summary of the various heat regulating mechanisms.

IR reflection is perhaps the most straight-forward and well known. Building heat gain is reduced by reflecting incoming solar irradiation, and this is the main basis of solar films, passive IR rejecting windows, and smart electrochromic windows *(5)*. Chapters 1–4 of this book describe this process in much detail. The effectiveness depends on how much IR radiation is reflected, and there is often a trade-off between window transparency and heat rejection. Another drawback of this technology is that the reflected heat stays in the building environment, and would end up heating the surrounding air. This is especially problematic if heat reflecting windows are

Energy Saving Coating Materials
http://dx.doi.org/10.1016/B978-0-12-822103-7.00005-4

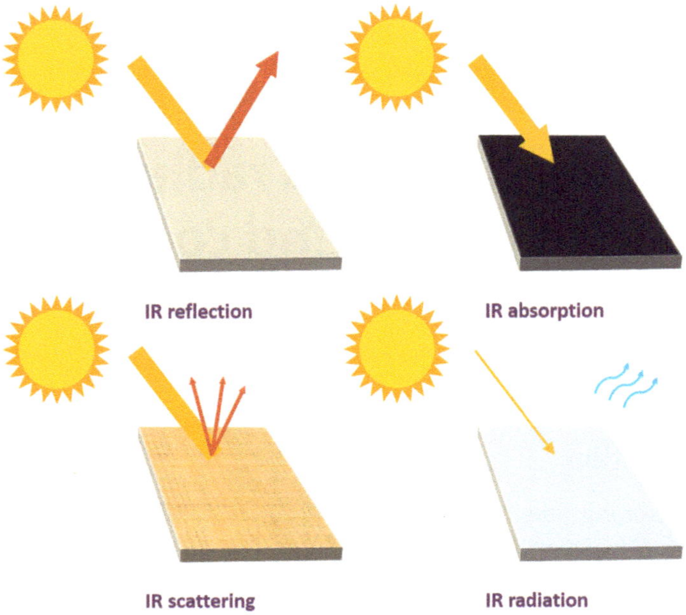

Figure 5.1 Working principles of materials that regulate infra-red radiation.

installed to directly face adjacent buildings, as the reflected IR radiation will end up increasing their building heat envelope instead.

The second process is IR absorption. By incorporating a material that absorbs much of the solar radiation, transmission into the building is greatly reduced. This is the main principle behind thermochromic and electrochromic windows *(6)* as described in Chapter 9, and also that of phase change materials for thermal energy storage in building applications *(7)*. While the transmittance reduction helps reduce the internal building temperatures, the absorbing material itself heats up, so some heat is still passed into the building via conduction. Effective heat dissipation strategies are necessary for this method to be effective.

The third process is the scattering of IR radiation. The effect is rather similar to IR reflection, since scattering is akin to a kind of diffused reflection as compared to specular reflection by a smooth mirror surface. IR scattering is particularly effective with particles of sizes that are on the order of the wavelength of visible and IR light *(8)*, and is the working mechanism of cool paint which will be described in detail in this chapter. Like heat rejecting windows, cool paint can reflect unwanted IR radiation away from the building, while mitigating the effect of heating up adjacent buildings due to the diffuse nature of the reflection.

The last process shown in Fig. 5.1 is radiative cooling, which makes use of the natural and passive black-body radiation of objects with nonzero absolute temperatures. Although it has been investigated since as early as the 1970s *(9)*, it is relatively unknown when compared to other heat mitigation strategies. Nevertheless, it is a true cooling process as compared to the others, which can be described as heat transfer processes. The concepts of radiative cooling and cool paint will be explained and elaborated extensively in this chapter, with the objective of raising the awareness of these lesser known concepts and encouraging more adoption and applications in building heat mitigation strategies.

2 Radiative cooling

2.1 Theoretical considerations

The Sun, Earth, and outer space can be modeled as black-bodies with absolute temperatures of approximately 5800 K, 300 K, and 3 K, respectively (Fig. 5.2). While the temperature of outer space is not at absolute zero due to the cosmic microwave background, it nevertheless represents extreme cold when compared to the Earth, and this cold darkness of the universe has been described as a renewable thermodynamic resource *(10)*, serving as the ultimate heat sink for the radiative heat output from the Earth. This is the basic mechanism of radiative cooling, which is passive cooling without energy inputs based on the natural black-body radiation of hot objects.

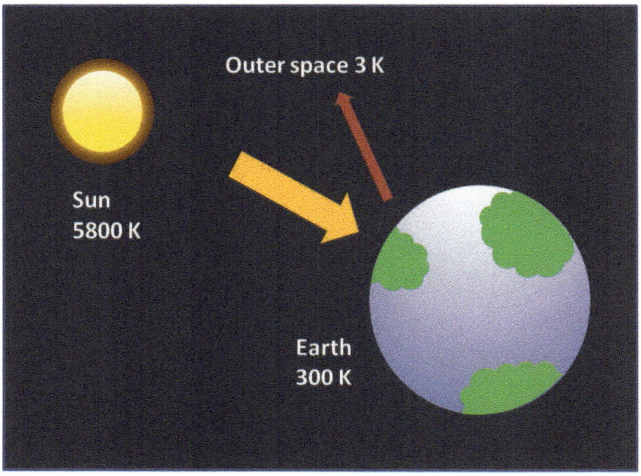

Figure 5.2 Illustration of the Sun, Earth, and outer space as black-bodies with their respective absolute temperatures. Radiative cooling works through the use of outer space as the ultimate heat sink, receiving the radiative heat output from the Earth.

The black-body radiation spectrum for an object at a temperature T may be derived from thermodynamics and quantum mechanics. It is given by Planck's Function $B_\lambda(T)$:

$$B_\lambda(T) = \frac{2hc^2}{\lambda^5\left(e^{hc/k\lambda T} - 1\right)}$$

where $h = 6.626 \times 10^{-34}$ Js is Planck's constant, $c = 3 \times 10^8$ m/s is the speed of light, λ is the wavelength, $k = 1.38 \times 10^{-23}$ J/K is Boltzmann's constant, and T is the absolute temperature *(8)*. Here, $B_\lambda(T)d\lambda$ may be interpreted as the emitted radiation intensity per unit solid angle contributed by the wavelength range λ to $\lambda + d\lambda$. Fig. 5.3A shows the spectral irradiation from the sun incident on the Earth, with and without atmospheric absorption (AM 1.5 and AM0 respectively, where AM stands for Air Mass). The AM0 spectrum corresponds approximately to a black-body at 5800 K, while the AM1.5 spectrum shows various absorption bands of the Earth's atmosphere. Meanwhile, the black-body spectra corresponding to common temperatures of terrestrial objects (in the range of 275–350 K) are plotted in Fig. 5.3(B). It should be noted that the black-body radiation of terrestrial objects peak in the spectral range of 6–14 μm.

When the Planck's Function is integrated over all wavelengths, a quantity known as the broadband flux $F(T)$ as a function of temperature is obtained, which is given by the Stefan–Boltzmann law:

$$F(T) = \sigma T^4$$

where $\sigma = 5.67 \times 10^{-8}$ W/m^{2}K^4 is the Stefan–Boltzmann constant *(8)*. This equation indicates that the total radiated power scales as the fourth power of the temperature.

Another quantity that is relevant to our understanding of radiative processes is the emissivity ε of a surface. So far, we have treated objects as ideal black-bodies, meaning they emit radiation according to the Planck's Function. In reality, objects and surfaces emit at a fraction of a black-body, and thus they may be more accurately referred to as gray-bodies. The fraction of radiation emitted is the emissivity ε, which is given by

$$\varepsilon_\lambda = \frac{I_\lambda}{B_\lambda(T)}$$

where ε_λ is the monochromatic emissivity at the wavelength λ, and I_λ is the actual intensity emitted *(8)*. Hence, for a black-body $\varepsilon_\lambda = 1$, while for

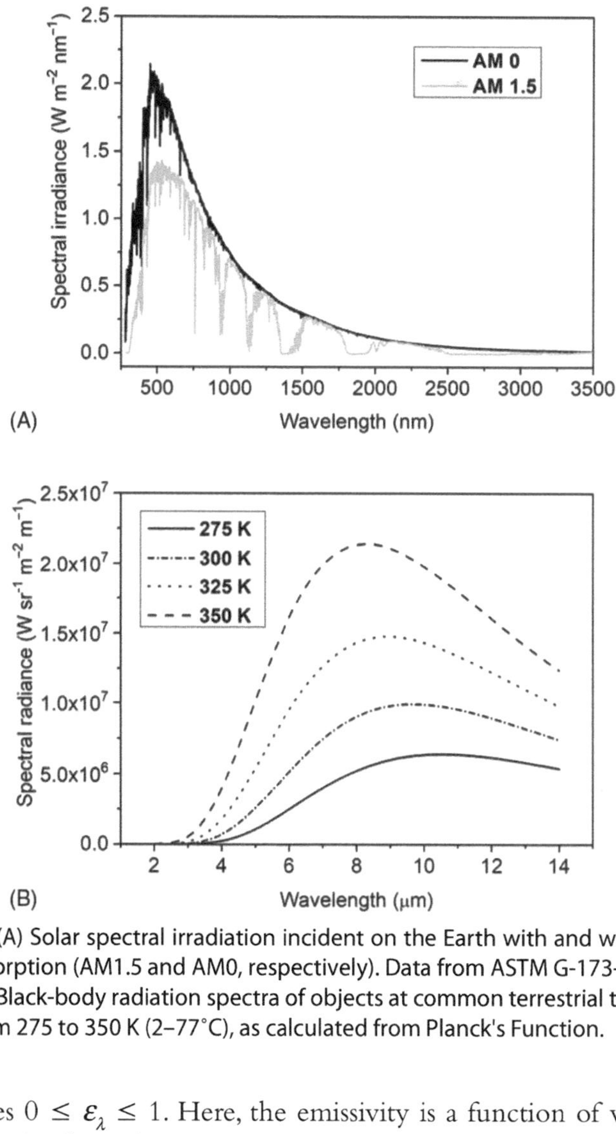

Figure 5.3 (A) Solar spectral irradiation incident on the Earth with and without atmospheric absorption (AM1.5 and AM0, respectively). Data from ASTM G-173-03 reference spectra; (B) Black-body radiation spectra of objects at common terrestrial temperatures ranging from 275 to 350 K (2–77°C), as calculated from Planck's Function.

gray-bodies $0 \leq \varepsilon_\lambda \leq 1$. Here, the emissivity is a function of wavelength, but it can also be a function of the viewing angles θ and φ. Moreover, the relationship between emissivity and absorptivity a may be expressed by Kirchoff's Law as follows:

$$\varepsilon_\lambda(\theta, \varphi) = a_\lambda(\theta, \varphi)$$

The emissivity is thus equal to the absorptivity, meaning that a highly emitting surface is also a highly absorbing one at the particular wavelength λ *(8)*.

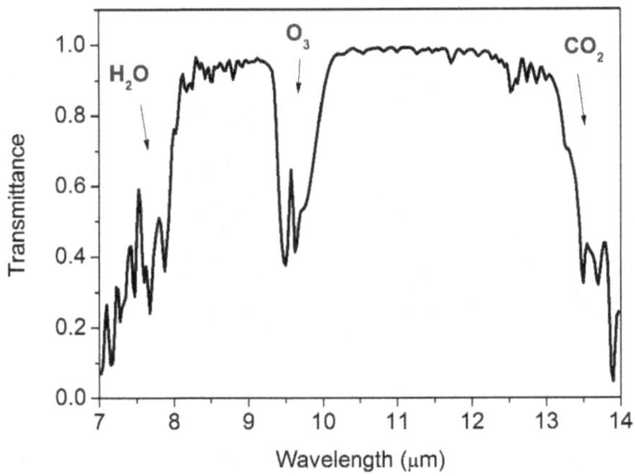

Figure 5.4 Spectral transmittance of the Earth's atmosphere at mid infra-red wavelengths *(11)*. Transmittance dips are due to absorption bands from H_2O, O_3, and CO_2, as indicated.

With the thermodynamic preliminaries out of the way, we are now ready to discuss the desired material properties for objects and surfaces that exhibit efficient radiative cooling. In order for the black-body radiation to escape to outer space, the object should emit in the infra-red window or the atmospheric transparency window, which is a spectral region in the wavelength range of 8–13 μm whereby the atmosphere is relatively transparent to radiation. This is illustrated in Fig. 5.4, which is a plot of atmospheric transmittance over Mauna Kea as measured by the Gemini observatory *(11)*. It can be seen that the atmosphere is largely transparent in the afore-mentioned wavelength range, with some dips in the transmittance corresponding to absorption by H_2O, O_3, and CO_2 gases. Coincidentally, this infra-red window also corresponds to the peaks in the black-body emission spectra of terrestrial objects at typical ground temperatures. Hence, it is absolutely essential for a surface to have high emissivity (as close to one as possible) in the infra-red window, in order for it to exhibit efficient radiative cooling.

On the other hand, the emitting surface should avoid absorbing the incoming solar irradiation as much as possible, to avoid heating up under the sun during daytime. We have seen previously that emissivity equals absorptivity, so it is best that the emissivity is zero outside the infra-red window. This is especially the case in the visible to near-IR spectral region, where the solar irradiation is significant (Fig. 5.3A). We have plotted the emissivity spectrum of the ideal selective emitter in Fig. 5.5. Such an ideal emitter would exhibit optimum radiative cooling performance, yet remain cool during the day when exposed to direct sunlight *(12)*.

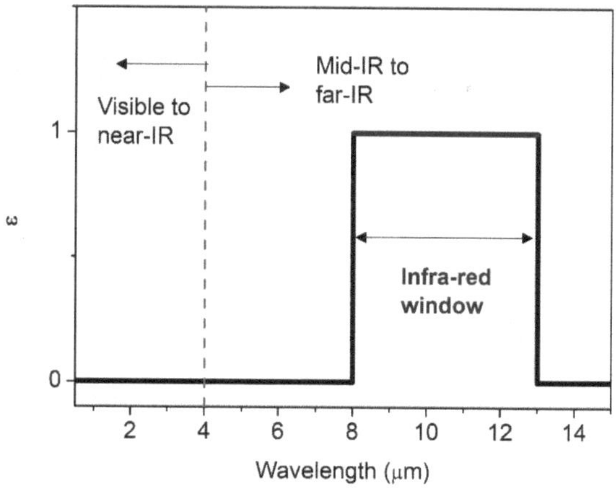

Figure 5.5 Emissivity profile of the ideal selective emitter for optimum radiative cooling performance. The emissivity should be maximum in the infra-red window, while minimum elsewhere especially in the visible to near-IR region to avoid absorbing energy from the incident solar irradiation.

It is instructive to calculate the cooling power P_{cool} that is possible due to the radiative cooling effect. To do this, we can integrate over the black-body spectral radiance and the emissivity across the infra-red window:

$$P_{cool} = \int d\Omega \cos(\theta) \int_{8\mu m}^{13\mu m} d\lambda B_{\lambda}(T) \varepsilon_{\lambda}(\theta, \varphi)$$

where $B_{\lambda}(T)$ is Planck's function and $\varepsilon_{\lambda}(\theta,\phi)$ is the monochromatic emissivity as defined previously. For a near-ideal selective emitter with $\varepsilon_{\lambda}(\theta,\phi) \approx 1$ in the infra-red window, it is possible to obtain a cooling power of around 150 W/m^2 for an emitter at 300 K *(13)*. This cooling power is significant; considering that the solar irradiance (AM0) is around 1360 W/m^2 *(14)*, the radiative cooling power can be over 10% that of the incident solar intensity.

We will conclude this section with a discussion on the effects of humidity and weather conditions. In Fig. 5.4 we have indicated some of the gases that are responsible for the spectral absorption lines of the atmosphere, including CO_2, O_3 and H_2O. The amount of CO_2 and other absorbing gases such as CO, N_2O, CH_4, and O_2 is roughly constant across the globe, while O_3 and H_2O can be highly variable *(8)*. In particular, the amount of water vapor in the air can have large effects on the atmospheric transmittance in the infra-red window; Granqvist et al. showed that a very humid atmosphere can reduce the transmittance by over 50% *(15)*, while a dry atmosphere is largely transparent in the infra-red window as indicated in Fig. 5.4. Another factor to consider is the cloud cover. Clouds are mostly

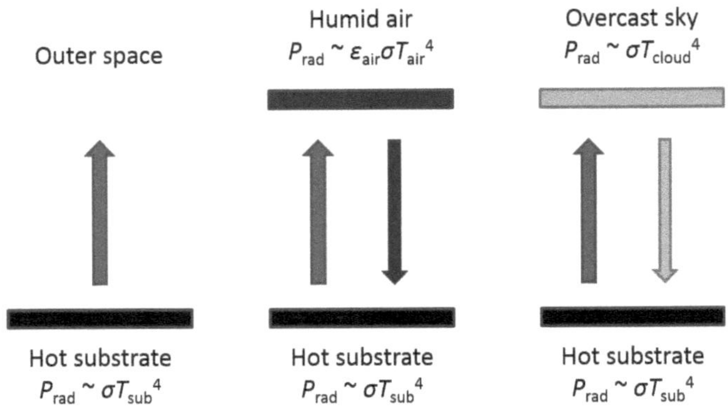

Figure 5.6 Radiative cooling under different environmental conditions. Ideal conditions are dry air and cloudless skies, so radiation can escape directly into outer space (left). Humid (middle) and cloudy (right) conditions reduce the cooling effect, since the substrate will reabsorb black-body radiation emitted from humid air and clouds.

comprised of liquid water and ice particles, which are highly opaque to infra-red radiation longer than 4 μm *(8)*. An overcast sky would therefore essentially eliminate the infra-red window.

The effect of the afore-mentioned environmental conditions may be summarized in Fig. 5.6, where we consider humid air and clouds as gray-bodies and black-bodies respectively, emitting black-body radiation back onto the hot substrate. To give an estimate of this effect, we take $T_{sub} = 300$ K, T_{air} and $T_{cloud} \sim 295$ K (temperature vs. height show a lapse rate of 6.5 K/km *(15)*, and we consider interaction with low-lying atmospheric air and clouds at a height up to ~1 km), and $\varepsilon_{air} \sim 0.5$. Plugging these values into the equations for radiated power (Fig. 5.6), we find that the drop in cooling power due to humid air and overcast skies are 47% and 93% respectively. These effects are significant, and indicate that radiative cooling would have a reduced (although non-zero) effect in a tropical hot and humid climate.

2.2 Experimental demonstrations

The potential cost savings from radiative cooling are huge; a modeling study by the Pacific Northwest National Laboratory (PNNL, USA) on the impact of radiative cooling concluded that cooling utility bill savings of 45%–68% can be achieved for five different cities across America *(16)*. The impact that radiative cooling can make to mitigate the effects of climate change is thus significant, and has motivated recent efforts to investigate and improve radiative cooling technologies. However, the report also mentioned challenges associated with implementation of the technology, such as

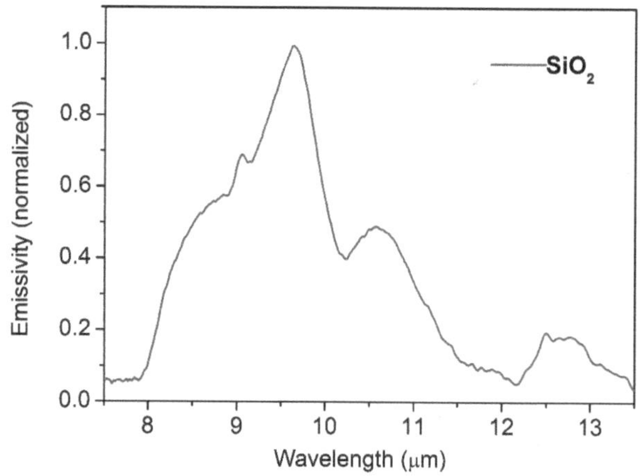

Figure 5.7 Normalized emissivity profile of a thin film silica sample.

difficulties with retrofitting, installation costs, and design complexity. They also came up with a maximum incremental cost from US$2.50 to $6.25 per square meter of rooftop heat exchanger area for a 5-year payback. It is thus prudent to come up with cost-effective setups that can be easily deployed on a large scale.

Before we survey recent efforts to implement radiative cooling technologies, it is instructive to briefly mention some of the popular materials that have been utilized. Recall from our previous discussions and Fig. 5.5 that an ideal radiative cooling material is one that emits strongly in the infra-red window, while having zero emissivity outside this window. A common and popular material that satisfies this criterion is silica, the main constituent of window glass. Silica emits strongly in the neighborhood of 10 μm, while being largely transparent in the visible and near-IR regions. A normalized emissivity profile of a thin film silica sample, as measured by Fourier-Transform Infra-Red spectroscopy (FTIR), is plotted in Fig. 5.7. It can be seen that silica exhibits broadband emissivity across much of the infra-red window, with a peak at around 9.5 μm. Other inorganic materials with strong emissivities in the infra-red window include SiC and Si_3N_4, both of which have been investigated for use in radiative cooling *(12,18,22)*. Organic materials also exhibit strong IR emission in general; an example is polyvinyl fluoride (PVF) with high emissivity also in the 10 μm region *(9,17)*.

Early studies to demonstrate radiative cooling have been based on the configuration illustrated in Fig. 5.8(A) (9,15–17). An emissive material is deposited as a thin film on a reflective substrate such as aluminum, and

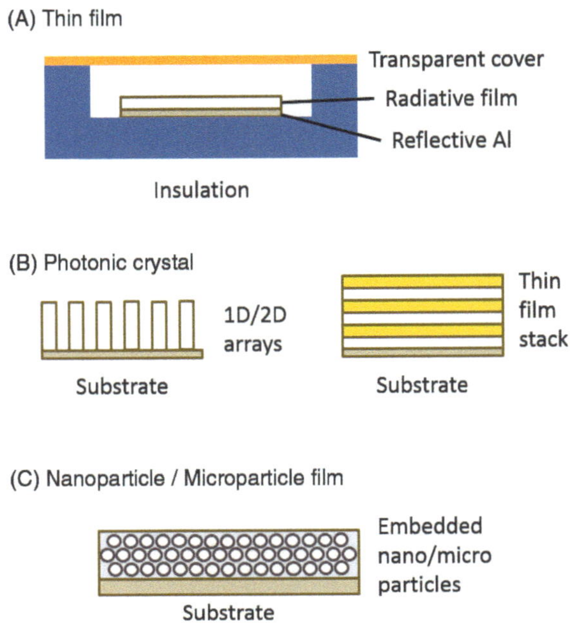

Figure 5.8 Approaches for implementation of selectively emissive surfaces for radiative cooling.

the radiating surface is insulated from the environment in order to achieve maximum cooling effect. PVF is a popular material of choice in these studies *(9, 17)*; in particular, Catalanotti et al. were able to demonstrate a cooling of 15°C below ambient with an insulated 12.5 µm thick PVF sheet that is coated with evaporated aluminum on one side *(9)*. Silica was another popular material of choice; Granqvist and Hjortsberg *(15)* were able to demonstrate a cooling of 13.8°C using just 2.0 µm thick amorphous SiO_x that was prepared by vacuum evaporation onto aluminum. In another study, Harrison and Walton *(16)* investigated the radiative cooling of TiO_2 white paint, with a coating prepared by painting onto an aluminum sheet. In this work, the authors also investigated the effect of different weather conditions; under clear skies, a cooling effect of up to 15°C was observed, while dense cloud cover reduces this to around 6°C.

After the afore-mentioned demonstrations in the 1970s and 1980s, research interest in radiative cooling seemed to have lay dormant for some time, until it was revived in the 2010s with a series of publications from Prof. Shanhui Fan of Stanford University *(10, 12, 13, 18, 19)*. Prof Fan's work is motivated toward producing selective emitters that more closely approach the emission profile of the ideal emitter (Fig. 5.5), which is not attained by that of natural materials (Fig. 5.7). He achieved this with metamaterials (Fig. 5.8(B)) that comprised of photonic crystals in 1D/2D *(18, 19)*, or via a stack of carefully engineered thin films *(10, 12)*; these engineered metamaterials were able to achieve close to 100% emissivity in the infra-red window by virtue of their unique photonic properties. By enclosing his emitter in a high vacuum chamber with a transparent ZnSe window, a record cooling performance of 42°C below ambient was demonstrated *(12)*.

Although photonic metamaterials have demonstrated impressive radiative cooling, their complicated structures require complex fabrication processes, and may lead to relatively high manufacturing costs. As such, there have been recent research efforts targeted towards low-cost and large-area deployment of radiative cooling emitters, which have been based on incorporation of nanoparticles or microparticles (Fig. 5.8C). Zhai et al. *(20)* demonstrated a cooling material based on a 50 μm thick polymer film embedded with ~8 μm diameter silica microspheres that can be produced with roll-to-roll manufacturing. The microsphere diameter was chosen to have maximum interaction with mid-IR radiation via Fröhlich resonances; the emitter showed emissivity close to 1 across the entire infra-red window, and a maximum cooling power of 140 W/m^2, which is close to the ideal value *(20)*. Recently, this technology has been scaled up in the form of a water-cooling module with the ability to cool water to 10.6°C below ambient at noon, which is one of the first demonstrations of a real-world application of radiative cooling *(21)*. In another effort, Bao et al. *(22)* fabricated a selective emitter by spray-coating a double layer of TiO_2 and SiO_2 particles; this approach is attractive since solution spray coating is ideally suited towards low cost and large area deployment. When tested outdoors under the sun, coated aluminum substrates have a temperature that is around 10°C cooler compared to uncoated substrates.

A summary of the various radiative cooling approaches that have been discussed is presented in Table 5.1, together with the materials that have been studied and the reported cooling performance. It can be seen that silica is the most popular material, and a cooling effect of 10°C has been routinely demonstrated. These results bode well for future large-scale deployment, with the first radiative cooling modules already being demonstrated *(21)*. A theoretical study by the US Department of Energy *(23)* showed that a proposed photonic radiative cooling system could save 45%–68% of cooling electricity costs across 5 different cities in the United States, which is highly significant. There is thus much potential for radiative cooling modules to make a great impact in energy savings and reduction of carbon emissions.

Table 5.1 Summary of radiative cooling approaches, materials utilized, and performance.

Approach	Material	Cooling performance	References
Thin film	PVF on Al	15°C below ambient	*(9)*
Thin film	SiO_x on Al	13.8°C below ambient	*(15)*
Paint coating	TiO_2 white paint	15°C below ambient	*(16)*
Thin film stack	SiO_2, HfO_2	5°C below ambient under sunlight	*(10)*
Thin film stack	Si_3N_4, Si	42°C below ambient under sunlight in high vacuum	*(12)*
2D photonic crystal	SiC, quartz	100 W/m^2 (theoretical)	*(18)*
1D photonic crystal	Silica	13°C below uncoated substrate	*(19)*
Embedded microspheres	Silica	140 W/m^2	*(20)*
Embedded microspheres	Silica	10.6°C water cooling below ambient	*(21)*
Nanoparticles	SiO_2, SiC, TiO_2	10°C below uncoated substrate	*(22)*

3 Cool paint

3.1 Theoretical considerations

Cool roofs with high solar reflectance (albedo) that reject a large fraction of the incident solar radiation (Fig. 5.9) have been proposed as a geo-engineering solution to combat global warming *(24)*. The impact of cool roofs is clear: rejection of solar irradiation would reduce building heat envelope, leading to improved indoor thermal comfort or reduction of air-conditioning bills. In fact, several building standards such as the ASHRAE Standard 90.1-2007 have prescribed cool materials for low-sloping roofs *(24)*. Cool paints have been at the forefront of the effort to transform regular roofs into cool roofs; by applying a coat of "cool paint" onto the roof, the solar reflectance has been observed to increase significantly *(25,26)*.

But what is the heat rejection mechanism of the cool paint? To answer this question, it will be necessary to look into the components of the paint. One of the most common pigments found in cool paint is TiO_2 *(26)*, which is a semiconductor with a large band gap of 3.2 eV. As such, it does not absorb light in the visible and near-IR, and it is transparent to most of the solar spectrum unless it is heavily doped (in which case it behaves like a transparent conducting oxide with IR rejection properties) *(27)*. No TiO_2 doping

Figure 5.9 Cool roofs that reflect a significant portion of the solar radiation can lower interior temperatures and save air-conditioning bills.

is mentioned in cool paint literature, so we can rule out IR reflection as a mechanism, even though the effect has always been described as an increase in solar reflectance. As a further clue, we note that the particle size in cool paint should be at least half the wavelength of light, in order for the cooling to be effective *(28)*. From these observations, we can conclude that the dominant cooling mechanism in cool paint is in fact light scattering, which may also be considered as a form of diffused reflection (Figs. 5.1 and 5.10).

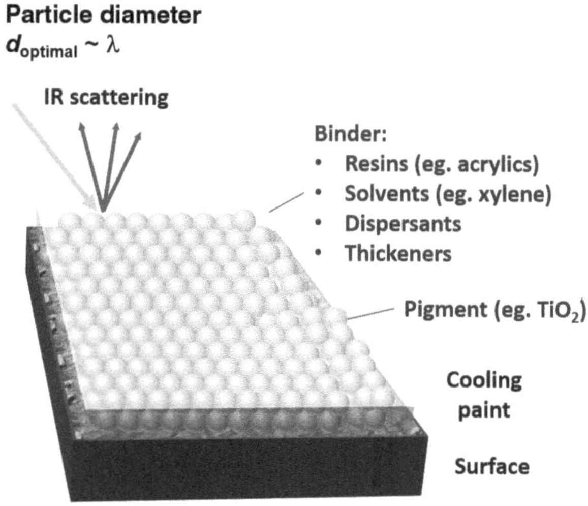

Figure 5.10 Schematic of the cooling mechanism via IR scattering of cool paint.

Table 5.2 Nature of particle interaction with radiation.

Size parameter	Interaction
$x > 2000$	Geometrical optics
$0.2 < x < 2000$	Mie scattering
$0.002 < x < 0.2$	Rayleigh scattering
$x < 0.002$	Negligible scattering

The main criterion that influences the nature of a particle's interaction with electromagnetic radiation is the nondimensional size parameter x, defined as *(8)*:

$$x \equiv \frac{2\pi r}{\lambda}$$

where r is the radius of the particle and λ is the wavelength of the radiation. The nature of the interaction can be determined based on the size parameter *(8)*, and this is summarized in Table 5.2.

From Table 5.2, it can be seen that Mie scattering is the correct interaction when considering the light scattering by particles in cool paint. The scattering is named after Gustav Mie, who completely solved Maxwell's equations corresponding to the interaction of light with particles of size corresponding to the wavelength *(29)*. The solution is based on slowly converging series that would need to be calculated by a computer; here, we will present the qualitative aspects using Mie scattering tables that have been published by Penndorf *(29)*.

A plot of the Mie Scattering coefficient K as a function of the particle size parameter x for refractive index $n = 1.5$ is shown in Fig. 5.11 *(29)*. It can be seen that maximum scattering occurs at around $x = 4.3$, which corresponds to $r \sim 0.7\ \lambda$. This is consistent with the previous observation that the particle size should be on the order of the wavelength of light *(28)*. Scattering ceases to be effective for $x < 1$ ($r < 0.2\ \lambda$). Hence, for effective scattering in the solar spectrum where 400 nm $< \lambda <$ 2000 nm, it is desirable for particle sizes to be at least 300 nm and preferably on the order of 0.5–2.0 μm. Finally, it may be noted that there are some oscillations (both major and minor) in the scattering coefficient plot; these may be attributed to resonances of magnetic dipole and multipoles *(29)*.

3.2 Materials and performance

In a typical composition of cool paint, pigments make up only about 25% of the paint *(26)*. The rest are composed of resins (~50%) that serve as the

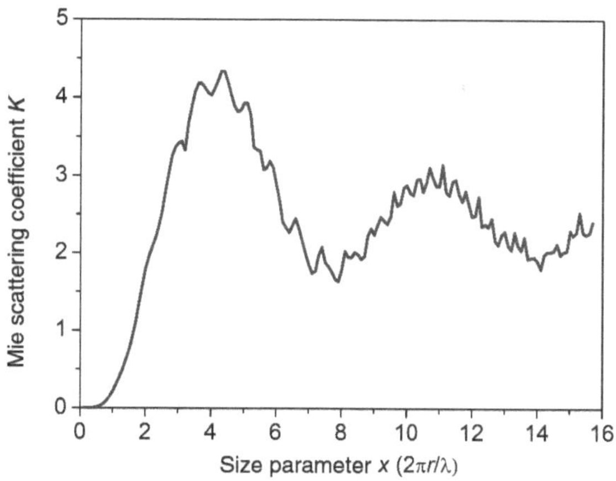

Figure 5.11 Mie scattering coefficient *K*, as a function of the particle size parameter *x*.

binder for the pigments, solvents (~25%) to allow paint coating in a liquid form, and thickeners, and dispersants (~1%) to ensure the paint is of the right viscosity and remains stable for long-term storage *(26)*. Nevertheless, the pigment is responsible for all of the scattering and hence the cooling effect of the paint, as there is no scattering from the rest of the components.

TiO_2 is the most common and effective pigment for use in cool paint *(26,28,30)*. TiO_2-based paints are white in color, and show high reflectance over 80% across essentially the entire visible and near-IR spectrum *(30)*. However, it would be truly boring indeed if all the buildings in a city are painted with white TiO_2 paints, although that would be the best case scenario from a cooling and energy saving perspective. What is needed are pigments that are able to give color to the paint, and yet reflect a greater portion of the solar spectrum compared to standard pigments; such pigments would enable color paints that also reduce the building heat envelope. To consider how this could be implemented, we revisit the solar AM 1.5 spectrum as indicated in Fig. 5.12, which has been divided into ultra-violet (<400 nm), visible (400–700 nm), and infra-red (>700 nm) regions. It can be seen that the amount of solar energy within the visible spectrum makes up just 42% of the total solar irradiation, while 53% of the energy can be found in the infra-red *(26)*. For a pigment to have a certain color, its spectral reflectance in the visible region needs to be a particular fixed profile, and cannot be modified. However, by choosing pigments to have spinel or rutile crystal structure, it is possible to obtain high reflectance in the infra-red *(30,31)*. Such pigments are sometimes referred to in the literature as complex inorganic color pigments *(30)*. The solar heat reflectance is still significant, since

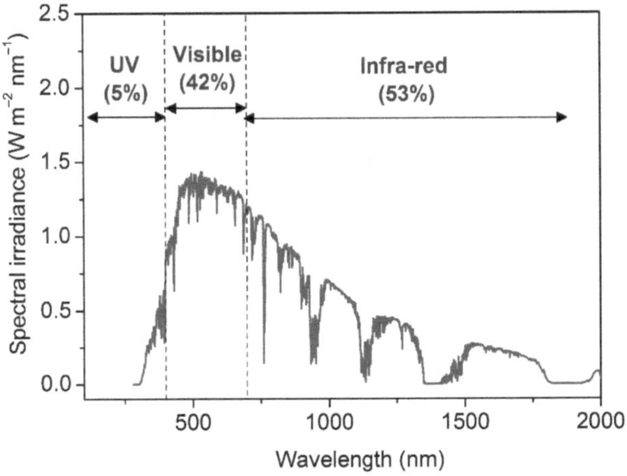

Figure 5.12 AM 1.5 solar spectrum, indicating the percentage of solar energy in the ultra-violet, visible, and infra-red regions. Data from reference *(26)*.

the infra-red region makes up 53% of the total solar irradiation. Hence, this is a viable strategy for designing "cool color pigments" that retain a certain color but exhibit higher solar reflectance than standard pigments.

Table 5.3 summarizes the reflectance data of selected pigments that have been reported in the literature *(26,31)*, including standard pigments as well as "cool pigments." As expected, pigments of the same color show similar reflectance in the visible region. However, the cool pigments exhibit significantly higher reflectance than the standard pigments in the infra-red region, and thus would be effective at reducing the solar heat gain compared to the standard pigments.

Table 5.3 Reflectance of selected pigments reported in the literature.

Color	Pigment	Visible reflectance	Infra-red reflectance	Reference
White	Titanium dioxide	90%	60%–80%	*(31)*
Standard yellow	Chrome antimony	10%–50%	30%–50%	*(26)*
Cool yellow	Chrome antimony	10%–60%	60%–80%	*(26)*
Cool yellow	Chrome antimony titanium	10%–70%	60%–80%	*(31)*
Standard brown	Zinc iron chromite	0%–15%	10%–15%	*(26)*
Cool brown	Zinc iron chromite	5%–25%	30%–60%	*(26)*
Standard brown	Iron oxide	0–25%	20%–30%	*(26)*
Cool brown	Iron oxide	0–30%	30%–60%	*(26)*
Standard black	Carbon	0–5%	0–5%	*(31)*
Cool black	Chromium iron oxide	0–5%	5%–30%	*(31)*

Table 5.4 Real-world testing on the effects of cool paint on roofs.

Location	Cooling effect	Reference
Poitiers, France	10–20°C reduction in roof surface temperature	*(25)*
Poitiers, France	1.5°C reduction in street facade temperature	*(32)*
Iraklion, Greece	20–30°C reduction in roof surface temperature	*(33)*
London, UK	8°C reduction in roof surface temperature, and 4°C reduction in internal ceiling temperature	*(34)*
Penang, Malaysia	5–8°C reduction in roof surface temperature	*(35)*
Burkina Faso (West Africa)	10–12°C reduction in inner roof temperature, and 4°C reduction in internal zone air temperature	*(36)*

Several studies *(25,32–35)* have reported real-world testing of the effects of cool paint, and are indicative of the actual impact that the cool coating can make. As indicated in Table 5.4, reduction in roof surface temperature with cool paint can be quite dramatic, with a report of roof surface temperature reduction as high as 30°C *(33)*. Of note are studies in Malaysia and West Africa demonstrating cooling of 8–12°C (35,36), demonstrating that cool paint can be useful in hot and tropical climates where cooling is most needed. The cooling of roof surfaces is extended to the cooling of internal temperatures as well, as evidenced by a study in London *(34)*. Street facades can also benefit from application of cool paint, although in this case the cooling of 1.5°C is not so dramatic *(32)*; this may be due to the use of cool brown paint as compared to white paint.

4 Conclusions

Both radiative cooling and cool paint are passive cooling technologies for outdoor surfaces, with proven cooling effect of 10°C or higher from outdoor studies. Cool paints are already commercially available, with cool pigments in the market for around 20–30 years to date. Radiative cooling technologies are still mostly demonstrated on a lab scale, although first demonstrations at large scale outdoor deployment are underway. They are both promising approaches for reducing building heat envelope gain and air-conditioning usage, and with climate change looming large, it is prudent to accelerate their usage and deployment with the purpose of mitigating the effects of global warming.

References

1. Coulomb, D.; Dupont, J. L.; Pichard, A. The Role of Refrigeration in the Global Economy. *29th Informatory Note on Refrigeration Technologies*, International Institute of Refrigeration: France, 2015.
2. Tembhekar, C. ACs Eat up 40% of City's Total Power Consumption. *The Times of India*, India, 2009.
3. Climate Change & Energy Efficiency. In *State of the Environment Report 2008 Singapore*, Ministry of Environment and Water Resources: Singapore, 2008.
4. Isaac, M.; van Vuuren, D. P. Modeling Global Residential Sector Energy Demand for Heating and Air Conditioning in the Context of Climate Change. *Energy Policy* **2009**, *37*, 507–521.
5. Dalapati, G. K.; Kushwaha, A. K.; Sharma, M., et al. Transparent Heat Regulating (THR) Materials and Coatings for Energy Saving Window Applications: Impact of Materials Design, Micro-Structural, and Interface Quality on the THR Performance. *Prog. Mater. Sci.* **2018**, *95*, 42–131.
6. He, T.; Yao, J. Photochromic Materials Based on Tungsten Oxide. *J. Mater. Chem.* **2007**, *17*, 4547–4557.
7. Zhou, D.; Zhao, C. Y.; Tian, Y. Review on Thermal Energy Storage with Phase Change Materials (PCMs) in Building Applications. *Appl. Energy* **2012**, *92*, 593–605.
8. Petty, G. W. *A First Course in Atmospheric Radiation*, 2nd ed.; Sundog Publishing: USA, 2006.
9. Catalanotti, S.; Cuomo, V.; Piro, G., et al. The Radiative Cooling of Selective Surfaces. *Solar Energy* **1975**, *17*, 83–89.
10. Raman, A. P.; Anoma, M. A.; Zhu, L., et al. Passive Radiative Cooling Below Ambient Air Temperature Under Direct Sunlight. *Nature* **2014**, *515*, 540–544.
11. Plotted from Mauna Kea Sky Transmission as Recorded by Gemini Observatory, with data downloaded from www.gemini.edu.
12. Chen, Z.; Zhu, L.; Raman, A., et al. Radiative Cooling to Deep Sub-Freezing Temperatures Through a 24-h Day-Night Cycle. *Nat. Commun.* **2016**, 7, 13729.
13. Kou, J.; Jurado, Z.; Chen, Z., et al. Daytime Radiative Cooling Using Near-Black Infrared Emitters. *ACS Photonics* **2017**, *4*, 626–630.
14. Kopp, G.; Lean, J. L. A new, Lower Value of Total Solar Irradiance: Evidence and Climate Significance. *Geophys. Res. Lett.* **2011**, *38* L01706.
15. Granqvist, C. G.; Hjortsberg, A. Radiative Cooling to Low Temperatures: General Considerations and Application to Selectively Emitting SiO Films. *J. Appl. Phys.* **1981**, *52* (6), 4205–4220.
16. Harrison, A. W.; Walton, M. R. Radiative Cooling of TiO_2 White Paint. *Solar Energy* **1978**, *20*, 185–188.
17. Berdahl, P.; Martin, M.; Sakkal, F. Thermal Performance of Radiative Cooling Panels. *Int. J. Heat Mass Transfer* **1983**, *26* (6), 871–880.
18. Rephaeli, E.; Raman, A.; Fan, S. Ultrabroadband Photonic Structures to Achieve High-Performance Daytime Radiative Cooling. *Nano Lett.* **2013**, *13*, 1457–1461.
19. Zhu, L.; Raman, A. P.; Fan, S. Radiative Cooling of Solar Absorbers Using a Visibly Transparent Photonic Crystal Thermal Blackbody. *Proc. Natl. Acad. Sci.* **2015**, *112*, 12282–12287.
20. Zhai, Y.; Ma, Y.; David, S. N., et al. Scalable-Manufactured Randomized Glass-Polymer Hybrid Metamaterial for Daytime Radiative Cooling. *Science* **2017**, *355*, 1062–1066.

21. Zhao, D.; Aili, A.; Zhai, Y., et al. Subambient Cooling of Water: Toward Real-World Applications of Daytime Radiative Cooling. *Joule* **2018**, doi: 10.1016/j.joule.2018.10.006.
22. Bao, H.; Yan, C.; Wang, B., et al. Double-Layer Nanoparticle-Based Coatings for Efficient Terrestrial Radiative Cooling. *Solar Energy Mater. Solar Cells* **2017**, *168*, 78–84.
23. Fernandez, N.; Wang, W.; Alvine, K., et al. *Energy Savings Potential of Radiative Cooling Technologies*. US Department of Energy: USA, 2015.
24. Akbari, H.; Matthews, H. D. Global Cooling Updates: Reflective Roofs and Pavements. *Energy Build.* **2012**, *55*, 2–6.
25. Bozonnet, E.; Doya, M.; Allard, F. Cool Roofs Impact on Building Thermal Response: A French Case Study. *Energy Build.* **2011**, *43*, 3006–3012.
26. Coser, E.; Moritz, V. F.; Kranzinger, A., et al. Development of Paints with Infrared Radiation Reflective Properties. *Polímeros* **2015**, *25* (3), 305–310.
27. Gardecka, A. J.; Goh, G. K. L.; Sankar, G., et al. On the Nature of Niobium Substitution in Niobium Doped Titania Thin Films by AACVD and its Impact on Electrical and Optical Properties. *J. Mater. Chem. A* **2015**, *3*, 17755–17762.
28. Bendiganavale, A. K.; Malshe, V. C. Infrared Reflective Inorganic Pigments. *Recent Pat. Chem. Eng.* **2008**, *1*, 67–79.
29. Penndorf, R. B. New Tables of Total Mie Scattering Coefficients for Spherical Particles of Real Refractive Indexes ($1.33 \leq n \leq 1.50$). *J. Opt. Soc. Am.* **1957**, *47*, 1010–1015.
30. Uemoto, K. L.; Sato, N. M. N.; John, V. M. Estimating Thermal Performance of Cool Colored Paints. *Energy Build.* **2010**, *42*, 17–22.
31. IR Reflecting Pigments. Heubach GmbH: Germany, 2014.
32. Doya, M.; Bozonnet, E.; Allard, F. Experimental Measurement of Cool Facades' Performance in a Dense Urban Environment. *Energy Build.* **2012**, *55*, 42–50.
33. Kolokotsa, D.; Diakaki, C.; Papantoniou, et al. Numerical and Experimental Analysis of Cool Roofs Application on a Laboratory Building in Iraklion, Crete, Greece. *Energy Build.* **2012**, *55*, 85–93.
34. Kolokotroni, M.; Gowreesunker, B. L.; Giridharan, R. Cool Roof Technology in London: An Experimental and Modelling Study. *Energy Build.* **2013**, *67*, 658–667.
35. Al-Obaidi, K. M.; Ismail, M.; Rahman, A. M. A. Passive Cooling Techniques Through Reflective and Radiative Roofs in Tropical Houses in Southeast Asia: A Literature Review. *Front. Arch. Res.* **2014**, *3*, 283–297.
36. Kabore, M.; Wurtz, E.; Coulibaly, Y., et al. Assessment on Passive Cooling Techniques to Improve Steel Roof Thermal Performance in Hot Tropical Climate. *Int. J. Energy Power Eng.* **2014**, *3*, 287–295.

CHAPTER 6

Understanding spectrally selective properties of solar absorbers

Atasi Dan[a], Bikramjit Basu[a,b]
[a]Materials Research Centre, Indian Institute of Science, Bangalore, Karnataka, India
[b]Interdisciplinary Centre for Energy Research, Indian Institute of Science, Bangalore, Karnataka, India

1 Introduction

Energy is the most fundamental and essential component to maintain a sustainable growth of human civilization. The progress of human civilization demands significant energy which leads to excessive use of fossil fuels namely coal, gas, oil, etc. It is predicted that with the increasing requirement, the consumption of fossil fuel might reach 123 million barrels per day by the year 2025 *(1)*. Such an indiscriminate and uncontrolled usage of fossil fuels has resulted in air pollution and global warming. Different reports suggest that CO_2 concentration was ~315 ppm at the end of the 1950s which now exceeds ~400 ppm *(2)*. Several other reports highlight global warming as one of the most concerned issues, which will raise the global average temperature about 1°C higher by the 2050s than that for 1960–1990, and as much as 6°C higher by the end of this century *(3)*. To address such difficulties, the utilization of diverse renewable energy sources has drawn significant attention. Among all sources of green energy, the inexhaustive nature of solar energy made it one of the reliable energy resources which can drastically change the energy supply scenario and possibly control environmental issues in near future.

Numerous methods have been introduced to convert the solar radiation into other form of energy which includes photovoltaic, solar cell *(4)*, solar-thermal, thermoelectric generator *(5)*, solar steam generator *(6)*, solar desalination equipments, photo-chemical and photo-catalytic reactors, etc. Among these, the popular way of utilizing solar energy is the employment of photovoltaics (PV), which has seen rapid growth in past couple of years. However, the concentrated solar power (CSP) systems need to complement photovoltaic technology to better integrate solar electricity

Energy Saving Coating Materials
http://dx.doi.org/10.1016/B978-0-12-822103-7.00006-6

into power grids. In CSP technology, the solar radiation collected by the receiver is transferred to a heat transfer fluid, which is heated up to produce steam. Then, solar-driven fluid evaporation is utilized to generate electricity using conventional turbine. The evaporation efficiency has been largely increased through the usage of solar absorber surfaces, which utilize the concept of interfacial heating. On the one hand, the amount of solar radiation absorbed by these selective absorbers plays a crucial role to determine the overall photothermal conversion efficiency. On the other hand, the solar thermal conversion is limited by the heat losses from the absorber which hinders the practical applications.

The spectrally selective absorbers on receiver tube involve the maximization of solar absorptance with a minimum heat loss in order to ensure high conversion of solar to thermal energy. Thus, the role of solar absorber on the receiver surface is to utilize solar radiation and to confine the energy, which suppresses heat loss to environment. A wide range of spectrally selective coating structures have been explored, which simultaneously satisfy the complex set of requirements of enhanced absorption with reduced thermal emission. Some coatings (V_2O_5, LaB_6 *(7)*, Fe_3O_4 *(8)*, β-$FeSi_2$ *(9)*) possess intrinsically unique selective properties as they have many available orbitals compatible with the wide energy range of solar spectrum. The key limitation of such coatings is that high temperature could lead to a failure, which hampers selective performance of the coating. Some of the semiconductor absorbers exhibit higher selectivity with a cut-off wavelength around 2 μm and enhanced resistance to thermal radiation but are not thermally stable compared to the rest of solar selective absorbers. Another special kind of selective absorber is the black pigmented particles, which have drawn attention as they can be easily deposited by spray coating or spin coating techniques on a large scale *(10,11)*. However, the formation of aggregates, agglomerates, and flocculates due to the presence of coarse particles in pigmented dispersion system causes non-uniform and less adhesive films. Another kind of absorbers, called cermet (Sn-Al_2O_3 *(12)*, Ni-Al_2O_3 *(13)*, Zr-ZrO_2 *(14)*, Cu-SiO_2 *(15)*, etc.) have been widely studied. In this class of absorber materials, metal nanoparticles are embedded in a ceramic host matrix to selectively absorb solar radiation, while having high transmission in the infrared region. Multilayer absorbers, for instance, W/WAlN/WAlON/Al_2O_3 *(16)*, TiB_2/TiB(N)/Si_3N_4 *(17)*, Al/NbMoN/NbMoON/SiO_2 *(18)*, W/AlSiN/AlSiON/SiO_2 *(19)*, etc. are fabricated in order to achieve highest absorption by minimizing surface reflections. Recently, there is a tremendous thrust for developing photonic crystals-based absorbers, which

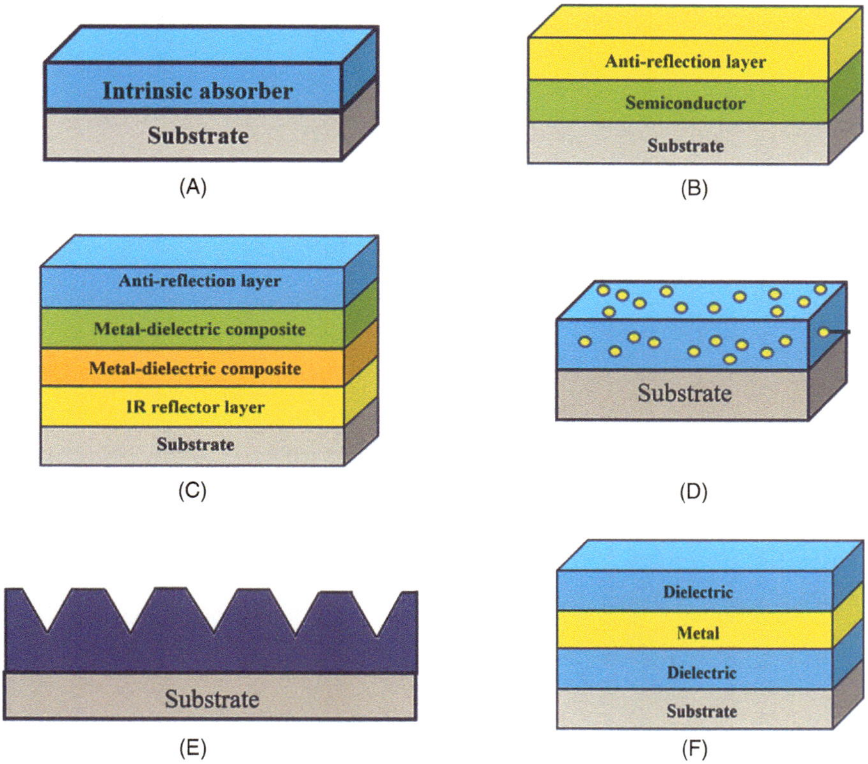

Figure 6.1 Spectrally selective absorber coatings: (A) Intrinsic absorber, (B) Semiconductor absorber, (C) Multilayer absorber, (D) Cermet absorber, (E) Textured surface, (F) Dielectric-metal-dielectric absorber. *Reproduced from Reference* (20) *with permission from Elsevier.*

has ample room for performance improvement in near future, owing to exceptional capabilities in terms of spectral selectivity with respect to wavelength, design, angle, and polarization. All different type of designs, shown in Fig. 6.1, provide alternate approaches for the spectrally selective absorbers. The performance of various solar selective absorbers is summarized in Table 6.1 *(20)*. Along with different designs, various factors such as morphology, surface roughness, different materials, irradiation of solar wavelength influence the performance of the solar selective absorbers.

In this chapter, we focus on two key questions: (1) how to evaluate the performance of spectrally selective solar absorbers and (2) how different properties dominate the selective behavior of thin films? Research activity is constantly progressing to optimize various deposition processes and conditions for simultaneously obtaining high solar absorptance and low thermal emittance.

Table 6.1 Different type of spectrally selective absorbers.

Type of absorbers	Description	Example	Solar absorptance (α)	Thermal emittance (ε)	Reference
Intrinsic absorber	Possess intrinsically selective properties having a cut-off around 2 μm	Fe_3O_4	0.74–0.93	0.11–0.62	*(21)*
		Transparent conductive oxide (TCO) coated on a metal substrate	0.71	0.08 (80°C)	*(22)*
Semiconductor absorbers	The bandgap energy of the material is tuned to the cut-off wavelength	CuO	0.93	0.16	*(23)*
		Amorphous Ge-CaF_2	0.76	0.05	*(24)*
Black absorbers	Absorbers with high-temperature binder applied stainless steel substrates to satisfy cost-effectiveness	Black chromium (Cr/Cr_2O_3)	0.90	–	*(25)*
		Black-pigmented polyolefin	0.95	0.93–0.95	*(26)*
		Black, high-temperature stable spinel type pigments (P3060, Ferro Co.) and rheological additives	0.89/0.93	0.2/0.36	*(27)*
Cermet absorbers	Metal-dielectric composites, composed of nanoscale metal particles in a dielectric or ceramic matrix	Sn–Al_2O_3	0.94	0.21 (100°C)	*(12)*
		Ag–Al_2O_3	0.93	0.04–0.05 (82°C)	*(28)*
		Ni:SiO_2	0.90–0.96	0.03–0.14	*(29)*
		Mo-Si_3N_4	0.926	0.017	*(30)*
		WTi-Al_2O_3	0.93	0.10	*(31)*
Multilayer absorbers	Multiple stacks of different layers usually fabricated using physical vapor deposition	$ZrO_x/Zr/ZrO_x/Al_xO_y$	0.93	0.11 (100°C)	*(32)*
		$MoSi_2$-Si_3N_4/ Si_3N_4/Al_2O_3	0.924 ± 0.005	0.463 ± 0.005 (600°C)	*(33)*
		W/W-WO_x/W-WO_x/WO_x	0.932	0.058	*(34)*
		AlTiO/AlTiON (20 at.%)/AlTiON (14 at.%)/AlTiN/TiN	0.94	0.047	*(35)*
Photonic absorbers	Periodic dielectric structures which may hinder the propagation of light for all directions within a frequency range	Mo crystal	0.99	0.17	*(36)*
		$Al/Si_3N_4/(Ti/Si_3N_4)^2$	0.94	0.07	*(37)*
		2D tantalum absorber	–	–	*(38)*
Textured surfaces	Tailoring light scattering and absorptance using wavelength-scale nanostructuring	CuO nanostructures	0.946	0.234 ± 0.019	*(39)*
		Fractal textured Cu surface	0.7	0.6	*(40)*
		W hemispherical shell array coated with an ultrathin Ge film	0.93	–	*(24)*

2 Solar absorptance and thermal emittance

Solar absorptance and thermal emittance are two main parameters in determining the performance of solar absorber coatings. The spectrally selective absorbers on receiver tube involves maximization of solar absorptance (≥0.95) in solar spectrum (0.25–2.5 μm) with a minimum heat loss, i.e., emittance (≤0.05) in the infrared spectrum (2.5–30 μm) in order to ensure high conversion of solar to thermal energy. Plank's law represented in Eq. (6.1) indicates the cut-off wavelength ($\lambda_{\text{cut-off}}$) at which the solar absorber should exhibit a transition from high absorptance to high reflectance.

$$E_{b\lambda}(\lambda, T) = \frac{2hc^2}{\lambda^5}\left[\exp\left(\frac{hc}{\lambda kT} - 1\right)\right]^{-1} \tag{6.1}$$

where T is the temperature, c is the speed of light (2.99×10^{-8} ms^{-1}), h is Planck's constant (6.63×10^{-34} Js), k is Boltzmann's constant (1.38×10^{-23} JK^{-1}), $E_{b\lambda}(\lambda, T)$ is the spectral radiance (Wm^{-3}sr^{-1}). The position of $\lambda_{\text{cut-off}}$ of a system depends on the temperature of thin film surface, the solar concentrations, thin film structure, thickness of coating, etc. In general, the desired $\lambda_{\text{cut-off}}$ should lie around 2 μm to achieve maximum selectivity, as shown in Fig. 6.2A *(20)*.

Considering the energy conservation, the solar absorptance (α) can be defined as the ratio of absorbed radiation by the surface to the incident

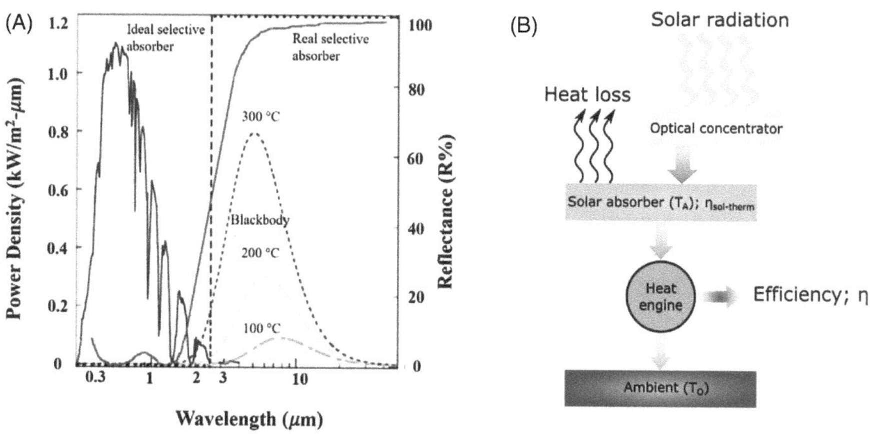

Figure 6.2 (A) The solar spectrum (*blue*); the blackbody radiation at 100, 200, and 300°C (*green, yellow, maroon*, respectively); The reflectance spectrum of an ideal (*black*) and real (*red*) solar selective absorber. *Reproduced from Reference* (20) *with permission from Elsevier.* (B) *schematic of photothermal conversion system.*

solar radiation. The solar absorber transforms the absorbed radiation into heat and transfers the heat to a medium (water or solar fluid). The absorbed energy is transferred via three methods; conduction, convection, and radiation. The heat loss via radiation is a critical factor for solar thermal systems. All objects having thermal energy radiates electromagnetic radiation, which is proportional to the fourth power of absolute temperature. The dependence can be expressed by the equation given by

$$P = \varepsilon \sigma A\left(T^4 - T_c^4\right) \tag{6.2}$$

where P is amount of thermal energy radiated per unit area, ε is the emittance of the object, A is the surface area, and T and T_c are the absolute temperature of the material and the environment, respectively. σ is the Stefan–Boltzmann constant, which is equal to 5.6697×10^{-8} Wm^{-2} K^{-4}. The heat loss from a coating surface at a certain temperature can be estimated by emittance, which is defined as the ratio of energy emitted by an object to energy emitted by a perfect black body at the same temperature and wavelength. The expression of absorptance and emittance can be given by following equations;

$$\alpha = \frac{\int_{0.25\mu m}^{2.5\mu m}\left[1 - R(\lambda)\right] G(\lambda) d\lambda}{\int_{0.25\mu m}^{2.5\mu m} G(\lambda) d\lambda} \tag{6.3}$$

$$\varepsilon(T) = \frac{\int_{25\ \mu m}^{2.5\ \mu m}\left[1 - R(\lambda)\right] E_{b\lambda}(\lambda, T) d\lambda}{\int_{25\mu m}^{2.5\mu m} E_{b\lambda}(\lambda, T) d\lambda} \tag{6.4}$$

where $G(\lambda)$ is the standard solar spectrum. The incident power density (W/m^2) can be calculated by integrating $G(\lambda)$ over entire solar wavelength range. As mentioned earlier, $E_{b\lambda}(\lambda, T)$ is the blackbody radiation. The value of absorptance and emittance can be calculated by measuring reflectance ($R(\lambda)$) in solar and infrared wavelength range.

Eventually, combining these two important factors (α and ε), we can calculate photothermal conversion efficiency of receiver (η_r), that is, sunlight-to-heat conversion factor using the equation mentioned below

$$\eta_r = \alpha - \frac{\varepsilon(T)\sigma\left(T^4 - T_c^4\right)}{CI} \tag{6.5}$$

In the above-mentioned equation, CI is the concentrated solar radiation.

In case of supplying electricity, the efficiency of solar-thermal system is also related to heat-to-electricity conversion process, which is dominated by Carnot efficiency (see Fig. 6.2B). Hence, the solar to electricity conversion efficiency (η) is estimated by the product of receiver efficiency (η_r) and Carnot efficiency (η_c) and represented as following:

$$\eta = \left(\alpha - \frac{\varepsilon(T)\sigma\left(T^4 - T_c^4\right)}{CI} \right)\left(1 - \frac{T_c}{T}\right) \tag{6.6}$$

From Eqs. (6.5) and (6.6), it is evident that the highest efficiency should be ultimately obtained by satisfying the contradictory requirement of high solar absorptance and low thermal emittance. Also, the temperature of the receiver surface and surrounding temperature would play a major role in enhancing the total efficiency.

3 Parameters influencing solar selectivity

This section intends to analyze the factors that influence the performance of spectrally selective absorbers. The solar thermal systems are desired to have high conversion efficiency. However, the theoretical efficiency cannot be fully achieved in practice. As with many other solar energy conversion technologies, photothermal system which absorbs solar radiation using receivers may also lose little energy to the environment. Therefore, such systems are never 100% efficient. The efficiency of solar thermal systems is limited by solar selectivity, which depends on a large number of factors, including substrate reflectance and surface roughness, metal volume fraction and compositional gradient, intrinsic absorption, particle size, destructive interference by multiple layers, film thickness, the presence of anti-reflection coating, refractive index, extinction coefficient, etc. These factors have been validated by experimental work and further utilized to systematically design different solar absorber thin films with the objective of optimizing selectivity of absorbers.

3.1 Thickness of absorber

Several researchers investigated the effect of thickness of absorber on the selectivity of thin films. The thickness of the coating can be modified in such a way that the destructive interference occurs in ultraviolet (UV) and visible region, where most of the solar spectrum lies. Interference effects can be used to tune the solar absorptance and thermal emission

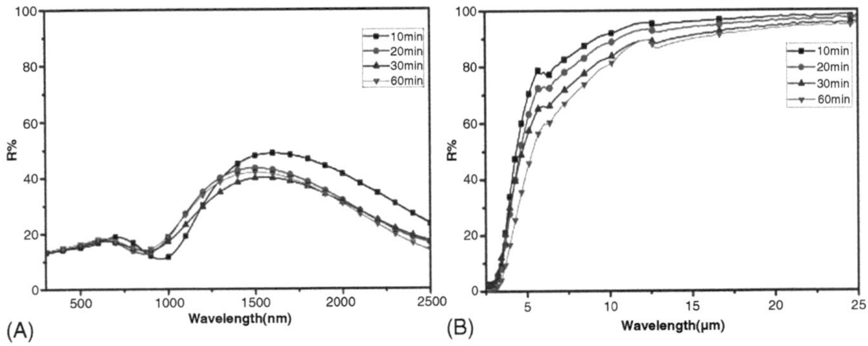

Figure 6.3 Reflectance spectra of Ni/C/NiO thin film deposited at different time (A) 0.3–2.5 μm and (B) 2.5–25 μm. *Reproduced from Reference* (22) *with permission from Springer Nature.*

by modifying the cut-off wavelength. In order to understand the effect of film thickness, Xiao et al. *(41)* prepared CuO_x thin films by varying reaction time. It was observed that the reflectance spectra degrade gradually due to the increase in reaction time, leading to an increase in film thickness. It is well known that, if the film thickness is increased, the absorption edge shifts toward longer wavelength, so that the absorptance increases, but at the same time, the emittance also increases. Khan et al. *(42)* recently prepared Ni/C/NiO nanocomposite thin film by a facile one step sol–gel method. They studied the effect of film thickness by varying the holding time, while annealing the film at 450°C in vacuum. Fig. 6.3 shows the reflectance spectra in solar spectrum and infrared region of the samples with different holding time. An increase of absorptance from 0.80 to 0.82 was observed with increasing the holding time from 10 to 30 min. However, it decreased to 0.78 as the holding time increased to 60 min, whereas the emissivity gradually increased from 0.11 to 0.22 as the holding time increased from 10 to 60 min. The main structure variations with various holding times introduced the decrease in thickness and enhancement of metallic Ni nanoparticles size. In a different account, it was observed that the interference peaks could be shifted by varying the film thickness of TiAlN thin film *(43)*. The appearance of the film also transformed from yellow to blue while adjusting the thickness. Liu et al. *(44)* developed a Cu/NbTiON (metallic–like)/NbTiON (dielectric–like)/SiON (anti-reflecting) thin film where it was observed that the emittance of the thin film increased while the thickness of the NbTiON (M) layer varied from 50 nm to 600 nm, keeping the film thickness of NbTiON(D) layer constant at 150 nm.

3.2 Size of the nanoparticles in host matrix

According to plasmonic theories, the size of the nanoparticles plays an important role on the optical properties of the selective absorber. The reason lies behind the excitation of surface plasmon resonance (SPR), originated from collective oscillation of conduction electrons of metal nanoparticles during interaction with electro-magnetic radiation, i.e., incident light. This effect is associated with wavelength-selective photon absorption, resonant Mie scattering together with a drastic enhancement of the local electromagnetic field in the vicinity of the nanoparticles *(45,46)*. A small change in the size of nanoparticles influences the interaction of the electromagnetic wave with fabricated thin film. For example, Liu et al. *(47)* used finite–difference time domain (FDTD) method to understand the effect of the size and shape of nanoparticles on the absorbing nature of spectrally selective absorber. Fig. 6.4 shows that the absorption of ellipsoid nanoparticles is lesser than spherical nanoparticles in the multilayer structure of tungsten nanoparticle layers and SiO_2 layers. This work involved a detail overview on the localized surface plasmon resonance (LSPR) to explain such absorptance property. In similar lines, the shape, size, and amount of Ag nanoparticles were proven to influence surface plasmon resonance absorption of Al_2O_3:Ag thin film. Wazwaz et al. *(48)* explored that the presence of a small amount of metals in Ni-Al_2O_3 thin film in the aluminum substrate might reduce the reflectance by about (30 ± 1)% than the highly

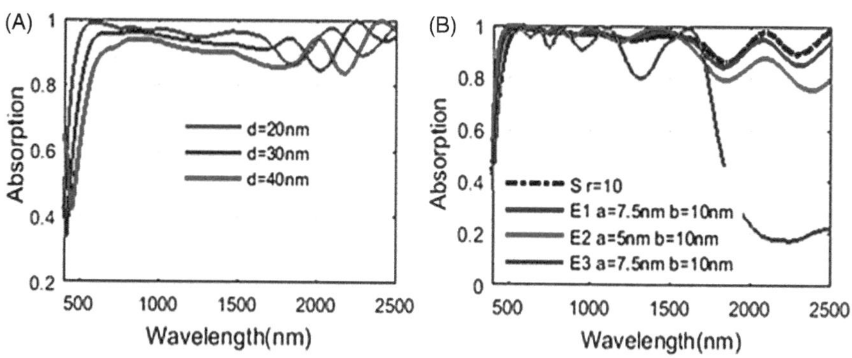

Figure 6.4 (A) Absorption spectra of metal nanoparticle absorber structure (NPA) varying with nanoparticle size. The absorber layer is composed of closely arranged nanoparticles of square array in cubic lattice embedded in SiO_2 layer. (B) Absorption pattern of the NPA structure for nanoparticles with different shapes. (S: sphere, E: ellipsoid), 'a' is the major axis semidiameter of the ellipsoid, and 'b' is the minor axis semidiameter of the ellipsoid. For E1 and E2, the electric filed is along the major-axis direction. For E3, the electric field is along the minor-axis direction. *Reproduced from Reference* (27) *with permission from Springer Nature.*

pure aluminum substrate could do. This is due to the fact that an increase in Ni content in the composite increases the absorption of solar energy.

3.3 Metal content in the film

Similar to size of metal nanoparticles, metal content in thin film is also considered as an influencing factor in tailoring the selective properties of a material. Changing the metal content in cermet-based coating is a highly efficient way of solar energy utilization. Wamae et al. *(12)* demonstrated that the solar absorptance property could be significantly increased with increasing Sn content in the pores of Al_2O_3, while the change of emissivity was insignificant. Owing to the obstruction in free electron movements, thermal conductivity of Sn-Al_2O_3 samples slightly decreased with increasing Sn content. As a result, instead of exchanging heat to the surrounding, Sn-Al_2O_3 transfers heat through Al substrate to the exchanger, which is preferable for solar thermal applications. The peak of surface plasmon resonance can be represented in terms of particle filling factor by the following expression:

$$\lambda_{SPR} = \lambda_p \left[\frac{2+q}{1-q} \varepsilon_{ext} + 1 \right]^{1/2} \tag{6.7}$$

where $\lambda_{SPR} = 2\pi c/\omega_p$ with $\omega_p = 9.2$ eV for Ag, q is the particle filling factor and ε_{ext} is the external dielectric constant surrounding the particle. It is evident from Eq. (6.7) that ε_{ext} and q determine λ_{SPR} of the system. An increase in q or ε_{ext} makes λ_{SPR} red shift. In this context, Xiao et al. *(49)* observed that prolonged annealing of Ag:Al_2O_3 film reduced the amount of Ag due to evaporation, which eventually leads to the degradation of solar absorptance. Another group demonstrated that solution–chemically fabricated Ni:Al_2O_3 coating with an optimum nickel content of 65% exhibited a solar absorptance of 0.93 and thermal emittance of 0.04 *(50)*.

3.4 Microstructure of the surface

Microstructure is a very important factor to determine the selectivity of solar absorber coatings. Except abnormal changes in grain size, several studies examined a number of different mechanisms involved with other microscale defects, which deeply influenced the properties of solar absorber coatings, such as widened interface, micro cracks and micro cavity. It was also documented that film morphology and microstructure could be tuned by judiciously controlling deposition process parameters, annealing temperatures, etc. Gong et al. *(51)* investigated the microstructure of the coating and reported that the absorptance and emittance of AlCrNO-based coating increased from 0.93 and 0.20 to 0.94 and 0.21, respectively, after annealing at 500°C for 100 h in air due to not only decrease in grain size from about 1.2 to 0.7 μm but also for the formation of macro-droplets, some holes, and cracks. With an objective to understand the influence of microstructure on selectivity, Wazwaz et al. *(48)* deposited a porous alumina top layer in Ni-Al_2O_3 film and found a reduction in the reflectance significantly, which indeed indicated an increase in solar absorptance. In a separate study, Fang et al. *(52)* demonstrated insignificant enhancement of thermal emittance (~0.05) in Ti_3SiC_2 by formation of microcracks, which were partially filled by amorphous SiO_2 in temperature range of 800–900 K. Al_2O_3-TiO_2 nano porous film, fabricated by Bayati et al. *(53)* revealed a high solar absorptivity of $\alpha = 0.99$ and low thermal emissivity of $\varepsilon = 0.06$. Their findings showed that porous structure (pore size: 60–550 nm) trapped the incident solar radiation by multiple reflections, resulting in high solar absorptance.

3.5 Number of layers

Analogous to other governing factors, a number of layers in multilayer coatings have a dramatic effect on performance of solar absorbers. By increasing the number of layers from one to several, the front surface reflection can be reduced, which improves the performance of the multilayer systems. In this regard, Liou and Lu *(54)* tested the spectrally selective properties of TiN_xO_y thin films with bilayer and gradient structures. The overall thickness for all types of coatings was kept at about 200 nm. The gradient of TiN_xO_y coating showed a better performance than that of bilayer coating with solar absorptance of 0.86 and thermal emittance of 0.11. The observed effect can be attributed to light trapping capacity of gradient stack due to versatile functionality. However, the effect of number of layers beyond a certain number is negligible. To understand the effect of number of layers on the selectivity, Farooq et al. *(55)* investigated on one, four, and ten layer systems on different material combinations such as W:Al_2O_3, Co:SiO_2, and Cr:SiO_2. It was observed that the highest absoprtance was obtained from the film with four layers. Several authors advocated that a thin oxide layer plays a crucial role in oxidation protection, because the amorphous oxide layer may delay the inward diffusion of oxygen ions and prevent accelerating oxidation.

3.6 Refractive index

If a thin film possesses a gradient in refractive index with the highest refractive index of the substrate followed by a steady decreasing trend toward the top layer, it absorbs maximum solar irradiation. This kind of coating architecture instigates destructive interference and suppresses front surface reflections. An elegant demonstration by Adsten et al. [73] showed that effective refractive index of the top layer of industrially sputtered nickel-nickel oxide is somewhat lower than that of the base layer, leading to a solar absorptance and thermal emittance of 0.94–0.96 and 0.13–0.15, respectively. The basic electromagnetic theory indicates that metal has a higher refractive index than that of oxide materials. In this context, Zhao et al. *(56)* introduced an innovative way of fabricating a thin film using DC magnetron reactive sputtering in a roll coater, where the properties of the thin films changed from metallic to metal oxide due to non–uniform oxidation along the sputtering zone. Indeed, a high selectivity with a solar absorptance and emittance of 0.91 and 0.05, respectively, was obtained by a single graded film without any anti-reflecting layer. Recent studies by our group on W/WAlN/WAlON/Al_2O_3 and TiB_2/TiB(N)/Si_3N_4 spectrally selective absorber showed that a gradation in refractive index from bottom to top layer, as depicted in Fig. 6.5A and B dominates the high absorptance value *(17,57)*. One typical observation is that the gradation in refractive index could change the individual layer from conductive to dielectric and would correspondingly change the optical properties from opaque, translucent, to transparent. The fabrication of graded refractive index coatings is progressing rapidly for achieving high selectivity and thermal stability. The examples include ZrO_x/Zr/ZrO_x/Al_xO_y *(32)*, Al infrared/Cr-N-O/SiO_2 *(58)*, W/WAlN/WAlON/Al_2O_3 *(57)*, Cu/$Zr_{0.2}AlN_{0.8}$/ZrN/AlN/ZrN/AlN/$Al_{34}O_{62}N_4$ *(59)*, Mo/ZrSiN/ZrSiON/SiO_2 *(60)*, Al/NbMoN/NbMoON/SiO_2 *(18)* and so on.

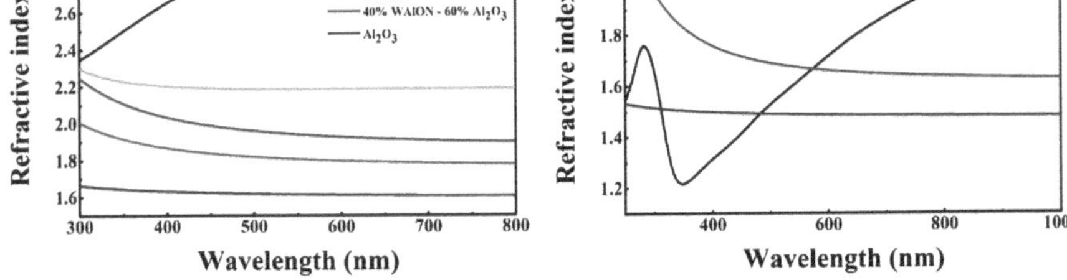

Figure 6.5 Gradation in refractive index in multilayer stack of (A) W/WAlN/WAlON/Al_2O_3 and (B) TiB_2/TiB(N)/Si_3N_4 thin films. *Reproduced from Reference* (17,57) *with permission from Elsevier.*

3.7 Effect of substrates

Having explained the influence of different other parameters on the selectivity of deposited film, it is also important to understand the effect of substrate on the selectivity of the coating. Several studies came up in recent years showing the role of nanoscale features on the substrate, leading to a change in optical characteristics. It was reported that when a solution of a chemical derived nickel–alumina coating was deposited on rough aluminum substrate, the solution could not mask rough surfaces, which caused scattering of light from the uneven places *(50)*. As a result, the destructive interference mechanism became less pronounced, compared to when a smooth surface was used as substrate. The film coated on rough surfaces possessed absorptance and emittance of 0.80 and 0.03, respectively, while the coating deposited on smooth surface had an absorptance and emittance of 0.82 and 0.04, respectively. In a separate study, Hu et al. *(61)* reported that an increase in surface roughness of Cu substrate enhanced the solar absorption of $SiO_2/Cr/SiO_2$ by about 2% by shifting the absorption band to a higher wavelength side. However, the surface-roughed dielectric-metal-dielectric coating possessed a slightly higher emittance than that of smooth surface. The deposition of one additional metallic layer on substrate could also improve selectivity by reducing the thermal emittance. For example, Sibin et al. *(62)* reported a reduction in thermal emittance of stainless steel from 0.12–0.13 to 0.03 by introducing a W layer below AlTiN/AlTiON/AlTiO absorber, deposited on SS substrate. The free electron in the pure W metal did not influence lattice vibration. As a result, thermal heat loss due to phonon oscillations decreased. Similar studies to reduce thermal emittance using infrared reflectors were carried out by Gelin et al. *(63)*. In this work, they reported that thermal emittance in spectrally selective absorber could be reduced from 0.12 to 0.06 by replacing nickel-chromium by copper-nickel due to its sensitivity toward magnetron sputtering process parameters. For a comparative study, Farooq et al. *(63a)* designed spectrally selective absorbers on aluminum and copper substrates, covered with sputtered or electroplated metal base layer. A metal sputtered base layer of 100 nm was optimized for solar selective coatings as the thin films with sputtered base layer exhibited higher absorptance along with higher emittance and coatings with electroplated base layered possessed comparatively lower absorptance and lower emittance.

3.8 Incident radiation

There are various factors, including latitude of the place, season, time of the day, etc. can determine the amount of solar radiation incidents on the receiver.

Therefore, the incident angle of solar radiation varies continuously. Moreover, a significant fraction, ≈10% of the incident light is constituted by diffuse sunlight from the atmosphere. In addition to exhibit spectral selectivity at near normal incidence, an ideal solar absorber should also possess consistent performance at various incidence angles to harvest most sunlight incident from oblique directions. The influence of the incident angles and polarization were extensively analyzed. It is concluded that reflectance is strongly correlated with incident light under different polarized conditions. In most of the cases, it was observed that the reflectance of the samples was very low even when the incidence angle increased up to 50°, indicating a wide angle absorptance of the multilayer coatings *(64)*. Hedayati et al. *(65)* prepared a thin film on a glass substrate composed of a optically thick metallic film (Au), followed by a thin dielectric layer (SiO_2) acting as the spacer layer and at the very top, a thin (20 nm) nearly percolated film of nanocomposite (Au/SiO_2). They found that solar absorption by the film remained around 90% at different angles of incidence, which indicated the promise of that system as a potential solar absorber. The angular absorption effect was also simulated on a 1-μm-thick ITO-coated W substrate. The study revealed that an isotropic absorption can be obtained up to an incidence angle as large as 40^c. However, beyond 40°, though spectrally selective properties were obtained, the absorption edge slightly shifted to shorter wavelengths *(22)*. In recent years, research on plasmonic metal nanoparticles is emerging for achieving high angular solar absorption as by trapping the solar radiation, plasmonic effect enhances the light-matter interaction. For example, plasmonic-nanoparticle-coated metal foils (PNFs) composed of Cu nanoparticles on Zn substrate exhibited an excellent wide-angle solar absorptance, with α ranging from 0.96 at 15°, to a peak of 0.97 at ≈35°, to 0.79 at 80° *(66)*.

4 Future outlook and conclusion

In this chapter, some essential questions are discussed, that is, what are all the possible factors that could influence solar selectivity and how those could affect photothermal conversion efficiency. A combined analysis of performance and thermal stability is necessary for identifying spectrally selective absorbers with greatest promise of widespread application *(67)*. When designing a solar collector which is expected to operate at any location, to be functional in a large operational temperature range, and under different climatic conditions, the manufacture ought to find out the most cost-effective solution. Since the bulk materials would not be applicable for most of the cases, a solution to utilize these potential materials is required. Bulk Ultra high temperature ceramics (UHTCs), if deposited as low dimensional materials in the form of a thin film, may intrigue a breakthrough in solar thermal systems. However, simple fabrication process for thin film deposition should be implemented for mass production. Also, the design of innovative thin film structure is essential for inducing high optical absorption in the system and preventing heat dissipation. Nevertheless, researchers believe that innovative approaches of nanotechnology will unambiguously provide an engineering platform to fabricate advanced thin film designs for light-to-heat conversion. There has been a tremendous concern about thermal stability of solar absorbers. Much of the direct and indirect evidences confirmed that degradation of optical selectivity at high temperature occurs in materials due to many consequences. So far, more studies are required to address many open questions in order to completely understand their impact on the performance of solar thermal systems. Some parameters, including mechanical properties, thermal conductivity, coefficients of thermal expansion, etc. could be extracted, measured, or approximated to estimate the long-term durability of materials in harsh operational conditions (Fig. 6.6) *(16)*. The explicit analysis of these factors helps identifying future

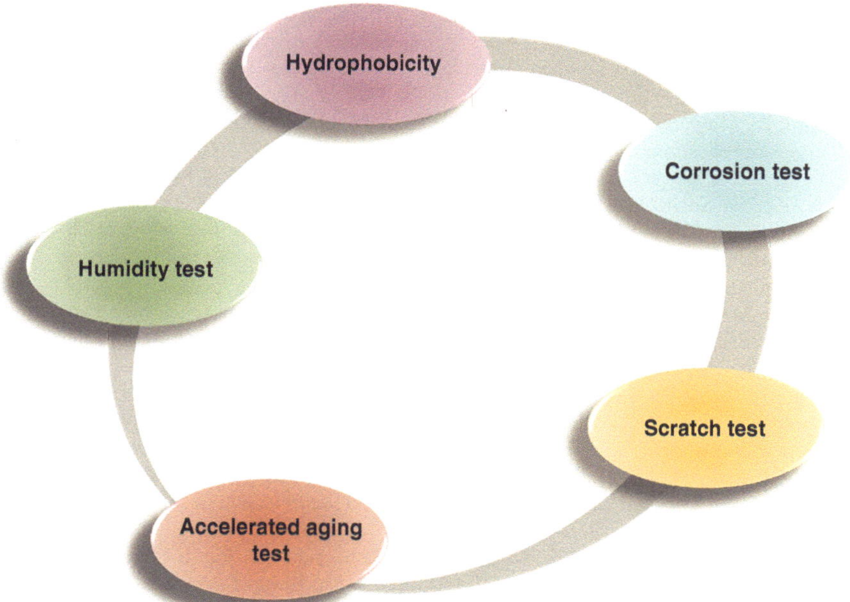

Figure 6.6 ***Various performance qualifying tests required to perform on spectrally selective absorber.***

barriers that could lead to less lifetime of receiver. The additional parameters, which are not directly related to selectivity such as toxicity of materials can be investigated, which has significant impact on environmental and life-cycle analyses. A better approach would be to perform an integrated study, including thin film design, performance, evaluation of robustness, environmental stability, flexibility in different regions and costs for a specific installation. Thus, it is imperative to conduct more generalized experiments to overcome various drawbacks in achieving highest photothermal conversion efficiency. Addressing such issues will undoubtedly accelerate the progress of solar thermal system over PV.

References

1. Doe, U. S.; Admin EI. Annual Energy Outlook 2005, with Projections to 2025. DOE/EIA 2005;383.
2. Cetin, M.; Sevik, H.; Isinkaralar, K. Changes in the Particulate Matter and CO_2 Concentrations Based on the Time and Weather Conditions: The Case of Kastamonu. *Oxidation Commun.* **2017**, *40*, 477–485.
3. Rackley SA. Carbon capture and storage: Butterworth-Heinemann; 2017.
4. Bisquert, J.; Qi, Y.; Ma, T.; Yan, Y. Advances and Obstacles on Perovskite Solar Cell Research from Material Properties to Photovoltaic Function. *ACS Energy Lett.* **2017**, *2*, 520–523.
5. Kraemer, D.; Jie, Q.; McEnaney, K., et al. Concentrating Solar Thermoelectric Generators with a Peak Efficiency of 7.4%. *Nat. Energy* **2016**, *1*, 16153.
6. Hu, X.; Xu, W.; Zhou, L., et al. Tailoring Graphene Oxide-Based Aerogels for Efficient Solar Steam Generation under One Sun. *Adv. Mater.* **2017**, *29*, 1604031.
7. Sani, E.; Mercatelli, L.; Meucci, M.; Zoli, L.; Sciti, D. Lanthanum hexaboride for solar energy applications. *Sci. Reports* **2017**, *7*, 718.
8. González, F.; Barrera-Calva, E.; Huerta, L.; Mane, R. S. Coatings of Fe_3O_4 Nanoparticles as Selective Solar Absorber. *Open Surf. Sci. J.* **2011**, *3*, 131–135.
9. Okuhara, Y; Kuroyama, T; Takata, M; Tsutsui, T; Noritake K. Solar selective absorbers based on semiconducting β-FeSi2 for high temperature solar-thermal conversion. 1 ed: AIP Publishing. p. 220005.
10. Gunde MKe; Orel, Z. C.; Hutchins, M. G. The Influence of Paint Dispersion Parameters on the Spectral Selectivity of Black-Pigmented Coatings. *Sol. Energy Mater. Sol. Cells* **2003**, *80*, 239–245.
11. Povacz, M.; Wallner, G.; Lang, R. Black-pigmented polypropylene materials for solar thermal absorbers–effect of carbon black concentration on morphology and performance properties. *Solar Energy* **2014**, *110*, 420–426.
12. Wamae, W.; Suriwong, T.; Threrujirapapong, T. Influence of Tin Content on Spectral Selectivity and Thermal Conductivity of Sn–Al_2O_3 Solar Selective Absorber. *Mater. Renew. Sustain. Energy* **2018**, 7, 2.
13. Boström, T. K.; Wäckelgård, E.; Westin, G. Anti-Reflection Coatings for Solution-Chemically Derived Nickel—Alumina Solar Absorbers. *Sol. Energy Mater. Sol. Cells* **2004**, *84*, 183–191.
14. Zhang, Q. -C.; Hadavi, M.; Lee, K.; Shen, Y. Zr–ZrO_2 Cermet Solar Coatings Designed by Modelling Calculations and Deposited by dc Magnetron Sputtering. *J. Phys. D: Appl. Phys.* **2003**, *36*, 723.

15. Garnich, F.; Sailer, E. Cu-SiO_2/Cu-Cermet Selective Absorbers for Solar Photothermal Conversion. *Sol. Energy Mater.* **1990**, *20*, 81–89.
16. Dan, A.; Basu, B.; Echániz, T.; de Arrieta, I. G.; López, G. A.; Barshilia, H. C. Effects of Environmental and Operational Variability on the Spectrally Selective Properties of W/WAlN/WAlON/Al_2O_3-based Solar Absorber Coating. *Sol. Energy Mater. Sol. Cells* **2018**, *185*, 342–350.
17. Dan, A.; Chattopadhyay, K.; Barshilia, H. C.; Basu, B. Shifting of the Absorption Edge in TiB_2/TiB (N)/Si_3N_4 Solar Selective Coating for Enhanced Photothermal Conversion. *Solar Energy* **2018**, *173*, 192–200.
18. Song, P.; Wu, Y.; Wang, L., et al. The Investigation of Thermal Stability of Al/NbMoN/NbMoON/SiO_2 Solar Selective Absorbing Coating. *Sol. Energy Mater. Sol. Cells* **2017**, *171*, 253–257.
19. Rebouta, L.; Sousa, A.; Capela, P., et al. Solar Selective Absorbers Based on Al_2O_3: W Cermets and AlSiN/AlSiON Layers. *Sol. Energy Mater. Sol. Cells* **2015**, *137*, 93–100.
20. Dan, A.; Barshilia, H. C.; Chattopadhyay, K.; Basu, B. Solar Energy Absorption Mediated by Surface Plasma Polaritons in Spectrally Selective Dielectric-Metal-Dielectric Coatings: A Critical Review. *Renewable and Sustainable Energy Reviews* **2017**, *79*, 1050–1077.
21. Fu, R.; Wu, X.; Wang, X., et al. Low-Temperature Hydrothermal Fabrication of Fe_3O_4 Nanostructured Solar Selective Absorption Films. *Appl. Surf. Sci.* **2018**, *458*, 629–637.
22. Shimizu, M.; Suzuki, M.; Iguchi, F.; Yugami, H. High Spectral Selectivity for Solar Absorbers Using a Monolayer Transparent Conductive Oxide Coated on a Metal Substrate. *J. Appl. Phys.* **2017**, *121*, 183103.
23. Márquez, A.; Blanco, G.; De Rapp, M. E. F.; Lamas, D. G.; Tarulla, R. Properties of Cupric Oxide Coatings Prepared by Cathodic Arc Deposition. *Surf. Coat. Technol.* **2004**, *187*, 154–160.
24. Thomas, N. H.; Chen, Z.; Fan, S.; Minnich, A. J. Semiconductor-based Multilayer Selective Solar Absorber for Unconcentrated Solar Thermal Energy Conversion. *Sci. Rep.* **2017**, 7, 5362.
25. Nunes, R. A. X.; Costa, V. C.; Sade, W.; Araújo, F. R.; Silva, G. M. Selective Surfaces of Black Chromium for Use in Solar Absorbers. *Mater. Res.* **2018**, 21.
26. Povacz, M.; Wallner, G. M.; Lang, R. W. Black-pigmented polypropylene materials for solar thermal absorbers–effect of carbon black concentration on morphology and performance properties. *Solar Energy* **2014**, *110*, 420–426.
27. Orel, Z. C.; Gunde, M. K.; Lencˇek, A.; Benz, N. The Preparation and Testing of Spectrally Selective Paints on Different Substrates for Solar Absorbers. *Solar Energy* **2001**, *69*, 131–135.
28. Barshilia, H. C.; Kumar, P.; Rajam, K. S.; Biswas, A. Structure and Optical Properties of Ag–Al_2O_3 Nanocermet Solar Selective Coatings Prepared Using Unbalanced Magnetron Sputtering. *Sol. Energy Mater. Sol. Cells* **2011**, *95*, 1707–1715.
29. Farooq, M.; Green, A. A.; Hutchins, M. G. High Performance Sputtered Ni:SiO_2 Composite Solar Absorber Surfaces. *Sol. Energy Mater. Sol. Cells* **1998**, *54*, 67–73.
30. Céspedes, E.; Wirz, M.; Sánchez-García, J. A., et al. Novel Mo–Si_3N_4 Based Selective Coating for High Temperature Concentrating Solar Power Applications. *Sol. Energy Mater. Sol. Cells* **2014**, *122*, 217–225.
31. Wang, X.; Gao, J.; Hu, H., et al. High-Temperature Tolerance in WTi-Al_2O_3 Cermet-Based Solar Selective Absorbing Coatings with Low Thermal Emissivity. *Nano Energy* **2017**, *37*, 232–241.
32. Khoza, N.; Nuru, Z. Y.; Sackey, J., et al. Structural and Optical Properties of ZrO_x/Zr/ZrO_x/Al_xO_y Multilayered Coatings as Selective Solar Absorbers. *J. Alloys Compds.* **2019**, *773*, 975–979.
33. Céspedes, E.; Rodríguez-Palomo, A.; Salas-Colera, E., et al. Role of Al_2O_3 Antireflective Layer on the Exceptional Durability of Mo–Si–N-Based Spectrally Selective Coatings in Air at High Temperature. *ACS Appl. Energy Mater.* **2018**.

34. Wang, W.; Wen, H.; Ling, S.; Li, Z.; Su, J.; Wang, C. Self-Doped W–WO_x Nanocermet Multilayer Films Fabricated by Single Tungsten Target Reactive Sputtering For Selective Solar Absorption. *J. Mater. Chem. A* **2018**, *6*, 15690–15700.
35. Heras, I.; Guillén, E.; Lungwitz, F., et al. Design of High-Temperature Solar-Selective Coatings Based on Aluminium Titanium Oxynitrides Al y Ti 1− y (O x N 1− x). Part 1: Advanced Microstructural Characterization and Optical Simulation. *Sol. Energy Mater. Sol. Cells* **2018**, *176*, 81–92.
36. Ollier, E.; Dunoyer, N.; Szambolics, H.; Lorin, G. Nanostructured Thin Films for Solar Selective Absorbers and Infrared Selective Emitters. *Sol. Energy Mater. Sol. Cells* **2017**, *170*, 205–210.
37. Li, B.; Qi, D.; Wang, X.; Wang, F.; Nie, Y.; Gong, R. Enhanced Spectra Selectivity of Solar Absorber Film With Ti/Si_3N_4 Photonic Structures. *Mater. Lett.* **2017**, *201*, 5–8.
38. Nam, Y.; Yeng, Y. X.; Lenert, A., et al. Solar Thermophotovoltaic Energy Conversion Systems With Two-Dimensional Tantalum Photonic Crystal Absorbers and Emitters. *Sol. Energy Mater. Sol. Cells* **2014**, *122*, 287–296.
39. Jeong, D.; Lee, J.; Hong, H., et al. Absorption Mechanism and Performance Characterization of CuO Nanostructured Absorbers. *Sol. Energy Mater. Sol. Cells* **2017**, *169*, 270–279.
40. Jain, R.; Pitchumani, R. Fabrication and Characterization of Multiscale, Fractal Textured Solar Selective Coatings. *Sol. Energy Mater. Sol. Cells* **2017**, *172*, 213–219.
41. Xiao, X.; Miao, L.; Xu, G., et al. A Facile Process to Prepare Copper Oxide Thin Films as Solar Selective Absorbers. *Appl. Surf. Sci.* **2011**, *257*, 10729–10736.
42. Khan, S.; Wu, Z.; Yuan, G., et al. Study of Annealing Effect Upon the Structural and Solar-Selective Properties of C/Ni/NiO Composite Coatings Prepared by Sol–Gel Method. *J. Sol–Gel Sci. Technol.* **2018**, 1–8.
43. Zhu, D.; Zhao, S. Chromaticity and Optical Properties of Colored and Black Solar–Thermal Absorbing Coatings. *Sol. Energy Mater. Sol. Cells* **2010**, *94*, 1630–1635.
44. Liu, Y.; Wang, C.; Xue, Y. The Spectral Properties and Thermal Stability of NbTiON Solar Selective Absorbing Coating. *Sol. Energy Mater. Solar Cells* **2012**, *96*, 131–136.
45. Granqvist CG. Energy-efficient windows: present and forthcoming technology. Materials science for solar energy conversion systems: Elsevier; 1991. p. 106-67.
46. Kelly, K. L.; Coronado, E.; Zhao, L. L.; Schatz, G. C. *The optical properties of metal nanoparticles: the influence of size, shape, and dielectric environment.* ACS Publications, 2003.
47. Liu, C.; Zhang, D.; Liu, Y., et al. Numerical Study of an Efficient Solar Absorber Consisting of Metal Nanoparticles. *Nanosc. Res. Lett.* **2017**, *12*, 601.
48. Wazwaz, A.; Al-Salaymeh, A. Theoretical and Measured Optical Properties of Ni-Al_2O_3 over Al Substrate Selective Absorber. *Sustain. Cities Soc.* **2015**, *16*, 13–23.
49. Xiao, X.; Xu, G.; Xiong, B.; Chen, D.; Miao, L. The Film Thickness Dependent Thermal Stability of Al_2O_3:Ag Thin Films as High-Temperature Solar Selective Absorbers. *J. Nanoparticle Res.* **2012**, *14*, 746.
50. Boström, T.; Wäckelgård, E.; Westin, G. Solution-Chemical Derived Nickel–Alumina Coatings For Thermal Solar Absorbers. *Solar Energy* **2003**, *74*, 497–503.
51. Gong, D.; Liu, H.; Luo, G., et al. Thermal Aging Test of AlCrNO-Based Solar Selective Absorbing Coatings Prepared By Cathodic Arc Plating. *Sol. Energy Mater. Sol. Cells* **2015**, *136*, 167–171.
52. Fang, Z.; Lu, C.; Guo, C., et al. Suitability of Layered Ti_3SiC_2 and Zr_3 $[Al(Si)]_4C_6$ Ceramics as High Temperature Solar Absorbers for Solar Energy Applications. *Sol. Energy Mater. Sol. Cells* **2015**, *134*, 252–260.
53. Bayati, M. R.; Zargar, H. R.; Talimian, A.; Ziaee, A.; Molaei, R. Characterization of Al_2O_3–TiO_2 Nano Porous Solar Absorbers Derived via MAO/Sol-Gel Hybrid Process. *Surf. Coat. Technol.* **2010**, *205*, 2483–2489.
54. Liou, Y. -C.; Lu, F. -H. Air-Based Sputtering Deposition of TiN_xO_y Films for Solar Selective Absorber Coatings Applications. *Thin Solid Films* **2018**, *660*, 733–740.

55. Farooq, M.; Hutchins, M. G. A novel Design in Composities of Various Materials for Solar Selective Coatings. *Sol. Energy Mater. Sol. Cells* **2002**, *71*, 523–535.
56. Zhao, S.; Ribbing, C. -G.; Wäckelgård, E. New Method to Optimize a Solar Absorber Graded Film Profile. *Sol. Energy* **2005**, *78*, 125–130.
57. Dan, A.; Biswas, A.; Sarkar, P., et al. Enhancing Spectrally Selective Response of W/WAlN/WAlON/Al_2O_3-Based Nanostructured Multilayer Absorber Coating Through Graded Optical Constants. *Sol. Energy Mater. Sol. Cells* **2018**, *176*, 157–166.
58. Liu, J.; Sun, Z. -Q.; Wang, H. Design and Characterization of Solar Absorbing Multilayer Stack Based on Al/Cr-NO/SiO_2 Layers. *Sol. Energy Mater. Sol. Cells* **2018**, *188*, 18–26.
59. Meng, J-p.; Liu, X-p.; Fu, Z-q.; Zhang, K. Optical Design of Cu/$Zr_{0.2}AlN_{0.8}$/ZrN/AlN/ZrN/AlN/$Al_{34}O_{62}N_4$ Solar Selective Absorbing Coatings. *Sol. Energy* **2017**, *146*, 430–435.
60. Ning, Y.; Wang, W.; Wang, L., et al. Optical Simulation and Preparation of Novel Mo/ZrSiN/ZrSiON/SiO_2 Solar Selective Absorbing Coating. *Sol. Energy Mater. Sol. Cells* **2017**, *167*, 178–183.
61. Hu, E. -T.; Guo, S.; Gu, T., et al. Enhancement of Solar Absorption by a Surface-Roughened Metal–Dielectric Film Structure. *Jpn. J. Appl. Phys.* **2017**, *56*, 112301.
62. Sibin, K. P.; John, S.; Barshilia, H. C. Control of thermal emittance of stainless steel using sputtered tungsten thin films for solar thermal power applications. *Sol. Energy Mater. Sol. Cells* **2015**, *133*, 1–7.
63. Gelin, K.; Boström, T.; Wäckelgård, E. Thermal Emittance of Sputter Deposited Infrared Reflectors in Spectrally Selective Tandem Solar Absorbers. *Sol. Energy* **2004**, *77*, 115–119.
63a. Farooq, M.; Green, A. A.; Hutchins, M. G. High performance sputtered Ni : SiO_2 composite solar absorber surfaces. Solar Ener. Mater. Solar Cells **1998**, *54*, 67–73.
64. Hu, E. -T.; Guo, S.; Gu, T., et al. High Efficient and Wide-Angle Solar Absorption with a Multilayered Metal-Dielectric Film Structure. *Vacuum* **2017**, *146*, 194–199.
65. Hedayati, M. K.; Javaherirahim, M.; Mozooni, B., et al. Design of a Perfect Black Absorber at Visible Frequencies Using Plasmonic Metamaterials. *Adv. Mater.* **2011**, *23*, 5410–5414.
66. Mandal, J.; Wang, D.; Overvig, A. C., et al. Scalable, "Dip-and-Dry" Fabrication of a Wide-Angle Plasmonic Selective Absorber for High-Efficiency Solar–Thermal Energy Conversion. *Adv. Mater.* **2017**, *29* 1702156.
67. Dan, A.; Chattopadhyay, K.; Barshilia, H. C.; Basu, B.; Thermal Stability of WAlN/WAlON/Al_2O_3-Based Solar Selective Absorber Coating. MRS Advances 2016;CJO2016:doi:10.1557/adv.2016.388.

CHAPTER 7

Metal oxides and metal thin films by atomic layer deposition (ALD), liquid-ALD, and successive ionic layer adsorption and reaction methods for THR applications

Hongfei Liu[a], Chandreswar Mahata[b]
[a]Institute of Materials Research and Engineering (IMRE), A*STAR (Agency for Science, Technology and Research), Singapore
[b]School of Electronics Engineering, Chungbuk National University, Cheongju, Republic of Korea

1 Introduction

Atomic layer deposition (ALD) is a unique technique that has been widely employed for conformally coating ultrathin films on flat or nanostructured substrates making use of the self-limited surface reactions of processors under certain conditions *(1,2)*. Ahvenniemi et al. has recently published an interesting review article, summarizing the earlier developments of ALD based on the quite completely collection of the publications *(3)*. In principle, gas-phase ALD is similar to chemical vapor deposition (CVD) but breaks the chemical reactions in time sequences by separately feeding the individual reaction precursors onto the growing surface in a periodic way. The growth temperature is also lowered in ALD as compared with that in CVD to eliminate the pyrolytic breakdown of precursors before their approaching the surface. The surface reactions in a half-cycle stop on their own at a full monolayer coverage and the growth saturates till the beginning of the next half-cycle. Doping and/or alloying of the films can be realized by periodically inserting an extra half-cycle of desired precursors, forming a supercycle ALD process. In this way, the ultrathin films or coatings can be precisely controlled in their thickness and composition. Another advantage of ALD is that the reactive species have very low kinetically energies at the growing surface and thus cause minor physical damages to the coating/substrate interfaces. This unique characteristic of ALD, together with its low-temperature process, makes it suitable for film growth on heat- and/or

Energy Saving Coating Materials
http://dx.doi.org/10.1016/B978-0-12-822103-7.00007-8

kinetic-energy-sensitive substrates, for example, organic and/or biological structures *(4–6)*.

Conventional ALD are based on gas-phase reactions, that is, the reaction species are fed in gas phase onto the surface of the substrate located in a chamber reactor. The gas-phase species can be evaporated from the source materials that are not necessarily gaseous. Recently, liquid-phase ALD (liquid-ALD) has been developed for the growth of ultrathin oxide and metal films on planar substrates and through-hole structures from source species in liquid-phase solutions *(7,8)*. Since neither vacuum chamber system nor gas-phase precursors, for example, metalorganic chemicals, are required in the liquid-ALD process, it can be a low-cost method for developing ultrathin film with novel and/or additional functions. It should be noted that liquid-inject ALD is also seen in the literature *(9–11)*. This technique, in terms of the reactions on the surface of the substrate, could be a mixing of gas- and liquid-phase ALD since vaporization from the injected liquid might not that complete. Successive ionic layer adsorption and reaction (SILAR) *(12,13)*, similar to liquid-ALD, is another liquid-phase thin film growth technique. During the growth of oxide, the substrate is dipped into the liquid salt solutions to adsorb metallic ions at the surface of the substrate. Those metal ions having loosely bonds on the surface can be further removed by washing the substrate in deionized water. Subsequently, the substrate is dipped into an oxygen source to complete the chemical reactions of the surface-absorbed metal ions. Although the periodic deposition processes of SILAR are similar to those of liquid-ALD, the SILAR technique has different setup and limit cationic and anionic ion source precursors.

The ALD, liquid-ALD, and SILAR processes have been, more or less, well established for growing ultrathin compound films including metal oxides, nitrides, and sulfides on various substrates *(14)*. On the other hand, the process for growing metals has also been under significant developing. Researches on Cu, Ag, Au, etc. *(15–19)* have been showing good models for the growth of ultrathin metals by ALD. Both the compounds and the metals are the dominant components when engineering their coatings on substrates for transparent heat regulation (THR) applications *(20–23)*. In the following sections, we will be presenting the fundamentals of the ALD, LS-ALD, and SILAR techniques. In general, these processes for ultrathin metal-oxide, -nitride, and -sulfide as well as metals can be designed and engineered for THR coatings. Morphological, structural, electrical, and optoelectronic properties of such ultrathin films as well as their characterization and analysis methodologies will also be discussed; so as to provide the readers a perspective on recent developments of ALD, LS-ALD, and SILAR toward their applications in THR.

2 ALD, liquid-ALD, and SILAR

Although ALD technology is, from the point of view of the chamber reactor and the reaction precursors, similar to conventional CVD, it has its own unique functions *(2)*. In this section, the fundamentals and new developments of ALD will be discussed, targeting at their applications in conformally coating of thin films for consequent THR fabrications.

2.1 ALD system and ALD process

Fig. 7.1A presents a schematic diagram of a typical gas-source ALD system. In this system precursors, carried by inert or less reactive gases (e.g., nitrogen), were fed into the reaction chamber through high-speed valves, which can complete the switches within 0.05 s. The ballast setup can be simply replaced by using a throttle valve (located in between the chamber and the pump) with its opening controlled by the pressure of the reactor. Similar to CVD or metalorganic CVD (i.e., MOCVD) that a load-lock chamber can be installed, attaching to the reaction chamber *(24–26)*, to save the running time as well as to protect the reaction chamber from contaminations.

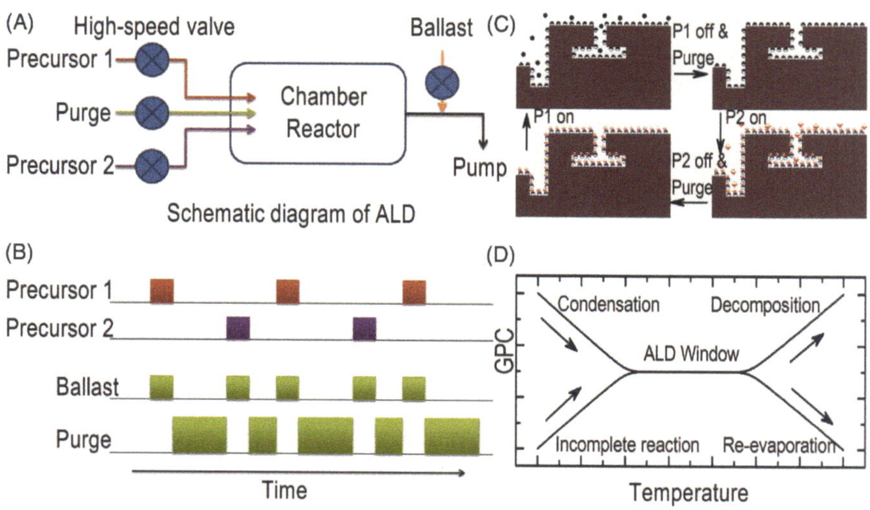

Figure 7.1 Schematic diagram showing the general (A) ALD set-up, (B) ALD process, (C) surface self-limited ALD reactions, and (D) ALD growth per cycle (GPC) as a function of temperature.

The growth setup can be horizontal as the schematic diagram shown in Fig. 7.1A, where the gaseous precursors flow in the direction parallel to the surface of the substrate in the reactor. It can also be vertical, where the precursors are fed in the surface normal direction of the substrate while the reaction by-products and/or the remaining precursors are pumped out from the behind of the substrate holder. Plasma generator can be installed to the chamber reactor to enhance the surface reactions of some less reactive precursors, which is known as plasma-enhanced ALD (PE-ALD) in the literature. Another function of the plasma generator is to in situ clean the surface of the substrate prior to the ALD process. Remote ozone generator can also be attached to the ALD reaction camber for producing O_3 from O_2 as the oxygen source precursors. To in situ monitor, the growth rate of the ultrathin films, ellipsometry, and/or quartz crystal microbalance (QCM) are, sometimes, installed into the ALD chamber.

Fig. 7.1B presents a typical ALD process. In general, the growth initiates, after the temperature settings for the substrate, the precursor bottles, and the feeding lines (from the precursor bottles to the reaction chamber) are reached, at the feeding of one precursor (e.g., precursor 1 in Fig. 7.1B). The ballast (or the throttle valve) is triggered simultaneously to stabilize the growth. After the pulse of precursor 1, the chamber is purged. The pulse of precursor 1 and its following purge form the first half-cycle of the ALD process. At the end of the first half-cycle, precursor 2 is fed and, meanwhile, the ballast (or the throttle valve) is triggered. After the pulse of precursor 2, the chamber is purged. The pulse of precursor 2 and its following purge form the second half-cycle of the ALD process. At the end of this half-cycle, a single layer coating of materials is basically completed. To promote the surface coverage of the reaction species, soaking can be inserted, in either or both the half-cycles, by switching off both the inlet- and outlet-valve in between the precursor pulse and its following purge.

General surface reactions during the ALD process are presented in Fig. 7.1C. During the feeding of precursor 1 (P1), reaction between P1 and the surface create reaction species chemically absorbed on the surface. The reaction by-products and the remaining P1 are purged after the feeding of P1. An introduction of soaking process, that is, just after the pulse of P1, tends to enhance the surface absorptions, which is essential for the conformal growth on nanostructured substrates, especially those with large aspect ratios. After the purge, precursor 2 (P2) is fed and the reactions between P2 and the surface absorbed reaction species form the coating material. The thickness of the coating material will be monotonically increased by repeating the process of P1-pulse/purge/P2-pulse/purge or P1-pulse/soak/purge/P2-pulse/soak/purge.

In a typical ALD process, the pulse length of the precursors, the time length of soaking, and the time length of purging must be optimized along with the growth temperature. The essential parameter is the growth-per-cycle (GPC) that measures the thickness of the film grown within one cycle. Fig. 7.1D presents the typical GPC that have been experienced in ALD process as a function of growth temperature. In terms of the growth temperature, three typical ranges can be seen in Fig. 7.1D. In the lower temperature range, the decrease of GPC with the increase in temperature has been attributed to the reduced condensation of the precursors at higher temperatures while the increase of GPC is due to the increased surface reactions. In the higher temperature range, the increase of GPC with temperature is a typical behavior of CVD growth where pyrolytic decomposition of the precursors occurred while the decrease of GPC is due to evaporation of the surface-absorbed reaction species at high temperatures. In between the low- and high-temperature ranges, the GPC is constant and this temperature range forms the typical ALD growth window. The temperature window of ALD growth, in comparison with those of the most advanced semiconductor epitaxy techniques, that is, molecular beam epitaxy (MBE) and MOCVD, *(27,28)* is fairly lower, which greatly suppress the undesired self-heating-induced atomic intermixing, that is, atomic interdiffusions induced by in situ thermal annealing of the grown materials and/or heterostructures in their following growth stages.

It is worth to note that spatial ALD has recently been extensively developed for large-scale and mass-production of ultrathin coatings, for example, for solar cell plates, to save the processing time. In this technique, all the precursors (e.g., 1 and 2 in Fig. 7.1) are kept feeding into the reaction chamber but at their own spatial zones *(29)*. During growth, the substrate is kept holding and transporting in sequence among the individual precursor zones. Because the precursors are continuously and individually fed in their own spatial zones, there is no need for purging the chamber, but transporting the substrate, when switching the growth from one precursor to the other. In this way, the absolute growth rate, when compared with that of the conventional ALD process, can be increased by a few orders of magnitude.

2.2 Liquid-ALD

Different from the surface reactions between gas-phase species and solid substrates of the conventional gas-phase ALD *(30–32)*, the reactions in liquid-ALD occur between liquid-phase species and solid substrates *(7,8)*. The emergence and development of liquid-ALD have been happening not long time ago and at the current stage, commercial liquid-ALD that can process compatible materials on different types of surfaces is still not available. For laboratory studies, a liquid-ALD setup as shown in Fig. 7.2 has recently been designed by Wu et al. *(7)*, where liquid-phase precursors 1 and 2 are supplied, separately and in sequence, to the microfluidic reactor by the individual peristaltic pumps. The residual precursors and/or reaction species can be removed from the surface by supplying solvent after individual precursor supplies while before the next half-cycle reactions. The typical liquid-ALD process cycle thus consists of precursor 1 → solvent → precursor 2 → solvent and the thickness of the deposited film is depended on the GPC as well as the number of cycles repeated. Examples of liquid-ALD chamber reactors designed for depositing thin layers on planar and through-hole-structured substrates are shown in Fig. 7.2 B and C, respectively. In Fig. 7.2B, the substrate is located at the center of the reactor where the surface is lowered by ~1.0 mm as compared to the surrounding areas so

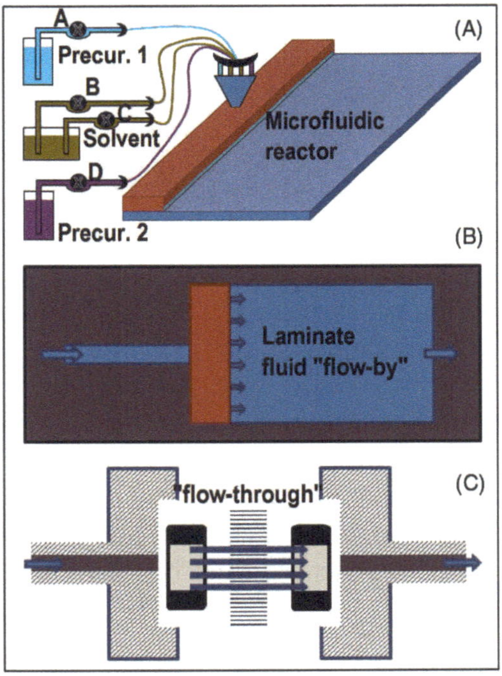

Figure 7.2 Schematic diagram (after Ref. *(7)*, ©2015 ACS) showing the typical liquid-ALD set-up for depositing ultrathin films on planar (A, B) and through-hole (C) substrates in the so-called flow-by and flow-through modes, respectively.

that the input liquid precursors can be uniformly distributed through micro channels onto the edge of the substrate and flow over the planar surface of the substrate (see the arrows indicated in the Fig. 7.2B), giving rise to the so-called "*flow-by* " growth. Likewise, the setup in Fig. 7.2C is used to deposit thin layers on the inner walls of the through holes in the substrate, which is named "*flow-through*" growth in the literature.

A simple comparison of the liquid-ALD with the conventional gas-phase ALD reveals that the both the liquid-ALD setups shown in Fig. 7.2B and C are less effective for conformally depositing thin films on porous substrates, especially nanostructured substrates with one-end-blocked holes/pores. This comparison indicates that to form conformal coating layers on complex nanostructure with dead-end nanoholes or pores of large aspect ratios is still a big challenge for current liquid-ALD techniques. This challenge may hinder further developments of liquid-ALD, especially for its applications in depositing thin layers on substrate with complex nanostructures.

On the other hand, pyrolytic breakdown of less thermally stable gas-phase precursors, for example, metal alkylamides and alkoxides, may occur in conventional gas-source ALD. In this regard, liquid injection techniques, which deliver the vapor direct to the point of use via injecting, have been proposed to guarantee the stability and make use of the less volatile precursors. To differ from the liquid-ALD, the ALD process based on the liquid injection techniques has been named liquid-injection ALD in the literature *(9,10)*. In practice, we believe that the growth on the surface is more or less a kind of mixing between the conventional gas-phase ALD and the liquid-ALD because of the incomplete vaporization of the injected liquid. In general, the growth temperature of liquid-ALD and liquid-injection ALD, due to the liquid-phase precursors, is usually at room temperature and lower than that of the conventional gas-phase ALD.

2.3 Successive ionic liquid adsorption and reaction (SILAR)

The successive ionic liquid adsorption and reaction (SILAR) method was developed in the 1980s for producing low-cost metal sulfide thin films for photovoltaic applications *(33–36)*. The earlier deposition procedure consisted of the successive soaking of the substrate in a cold complex solution C (source of cationic ions) and a hot solution A (source of anionic ions) to deposit polycrystalline and/or epitaxial thin films of water-insoluble compound CA. This process was improved further by inserting procedures to wash away those loosely bonded ions as well as the residual reaction species

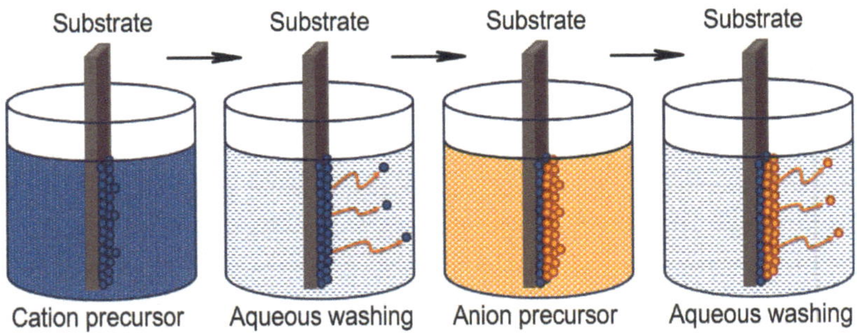

Figure 7.3 Schematic diagram showing the typical growth procedures of SILAR for coating ultrathin films on planar substrates.

and, sometimes, dry the surface when switching the soakings from one solution to the other. These additional procedures are designed to dense the films, smoothen the surfaces, and minimize the cross contaminations of the liquid source solutions so as to sustain the stabilized growth periodically. Thus, the individual SILAR growth cycles consist of soaking in cationic ion precursor, washing in deionized water and drying, soaking in anionic ion precursor, and washing in deionized water and drying (see Fig. 7.3 for a schematic diagram). Compared to the conventional gas-phase ALD, liquid-ALD, and liquid-injection ALD methods discussed above, SILAR is simpler and more cost-effective for the growth of compound thin films; it can also be used for large-area and mass productions.

3 ALD, liquid-ALD, and SILAR for metal oxides

3.1 ALD of Metal Oxides

ALD of metal oxide thin films has been attaining increasing interest for microelectronics due to the process conformality and precise control of thickness on different kind of substrates. Aluminum oxide (Al_2O_3) is one of the most important and widely studied oxide materials by ALD *(37–42)*. Al_2O_3 thin films grown by thermal ALD have excellent dielectric properties with good thermal and chemical stability on various surfaces, making them attractive for applications in electronic and optoelectronic devices. The enhanced conformity of ALD Al_2O_3 than those produced by other methods, for example, CVD *(24)*, sputtering *(43)*, etc., depends on the sequential self-limiting surface chemical reactions. Previous studies show that Al_2O_3 ALD can be achieved with various precursors at their respective ALD temperature windows *(44,45)*. Trimethylaluminum (TMA) is a typical ALD precursor for supplying alumina species, which react with hydroxyl group (–OH) of H_2O on the growing surface *(37)*. The chemical reaction was described and split into two half reactions:

$$Al - OH^{\star} + Al(CH_3)_3 \rightarrow Al - O - Al(CH_3)_2 {}^{\star} + CH_4 \quad (7.1)$$

$$Al - CH_3 {}^{\star} + H_2O \rightarrow Al - OH^{\star} + CH_4 \quad (7.2)$$

In this reaction process, TMA was introduced into the reaction chamber, which react and form chemical bonds with –OH formed on the surface in the prior half reaction. After multiple-cycle reactions, the reactions complete at a certain targeted film thickness. Excess TMA and the CH_4 by-products were purged and pumped away with a constant N_2 flow. Similar reactions occur when –OH reacts with surface TMA and, at the end, excess H_2O molecule and CH_4 by-products were pumped away by N_2. It has been confirmed that during ALD growth of Al_2O_3 thin films, the growth per cycles (GPC) can be achieved at 1.1 Å with a dielectric constant of ~7.6 at growth temperature of 177°C (see the results shown in Fig. 7.4A) *(37)*. It was also found that the dialectic constant of Al_2O_3 was varied depending on the film thicknesses as well as the deposition temperatures as shown in Fig. 7.4A andB, respectively.

Low-temperature depositions of Al_2O_3 have also been explored by ALD with desired growth conditions *(38)*. It is found that with the growth temperature decreases from 177 to 33°C, the growth time to complete one cycle ALD Al_2O_3 monotonically increased from 12 to 203 s. It is

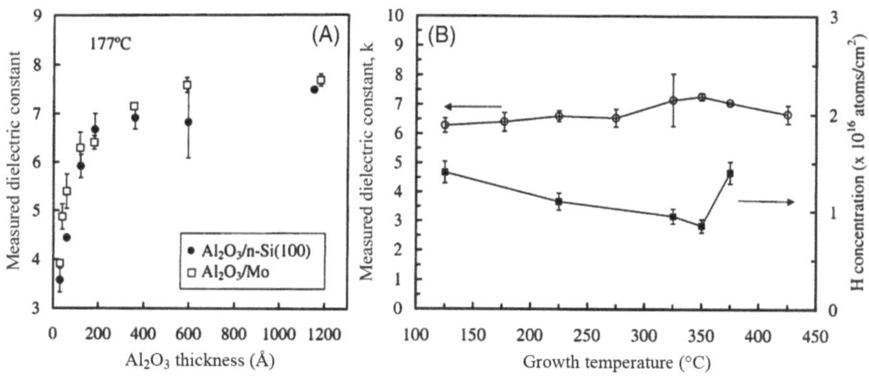

Figure 7.4 (A) Dielectric constant variations of ALD Al_2O_3 thin films as a function of film thicknesses at the growth temperature of 177°C; (B) dielectric constants and residual hydrogen concentrations of ~12.3-nm-thick Al_2O_3 films as a function of ALD temperatures on Mo-coated Si (1 0 0) substrates (Ref. *(37)*, ©2002, Elsevier).

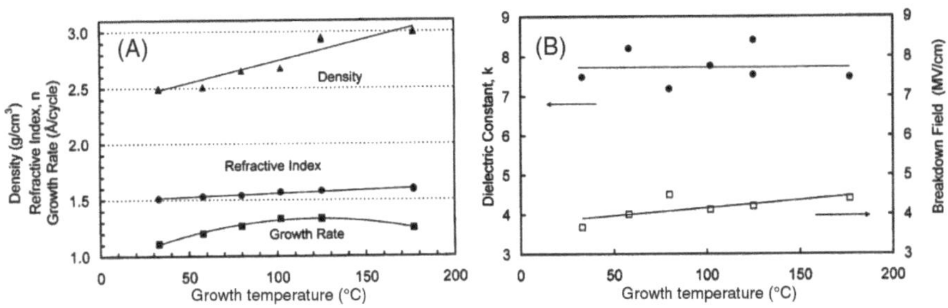

Figure 7.5 (A) The refractive indexes, growth rates, and material densities obtained from 300-cycle ALD Al_2O_3 thin films grown on Si substrates at various temperatures; (B) dielectric constants and breakdown fields of the ALD Al_2O_3 thin films depending on the growth temperatures (Ref. *(38)*, ©2004, ACS).

important to note that for low temperature ALD process the purge time should be increased to avoid chemical vapor deposition (CVD) of Al_2O_3. A comparison made for a 300-cycle ALD process of Al_2O_3 at 33 and 177°C shows that the film thickness is about 33.2 nm achieved by the growth at lower temperature, that is, 33°C, which is thinner than that of ~37.5 nm grown at higher temperature, that is, 177°C. Fig. 7.5A presents the material densities, the film refractive indexes, and the growth rates of ALD Al_2O_3 as a function of growth temperatures. The density measurements were carried out by using an in situ quartz crystal microbalance (QCM). One sees in Fig. 7.5A that the measured yields of ALD Al_2O_3 were 1.11 and 1.25 Å/cycle at 33 and 177°C, respectively. However, the surface roughness of the Al_2O_3 thin films was quite similar with one another and does not show any apparent dependence on the growth temperatures. The electrical characterization revealed that the breakdown voltage of the ALD Al_2O_3 thin films is a bit lower for those with lowered growth temperatures, although the dielectric constant of the ALD Al_2O_3 thin films is consistent in the range of ~7.7 ± 0.4 over the entire temperature range as shown in Fig. 7.5B.

To obtain ALD Al_2O_3 thin films with better properties and reduced leakage current density for electronic device applications, ozone was used as the oxidant *(46–48)*. It is believed that the residual unreacted hydroxyl groups in the ALD Al_2O_3 thin films play important roles, affecting the material and dielectric properties. In this light, high quality dielectric films can be produced by employing O_3, to replace H_2O, as the oxygen precursor in typical ALD process *(49)*. The material density of the Al_2O_3 thin films grown by ALD at 150–300°C using O_3 precursor is similar to those grown at 300°C but using H_2O precursor. The highest growth rate, when using O_3

precursor, was found to be 2.1–2.2 Å/cycle at low temperature (150°C). It is interesting to see that at a higher temperature (i.e., 300°C), the ALD growth rate is reduced by 15% when using O_3 in place of H_2O as the oxygen precursor.

The basic mechanism for ALD Al_2O_3 using TMA and O_3 precursors is presented in Fig. 7.6 *(47)*. In the process after an initial surface covering by $AlCH_3$*, the oxygen insertion produces $AlCH_2OH$* and ethylene (i.e., reaction by-product). The progress of this process, as well as its combination with adjacent species, form AlOH*, occupying the surface-active sites, and leave behind ethylene (see Fig. 7.6). In a parallel process, Al–C bonds react with oxygen atoms that produce methoxy groups; these adjacent methoxy groups combine and release CH_4 and H_2 gaseous by-products. As a consequence, the carbonate groups are formed after oxidizing the formats. Further releases of CO and CO_2 from the decomposing formate and carbonate groups leave behind hydroxyl groups on alumina and reaction by-product of CH_4. By reaping these cycles of TMA and O_3 with sequential reactions on the surface of the substrate Al_2O_3 can be deposited with controlled thickness *(47)*.

It has also been found that isopropyl alcohol (IPA) can effectively react with TMA for ALD Al_2O_3. With the IPA working as the oxygen source, the interfacial growth can be prevented during the ALD process *(50)*. The growth rate of ALD Al_2O_3 with TMA and IPA at 250°C is about 0.8 Å/cycle. This growth rate is lower than those using H_2O oxidant due to the reduced amount of secondary adsorbed chemical groups present in the process with IPA than H_2O since IPA* is larger molecule than H_2O*.

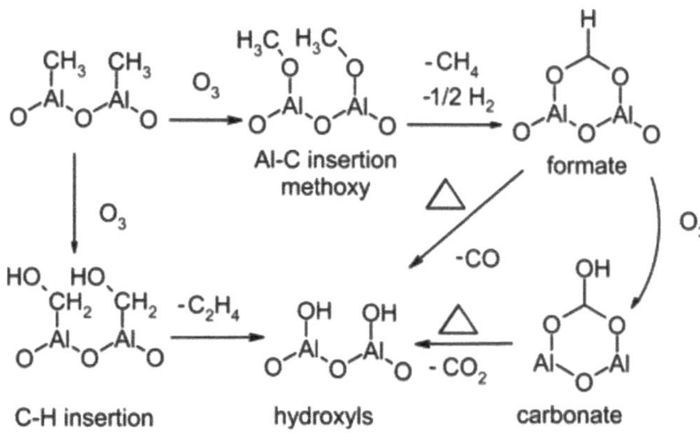

Figure 7.6 The reaction mechanism during ALD process of Al_2O_3 using O_3 and TMA as the oxygen and aluminum precursors, respectively (Ref. *(47)*, ©2008, ACS).

Dimethylaluminum isopropoxide [DMAI $(CH_3)_2AlOCH(CH_3)_2$] precursor has also been studied for Al_2O_3 deposition with H_2O oxidant *(51)*. In this study, differential thermal analysis (DTA) revealed an endotherm centered at 124°C due to the decomposition of the DMAI precursor. As the DMAI precursor temperatures ranged from 85 to 90°C, the growth rate of Al_2O_3 was constant at about 1.06 Å/cycles. The reaction was similar to TMA and can be explained as *(51)*:

$$2Me_2AlO^iPr + H_2O \rightarrow Al_2O_3 + 4CH_4 + 2HO^iPr \qquad (7.3)$$

Using this precursor, the growth of the interfacial layer can be suppressed within a monolayer thickness.

Dimethylaluminum chloride (DMACl) has also received attention as an alternative aluminum source precursor for ALD growth of Al_2O_3 thin films. The residual chlorine in the thin films could provide comparable passivation and thermal stability of the resultant Al_2O_3 *(52)*. As shown in Fig. 7.7, by employing DMACl and ozone, the ALD process GPC was obtained ≥0.76 Å/cycle at the temperatures ranging from 150 to 350°C. The GPC is apparently larger at higher growth temperatures. Also shown in Fig. 7.7A is that the refractive index of the Al_2O_3 thin films increases from 1.56 to 1.64 as the growth temperature is increased from 150 to 350°C. It should be noted that the impurity concentration of the Al_2O_3 thin films, typically carbon and chlorine, produced by using ozone is higher than those produced by using water (see Fig. 7.7B). This observation has been attributed to the less complete reactions of DMACl with ozone as compared with those reactions of DMACl with water.

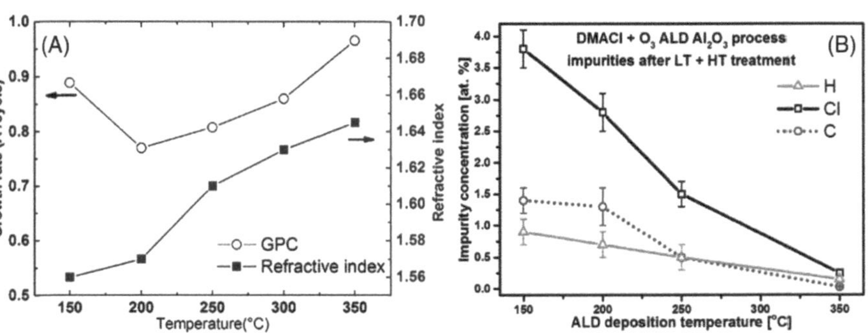

Figure 7.7 (A) Growth rate and refractive index of Al_2O_3 thin films grown by ALD using DMACl and ozone precursors at various temperatures; (B) typical impurities and their concentrations as a function of ALD growth temperatures of the Al_2O_3 thin films grown by using DMACl and ozone precursors (Ref. *(52)*, ©2017, Wiley).

From the point of view of both the refractive index and the impurity level, it is quite clear that the Al_2O_3 thin films grown by ALD using DMA-Cl and ozone precursors at higher temperatures have improved properties. Plasma enhanced atomic layer deposition (PEALD) has also been explored with TMA and O_2 as the aluminum and oxygen source precursors *(53)*. The magnetic field strength of the remote plasma generator plays a certain role in the ALD process, especially the GPC.

3.2 Liquid injection ALD of metal oxides

Liquid injection ALD (LI-ALD) techniques make use of vaporizations of precursors, which is suitable for precursors with less thermal stability. This ALD process can effectively minimize the thermal decompositions of alkylamides and alkoxides. In this light, the LI-ALD technique can be used for a wide variety of ALD precursors for producing thin metal oxides. Recent research results show that excellent physical properties of HfO_2 can be obtained by LI-ALD using $[Cp_2Hf(CH_3)_2]$ and H_2O as the Hf- and O-source precursors *(54)*. Post-growth thermal annealing was also introduced to reduce the material defects. Electrical characterizations, for example, via forming metal-oxide-semiconductor (MOS) capacitors, suggest that the negative fixed oxide charge can be significantly reduced by increasing the annealing temperature *(54)*.

By using $[Hf(mmp)_4]$ as the reaction precursor in LI-ALD process of HfO_2, it has been found that the ALD process is quite sensitive to the growth temperatures. At low temperatures, the growth rate is constant; however, it tends to increase in the temperature range of 275–350°C due to the thermal decomposition of the precursors (see Fig. 7.8A) *(11)*.

This study also revealed that the volume of the precursor injection per cycle plays an important role in controlling the growth rate of the LI-ALD process (see Fig. 7.8B) *(11)*. The growth rate monotonically increases with the increase in the volume of the injected precursors. The continuously increasing in the growth rate with injected volume of the precursors shows an absence of self-limiting in the studied conditions at 275°C. However, a lowering of the growth temperature to 225°C leads to a weak self-limiting growth (see Fig. 7.8B). This phenomenon can be attributed to the creation of reactive surface sites for nonpyrolytic precursors.

Recently, growth and characterization of anatase TiO_2 thin films have been demonstrated by direct LI-ALD using titanium tetraisopropoxide [TTIP, $Ti(OCH(CH_3)_2)_4$] as the Ti-source precursor at a deposition temperature of 280°C. In this study, the Ti-precursor was dissolved in anhydrous

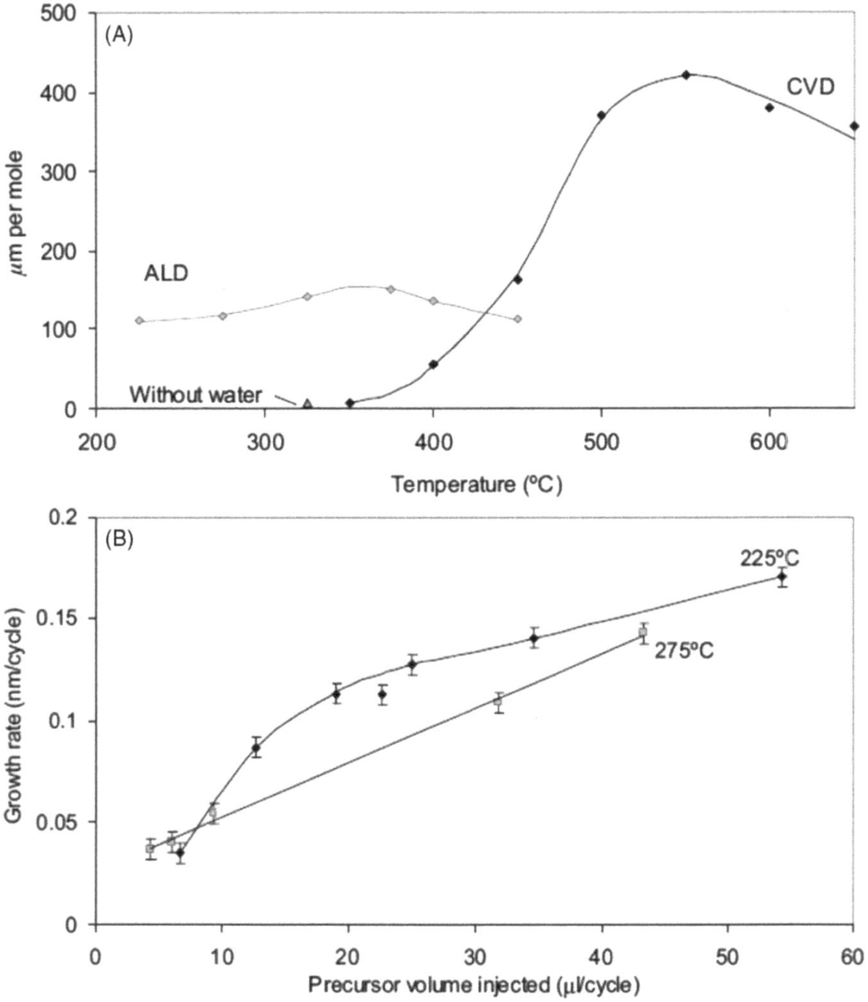

Figure 7.8 (A) Temperature-dependent LI-ALD growth rates of HfO_2, together with their comparisons with CVD; (B) growth rate variations with respect to the injected precursor volume per growth cycle in the LI-ALD process of HfO_2 (Ref. *(11)*, ©2005, Wiley).

cyclohexane (C_6H_{12}) to realize of the LI-ALD process *(55)*. It is shown in Fig. 7.9A that the growth rate monotonically increases with the growth temperatures ranging from 220 to 350°C; however, the increase was negligible at about 280–290°C, manifesting itself as an ALD window. It is most likely that the surface reactions of the precursors are incomplete at lower temperatures (i.e., below the ALD window); an increase in the growth temperature enhances the reactions, which in turn leads to the increased growth rates. In contrast, at the temperatures above the ALD window, thermal

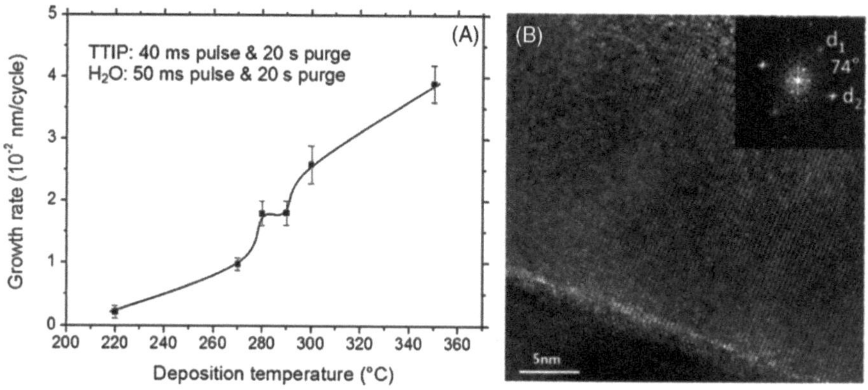

Figure 7.9 LI-ALD growth of TiO_2 thin films using TTIP and H_2O as the Ti- and O-source precursors, respectively. (A) ALD growth rates as a function of growth temperatures; (B) HRTEM image of an ALD-TiO_2 layer. The inset presents the zone-axis aligned diffraction pattern (Ref. *(55)*, ©2014, Elsevier).

decomposition occurred that results in an increased growth rate, which is the same as the mechanism of CVD growth. The effect of the Ti-precursor (i.e., TTIP) pulse duration on the LI-ALD growth rate was further studied in the ALD window. It was found that the growth rate increases with the TTIP-pulse durations and saturates when the pulse duration approaches the range of 35–47 ms. Likewise, an optimized pule duration of 50–100 ms was found for the LI-ALD growth based on the variations of the growth rate. As a result, to realize the self-limit surface saturation of the LI-ALD growth, a minimum H_2O pulse of 50 ms is necessary. Linear relationship between the TiO_2 film thickness and the LI-ALD cycles was observed as desired and the growth rate was 0.018 Å /cycle. Anatase phase TiO_2 was further confirmed by the crystallographic structure characterizations using high-resolution transmission electron microscopy (HRTEM). Typical results are shown in Fig. 7.9B, where one sees that along the zone axis the diffractions at d_1 and d_2 are 0.354 and 0.358 nm away from the zone center, respectively. The in-plane rotation from the azimuth of d_1 to that of d_2 is about 74° *(55)*.

Titanium dipivaloylmethanato di-isopropoxide [Ti(O-*i*-Pr)$_2$(DPM)$_2$, $Ti(OC_3H_7)_2(C_{11}H_{19}O_2)_2$], having a higher decomposition temperature and noncorrosive character, has also been used as Ti-source precursor for producing TiO_2 thin films by LI-ALD *(56)*. In this LI-ALD investigation, the Ti-precursor was dissolved in ethylcyclohexane and H_2O, and the growth temperature was varied in large range. The density of the Ti layer in an individual ALD cycles, grown at 365°C, has been monitored as a function of Ti-precursor pulse durations. The monitored results have been further compared with those of the process by using TTIP as the Ti-precursors and shown in Fig. 7.10A. One sees that the TiO_2 growth using Ti(O-*i*-Pr)$_2$(DPM)$_2$ shows a clearer saturation behavior, which indicates an absence of thermal decomposition of the precursors and the growth is realized in ALD, rather than CVD, reactions.

In principle, the growth rate of TiO_2 thin films by LI-ALD using different Ti-source precursors depends on their ligand structures. In this regard, it is reasonable that even at high temperatures the growth of TiO_2 using Ti(O-*i*-Pr)$_2$(DPM)$_2$ precursor has a reduced growth rate as compared with that using TTIP precursor. Presented in Fig. 7.10B are the film thicknesses of TiO_2 grown by using Ti(O-*i*-Pr)$_2$(DPM)$_2$ as a function of LI-ALD cycles; their best linear fittings revealed a growth rate 0.033 nm/cycle. Due to the large ligands present in the Ti(O-*i*-Pr)$_2$(DPM)$_2$ precursor, the growth rate is apparently lower when compared with those of other Ti-source precursors reported in the literature for ALD growth of TiO_2 thin films.

A typical HRTEM image taken from the cross-section of a TiO_2 thin film grown by LI-ALD at 365°C using Ti(O-*i*-Pr)$_2$(DPM)$_2$ Ti-source precursor is shown in Fig. 7.11A, which clearly shows the polycrystalline property with irregularly oriented grains formed due to the high-temperature growth. The grain structures of the TiO_2 thin films are also probed by AFM from the surface. Fig. 7.11B shows a typical AFM images taken from a TiO_2 film with the thickness of about 25–32 nm. One can see that the grains are randomly distributed on the surface; the root-mean-square (RMS)

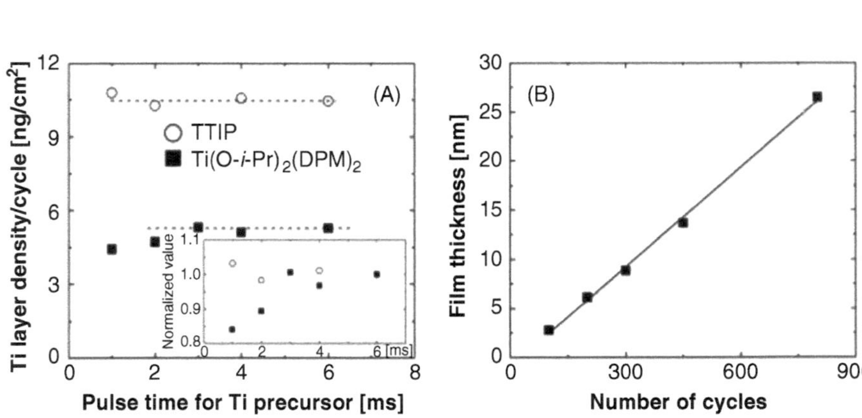

Figure 7.10 LI-ALD growth of TiO_2 thin films using TTIP and Ti(O-*i*-Pr)2(DPM)$_2$ as the Ti-source precursors. (A) Comparisons of the layer density in an individual ALD cycles with the changes in pulse time of the Ti(O-*i*-Pr)$_2$(DPM)$_2$ and TTIP precursors; the growth temperatures are 365 and 265°C, respectively; (B) film thicknesses of TiO_2 as a function of ALD cycles for the growth using the Ti(O-*i*-Pr)$_2$(DPM)$_2$ precursor (Ref. *(56)*, ©2009, ECS).

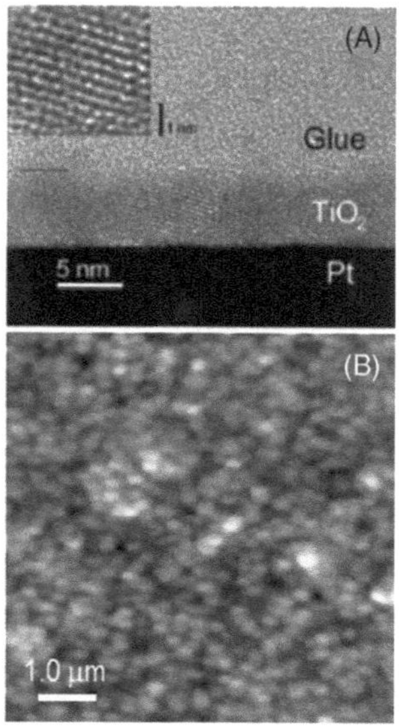

Figure 7.11 Structural and morphological studies of TiO_2 thin films grown by LI-ALD using Ti(O-*i*-Pr)2(DPM)$_2$ as the Ti-source precursor. (A) HRTEM image of a TiO_2 thin film grown at 365°C;(B) AFM image of a TiO_2 thin film with a thickness of 25--32 nm (Ref. *(56)*, ©2009, ECS).

roughness of the TiO_2 film is <2 nm, indicating a very smooth surface of the LI-ALD film using Ti(O-*i*-Pr)$_2$(DPM)$_2$ as the Ti-source precursor *(56)*.

RuO_2 thin thins have been recently drawing huge attentions for fabricating resistive random-access memory (RRAM) devices. The RRAM devices can be realized by controlling the distribution of defects in the oxide films, for example, a redistribution of oxygen vacancies by applying a voltage along the film thickness direction, so as to softly breakdown the oxide film via forming a conducting filament. In a recently reported work, LI-ALD growth of RuO_2 thin films was described with *iso*-octane dissolved metal-organic bis(2,2,6,6-tetramethyl-3,5-heptanedionato)(1,5-cyclooctadiene)ruthenium [Ru(thd)$_2$(cod)] as the Ru-source precursor *(57)*. Grazing incidence X-ray diffraction (GIXRD) reveals that the LI-ALD grown RuO_2 thin films are dominated by rutile phase RuO_2 (see Fig. 7.12A). Surface roughness as measured by AFM for a 17-nm-thick LI-ALD RuO_2 is about 1.1 nm (see Fig. 7.12B).

The best fitting to the plotting of the RuO_2 film thicknesses as a function of ALD cycles revealed a growth rate of 0.66 nm/cycle for the LI-ALD growth. Compared to other ALD techniques reported in the literature, this growth is a bit higher and the high growth rate has been attributed to an

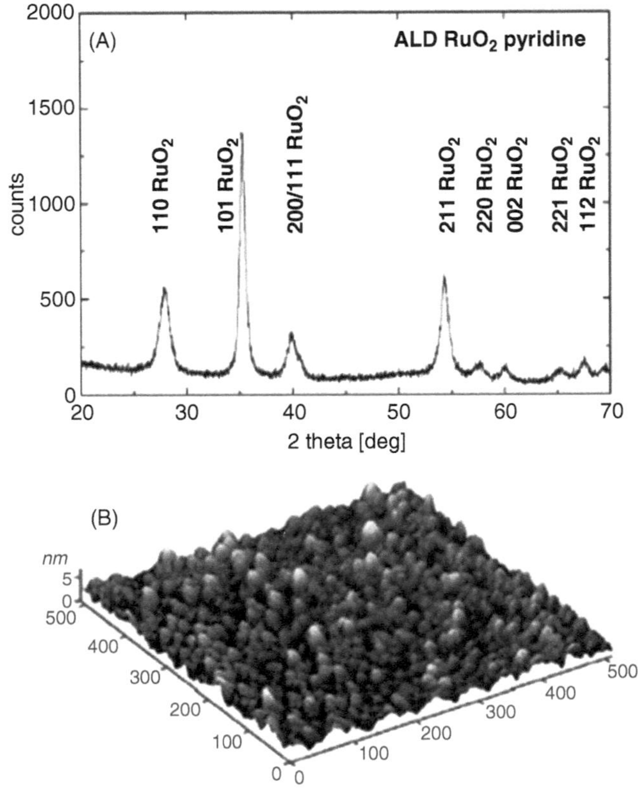

Figure 7.12 LI-ALD growth of RuO_2 thin films using $[Ru(thd)_2(cod)]$ as the Ru-source precursor. (A) GIXRD pattern of a RuO_2 thin film deposited by LI-ALD at 290°C; (B) AFM image recorded from a 17-nm-thick RuO_2 film (Ref. *(57)*, ©2010, Elsevier).

enhanced adsorption of oxygen on the surface of grown RuO_2. This LI-ALD was also tested on three-dimensional (3D) prepatterned trenches with the aspect ratio ranging from 2 to 12; the results show 100% conformal growth of RuO_2 on such 3D structures.

3.3 SILAR of metal oxides

Successive ionic layer adsorption and reaction (SILAR) method is attractive for growing metal oxide films because of its simple procedures and ideal for producing uniform thin films. The method consists of immersion and rinsing processes of the substrate in cations and anions solutions, which enables chemical reactions between solid phase and ions on the surface of the substrate in the solutions.

Recently, TiO_2 thin films were grown by SILAR method on glass substrates *(58)*. The Ti-source precursor used for the TiO_2 deposition was a mixture of 6 mL 15% $TiCl_3$ solution (P.S. in about 10% HCl) and 50 mL deionized water, 0.2 M urea solution was added to maintain the pH value of the Ti^{3+} cations contained solution at 0.5. The SILAR process consists of three steps. In the first step, the cleaned substrate was immersed in the precursor solution; in the second step, the substrate with Ti cations adsorbed on the surface was immersed in deionized water at 80°C to realize the reducing reactions on the surface; and in the third step, the substrate with deposited TiO_2 was again immersed deionized water but at room temperature. Since the $TiCl_3$ is the source of Ti^{3+} and the hot water (at 80°C) is a source of oxygen, after the first immersion the Ti^{3+} was adsorbed reduced to peroxotitanium on the surface of the substrate. In the second immersion, the peroxotitanium reacts with hot water and converts into TiO_2. The immersion in deionized water at room temperature is to remove the loosely adsorbed ions, which in turn enhances the film density. XRD studies show that after 150 cycles, the TiO_2 films are amorphous and do not show any diffraction peaks *(58)*.

4 ALD of metals

Among the various deposition techniques, ALD has gained the most considerable interest for producing metal depositions due to the self-limit nature of the ALD process, which makes the deposition of metals with a uniform thickness on 3D structures possible *(59)*. Pits, voids, and/or through holes that can be frequently observed in metal coatings by using physical vapor deposition methods (e.g., sputtering) can be avoided in the ALD process due to the self-limit surface reactions. ALD has been recognized as an ideal solution for depositing metal layers with smooth surface on planar and stepped structures.

Metal acetylacetonates ($M(acac)_2$, M = metal) precursors have been studied for ALD growth of metallic thin films such as Ni, Cu, Pt, etc. *(60)*. Utriainen et al. has described two different ALD processes for the metallic thin films *(60)*. One is controlled by the direct redox reactions of H_2:

$$\mathrm{M(acac)_2(g) + H_2(g) \rightarrow M(s) + 2H(acac)(g).} \tag{7.4}$$

The other is initiated with oxidization of the acetylacetonates by H_2O and followed by the redox reactions of H_2:

$$\mathrm{M(acac)_2(g) + H_2O(g) \rightarrow MO(s) + 2H(acac)(g),} \tag{7.5}$$

$$\text{and } \mathrm{MO(s) + H_2(g) \rightarrow M(s) + H_2O(g).} \tag{7.6}$$

When using $Pt(acac)_2$ as the Pt-source precursor, metallic Pt can be deposited at 155°C while for $Cu(acac)_2$, onset of partial decompositions has been observed at 250°C. In comparison, the decomposition of Ni(acac)$_2$ is observed at the temperatures above 280°C. Since the thermal decomposition of source precursors is undesired in ALD process, it was concluded that Pt(acac)$_2$ is not suitable for ALD growth of Pt metals while Cu(acac)$_2$ is limited to the temperatures lower than 250°C. Thermogravimetric (TG) analysis for the M(acac)$_2$ shown in Fig. 7.13 confirms the volatilizations studied by thermoanalyzer at low pressure of 2 mbar *(60)*. However, the ALD growth of metals using such metal acetylacetonates precursors is largely depending on the substrate materials and preferably on metallic substrate.

In the ALD process for metallic Ni with H_2O involved, NiO thin film was first deposited from the reactions during the Ni(acac)$_2$ and O_3 pulses. The poor reactivity between Ni(acac)$_2$ and H_2 leads to a negligible growth as compared with the reactions between Ni(acac)$_2$ and O_3. Further reduction of NiO to metallic Ni was performed at 260 °C and the produce of metallic Ni has been confirmed by XRD *(60)*.

ALD growth of metallic Al has been studied by using TMA and hydrogen plasma in separated half-cycles *(61)*. Thermal decompositions of the precursor have been studied at the temperatures ranging from 100 to 400°C on p-type Si, SiO_2, and TiN substrates, respectively. The ALD process was carried out at 250 °C as the thermal decomposition of the TMA precursor occurred at the temperatures above 300°C. However, no deposition could be obtained with sequential pulses of TMA and H_2 as hydrogen molecule

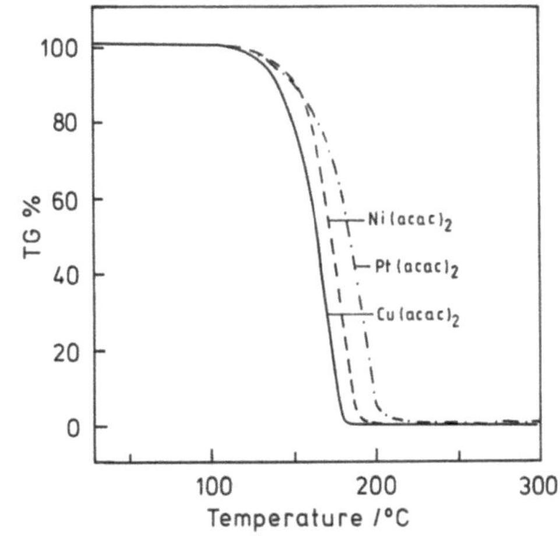

Figure 7.13 TG curves of M(acac)$_2$ (M = Ni, Pt and Cu) proposed for ALD growth of Ni, Pt and Cu metallic thin films at a heating rate of 10°C/min in vacuum (Ref. *(60)*, ©2000, Elsevier).

cannot reduce TMA at this temperature (i.e., 250°C). In comparison, an introducing of hydrogen plasma (at a power of 200 W) replacing the H_2 pulse leads to a successful growth of metallic Al. As shown in Fig. 7.14A, growth rate increases with the TMA pulse duration and saturates when the pulse time increases to above 3 s. The saturated and constant growth rate is ~0.15 nm/cycle. A test for the ALD growth of Al thin films on

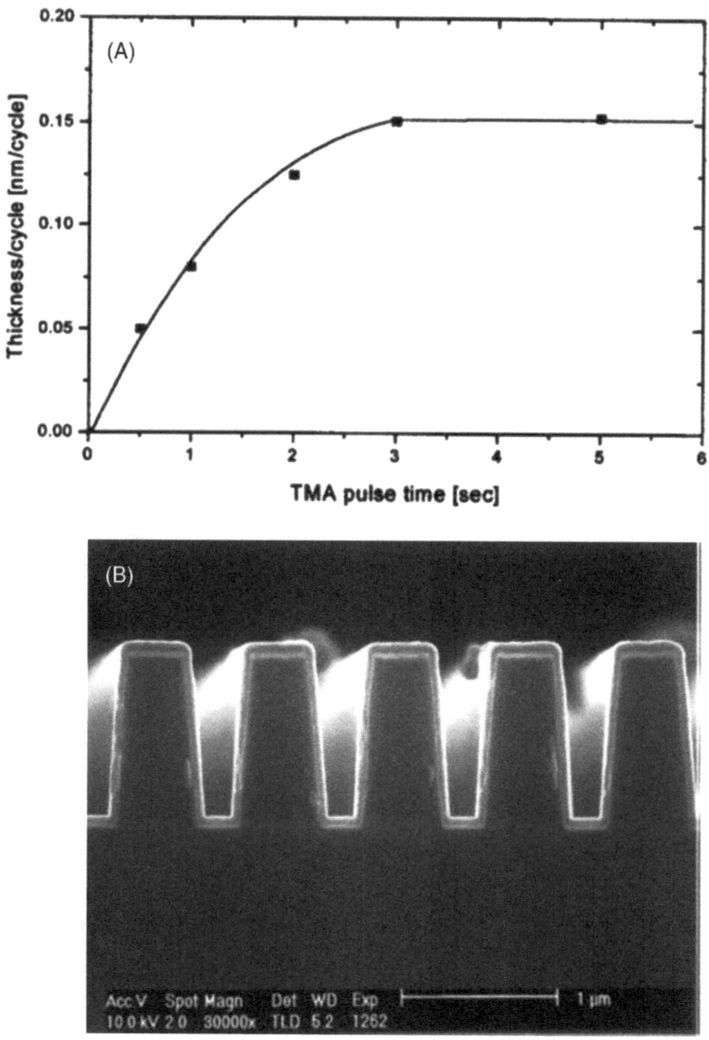

Figure 7.14 ALD growth of Al using TMA and H_2 plasma. (A) Dependence of Al film growth rate with respect to the TMA pulse time; (B) cross-sectional FESEM image of a 50-nm-thick Al film deposited by ALD at 250°C on a 3D patterned structure with the aspect ratio of 5:1 (Ref. *(61)*, ©2002, ECS).

3D structures has been further demonstrated. Fig. 7.14B shows a cross-sectional image taken by field-emission scanning-electron microscopy (FE-SEM) from a 50-nm-thick ALD-coated Al on a surface patterned substrate. One sees that the Al film has been conformally coated on the top- and bottom-surface as well as the sidewalls of the trenches with a uniform thickness *(61)*.

It has been mentioned above that the low thermal stability of $Pt(acac)_2$ limited its applications as a precursor for ALD growth of Pt metals *(60)*. Alternatively, (methylcyclopentadienyl)trimethylplatinum [$MeCpPtMe_3$, $(CH_3C_5H_4)Pt(CH_3)_3$] has been studied for ALD growth of Pt *(62)*. In this study, Pt films were deposited on 2-nm-thick Al_2O_3, borosilicate, and silicon with the precursor pulse time varied from 0.2 to 1.5 s and the growth temperature changed from 300 to 350°C. The upper temperature limit of the deposition was found at 300°C as thermal decomposition can be observed at 350°C. Although partial thermal decomposition cannot be completely avoided, after 0.2 s pulse time for $MeCpPtMe_3$ growth rate increased and saturated at 0.48 Å/cycle. SEM and AFM measurements show that the deposited Pt films are quite uniform and the surface roughnesses are 4 and 17 nm for the films with the thicknesses of 50 and 110 nm, respectively *(62)*.

Below 250°C, the reactions between $MeCpPtMe_3$ precursor and O_2 is relatively weak that leads to a low growth rate. Recent studies indicate that using O_3, in place of O_2, as the oxidant for $MeCpPtMe_3$, Pt thin films can be obtained at 100°C. This low temperature ALD process might be suitable for thermally fragile substrates. Further studies of the ALD process at the temperatures ranging from 60 to 300°C for 60 cycles show that the growth rate is about 0.45 Å/cycle *(63)*. The growth rate was also studied by changing the pulse time of the $MeCpPtMe_3$ and O_3 precursors. The results show that the pulse times of the $MeCpPtMe_3$ and O_3 precursors to approach the saturate growth rate are 1 and 3 s, respectively (see Fig. 7.15A and B) *(63)*.

X-ray photoelectron spectroscopy (XPS) and XRD were employed to validate the growth of Pt metals by the ALD process. The XPS $Pt4f_{7/2}$ core level probed from an ALD film grown for 200 cycles at 100°C exhibits a peak at 71.0 eV, which confirms the metallic character of the resultant Pt film. XRD results also indicate the metallic phase Pt with a dominant [1 1 1] orientation. The bulk resistivity of the ALD grown Pt thin films was found decreasing with the increase in the film thickness. A typical resistance of 11.9 μΩ cm was obtained for a 24-nm-thick Pt film grown at 150°C on a Si substrate with SiO_2 native oxide (~4 nm thick) *(63)*.

ALD growth of metallic Cu has also been studied. Continuous Cu thin films with high conductivity have been reported by using $Cu(^sBu\text{-}amd)_2$ as

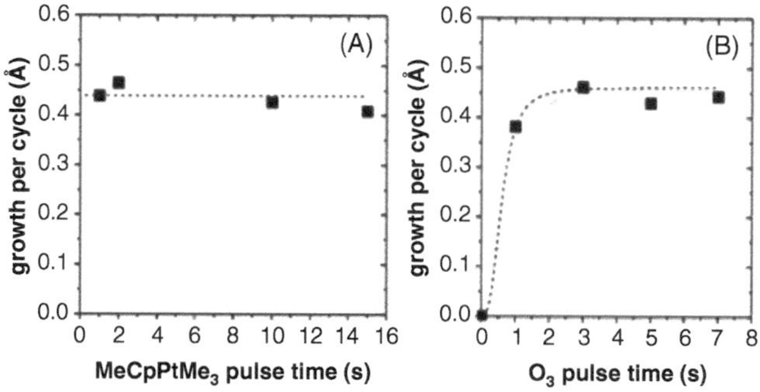

Figure 7.15 ALD growth rate for Pt films using $MeCpPtMe_3$ and O_3 precursors as a function of (A) $MeCpPtMe_3$ pulse time while keeping the O_3 pulse time fixed at 8 s; (B) O_3 pulse time while keeping the $MeCpPtMe_3$ pulse time fixed at 15 s (Ref. *(63)*, ©2013, ACS).

the ALD source precursor *(64)*. Thermal gravimetric (TG) analysis shows that among the low-melting Cu amidinates $Cu(^{s}Bu\text{-}amd)_2$ has a melting point of 77°C and the fastest vaporization rate of 38.8 nmol/(cm^2 min).

In ALD processes with the cycle number fixed at 1000, by increasing the Cu precursor exposure, the growth rate of Cu increases monotonically and eventually saturated at about 0.9 Å/cycle (see Fig. 7.16A). It should be noted that the precursor exposure, which is the overall effect of pressure and pulse time, was monitored. This is because the chamber pressure was varied when switching from one to another precursor pulse in adjacent half cycles. The H_2 exposure was about one order of magnitude larger than that of the Cu precursor exposure, which led to completely saturated ALD growth rates in the studied exposure range (see Fig. 7.16B). Also found is that on

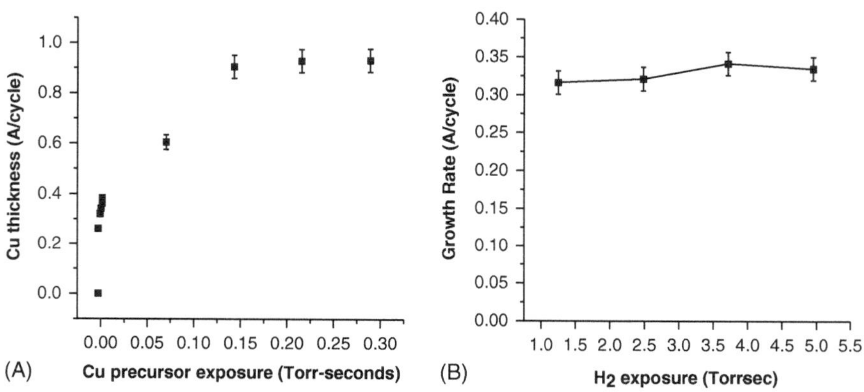

Figure 7.16 Growth rate in GPC of the ALD process of Cu on ALD Al_2O_3 as a function of (A) Cu precursor exposure; (B) H_2 exposure (Ref. *(64)*, ©2006, ECS).

the surface of oxides (i.e., ALD-grown-Al_2O_3, -HfO_2, and thermal-SiO_2), the initial growth rate was as high as ~2 Å/cycle; however, after the surface of the oxides was covered by the deposited Cu, the growth rate decreased to 0.4–0.5 Å/cycle. In contrast, the ALD growth rate of Cu on Ru substrate was constant *(64)*. Through morphological characterizations using SEM it was found that the initial deposited Cu agglomerated on the surface of oxides with large grains (~100 nm). Electrical resistivity measurements of the ALD Cu have been carried out for those nucleated well on Co and Ru surfaces; for a 4-nm-thick Cu, the sheet resistance was found to be ~50 Ω/cm, which is good enough for applications in electroplating process of microelectronics.

It has been demonstrated that aluminum hydride complexes have good thermal stability and are powerful reducing agents for application in ALD process. Recently, *tert*-butyl-substituted amido-amine ligands have been used to produce monomeric aluminum dihydride complexes with high thermal stability; and, as a result, dimethylaluminum hydride (Me_2AlH) precursor was found volatile and thermally stable at low temperatures, which is desired for applications in ALD process *(65)*.

To growth Al thin films by ALD using the above mentioned $MeAlH_2$ Al-source precursor the purging time was tuned between 20 and 10 s to realizing the self-limit surface reactions. Fig. 7.17A presents the growth rates of the thin films grown for 200 cycles and their respective sheet resistivity as a function of the growth temperatures ranging from 100 to 180°C. It is seen that the growth rate decreases while the resistivity increases with the increase in the growth temperatures. The low resistivity of the Al thin films

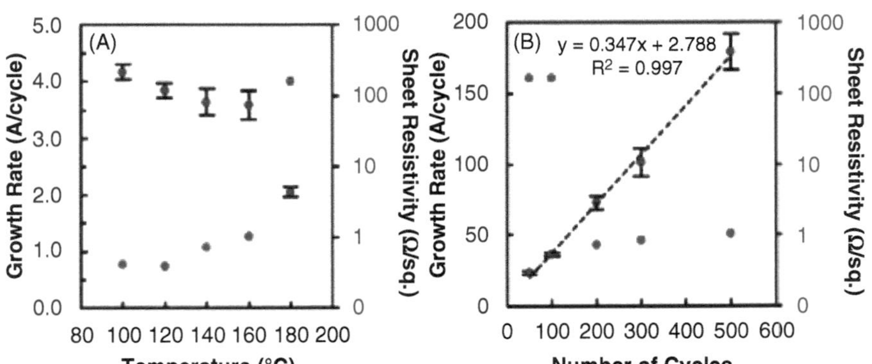

Figure 7.17 (A) Temperature-dependent growth rate of the ALD process for 200 cycles; (B) film thicknesses and their respective resistivity produced by the ALD process for certain cycles (Ref. *(65)*, ©2018, ACS).

grown for 200 ALD cycles at lower temperatures is quite close to the bulk value of metallic Al. The film thickness of the ALD Al also increases linearly with number of ALD cycles. However, the electrical resistivity exhibits a sharp drop (more than two orders of magnitude) when increasing the ALD cycles from 100 to 200 and keeps constant at 0.73–1.02 Ω/cm with further increases in the thickness (see Fig. 7.17B).

Morphological comparisons of the ALD grown Al thin films have further made by SEM. Fig. 7.18A–D presents the SEM images taken from the surface of the ALD Al thin films grown for 50, 100, 200, and 300 cycles, respectively. One sees that the Al grows in separated islands when the ALD cycles are <100. When the ALD cycles increase to 100, onset of islands connecting occurs and continuously connected chains formed at 200 cycles. These observations provide direct evidence for the resistivity variations as a function of the ALD cycles (see Fig. 7.17B).

5 Potentials of ALD, liquid-ALD, and SILAR techniques for THR

It is now quite clear that the thin film techniques, that is, ALD, liquid-ALD, and SILAR, discussed above have the capability in controlling the thin film growth in atomic thickness. With dedicated design, they can be used for growing not only metal and metal oxide thin films but also other films like

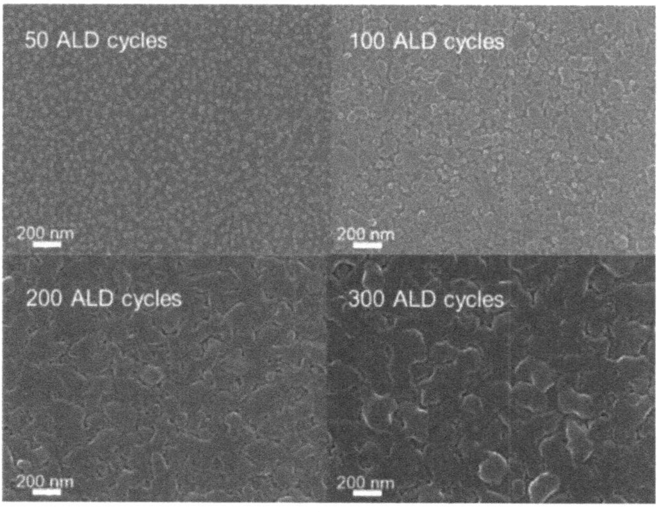

Figure 7.18 Morphological comparisons by SEM for ALD Al thin films grown at 140°C for different thicknesses (i.e., ALD cycles) (Ref. *(65)*, ©2018, ACS).

nitrides and disulfides. They are also suitable for growth heterostructures and/or multiple layers, for example, the dielectric/metal/dielectric structures for applications in transparent heat regulations (THR). To fabricate the dielectric/metal/dielectric-based multiple layered THR structures *(20)*, oxides (e.g., SnO_2, ZrO_2, $BaSnO_3$, TiO_2, ZnO, WO_3, etc.), nitrides (e.g., AlN), and sulfides (e.g., ZnS) thin films have been demonstrated as effective antireflection layers and all these dielectrics can now be produced by ALD, liquid-ALD, and/or SILAR process. For the metals, as we have mentioned above that Ni, Cu, Pt, Al, Au, Ag, etc. can now be produced by ALD, liquid-ALD, and/or SILAR process although some of them need more improvement in their qualities by, for example, synthesis more suitable source precursors. By further developing the dedicated process, ALD, liquid-ALD, and SILAR do have the potential for producing few-atoms-thick layers with the desired uniformity, conformality, and purity. As these thin film techniques entering the industry, for example, for large-scale THR fabrications, they have to be able to be scaled up. Currently, general commercial gas-source ALD systems can be used for process on 200-mm-diameter substrates, updating the capability to 300-mm-diameter is on the way. For the liquid-ALD and SILAR techniques, the scaling up is much easier and readily available for industrializations. Since the ALD, liquid-ALD, and SILAR processes are strongly dependent on the precursors, developing and exploring of novel suitable precursors dominated the new developments of these thin film techniques. A great of additional investigations in this area are desired to enhance their growth rate and overall throughput.

References

1. Sheng, J.; Lee, J. -H.; Choi, W. -H., et al. Review Article: Atomic Layer Deposition for Oxide Semiconductor Thin Film Transistors: Advances in Research and Development. *J. Vac. Sci. Technol. A* **2018**, *36* (6), 060801.
2. Liu, H. Recent Progress in Atomic Layer Deposition of Multifunctional Oxides and Two-Dimensional Transition Metal Dichalcogenides. *J. Mol. Eng. Mater.* **2016**, *4* (2), 1640010.
3. Ahvenniemi, E.; Akbashev, A. R.; Ali, S., et al. Review Article: Recommended Reading List of Early Publications on Atomic Layer Deposition—Outcome of the "Virtual Project on the History of ALD". *J. Vac. Sci. Technol. A* **2017**, *35* (1), 010801.
4. Sharma, K.; Hall, R. A.; George, S. M. Spatial Atomic Layer Deposition on Flexible Substrates Using a Modular Rotating Cylinder Reactor. *J. Vac. Sci. Technol. A* **2015**, *33* (1), 01A132.
5. Jarvis, K. L.; Evans, P. J. Growth of Thin Barrier Films on Flexible Polymer Substrates by Atomic Layer Deposition. *Thin Solid Films* **2017**, *624*, 111–135.
6. Maydannik, P. S.; Kääriäinen, T. O.; Cameron, D. C. An Atomic Layer Deposition Process for Moving Flexible Substrates. *Chem. Eng. J.* **2011**, *171* (1), 345–349.

7. Wu, Y.; Döhler, D.; Barr, M., et al. Atomic Layer Deposition from Dissolved Precursors. *Nano Lett.* **2015**, *15* (10), 6379–6385.
8. O'Neill, B. J.; Jackson, D. H. K.; Crisci, A. J., et al. Stabilization of Copper Catalysts for Liquid-Phase Reactions by Atomic Layer Deposition. *Angew. Chem. Int. Ed.* **2013**, *52* (51), 13808–13812.
9. Avril, L.; Decams, J. M.; Imhoff, L. Pulsed Direct liquid Injection ALD of TiO_2 Films Using Titanium Tetraisopropoxide Precursor. *Phys. Procedia* **2013**, *46*, 33–39.
10. Hoffmann-Eifert, S.; Watanabe, T.; Hwang, C. S.; et al. Liquid Injection Atomic Layer Deposition of Perovskite-type Multi-component Oxide Thin Films for Ferroelectric and Higher-k Three Dimensional Capacitor Structures. In 17th IEEE International Symposium on the Applications of Ferroelectrics, 23–28 February. 2008; pp 1–2.
11. Potter, R. J.; Chalker, P. R.; Manning, T. D., et al. Deposition of HfO_2, Gd_2O_3 and PrO_x by Liquid Injection ALD Techniques. *Chem. Vapor Depos.* **2005**, *11* (3), 159–169.
12. Sankapal, B. R.; Mane, R. S.; Lokhande, C. D. Successive Ionic Layer Adsorption and Reaction (SILAR) Method for the Deposition of Large Area ($\sim$10 cm^2) Tin Disulfide (SnS_2) Thin Films. *Mater. Res. Bull.* **2000**, *35* (12), 2027–2035.
13. Dubal, D. P.; Holze, R. A Successive Ionic Layer Adsorption and Reaction (SILAR) Method to Induce Mn_3O_4 Nanospots on CNTs for Supercapacitors. *New J. Chem.* **2013**, *37* (2), 403–408.
14. Barry, A.; Jonathan, G.; Alain, K. Understanding ALD, MLD and SAMs as They Enter the Fab. *Solid State Technol.* **2018**, *61* (1), 3.
15. Lee, B. H.; Hwang, J. K.; Nam, J. W., et al. Low-Temperature Atomic Layer Deposition of Copper Metal Thin Films: Self-Limiting Surface Reaction of Copper Dimethylamino-2-propoxide with Diethylzinc. *Angew. Chem. Int. Ed.* **2009**, *48* (25), 4536–4539.
16. Tobias, T. Atomic Layer Deposition of Copper, Copper(I) Oxide and Copper(I) Nitride on Oxide Substrates. ACTA Universitatis Upsaliensis: Uppsala, Sweden, 2004.
17. Mäkelä, M.; Hatanpää, T.; Mizohata, K., et al. Studies on Thermal Atomic Layer Deposition of Silver Thin Films. *Chem. Mater.* **2017**, *29* (5), 2040–2045.
18. Maarit, M. Studies on Atomic Layer Deposition of Gold and Aliver Thin Films. University of Helsinki: Helsinki, 2018.
19. Griffiths, M. B. E.; Pallister, P. J.; Mandia, D. J., et al. Atomic Layer Deposition of Gold Metal. *Chem. Mater.* **2016**, *28* (1), 44–46.
20. Dalapati, G. K.; Kushwaha, A. K.; Sharma, M., et al. Transparent Heat Regulating (THR) Materials and Coatings for Energy Saving Window Applications: Impact of Materials Design, Micro-structural, and Interface Quality on the THR Performance. *Progr. Mater. Sci.* **2018**, *95*, 42–131.
21. Ko, R. H. H.; Khalatpour, A.; Clark, J. K. D., et al. Ultrasmooth Ultrathin Ag Films by AlN Seeding and Ar/N2 Sputtering for Transparent Conductive and Heating Applications. *APL Mater.* **2018**, *6* (12), 121112.
22. Dmitrieva, E. A.; Mukhamedshina, D. M.; Mit, K. A.; et al. The Influence of Isothermal Annealing on the Optical and Electrical Properties of Thin SnO2 Films Doped with Fluorine. Rec. Contrib. Phys. 2018, 50 (2), 68–75.
23. Nazemiyan, M.; Jalili, Y. S. Record Low Temperature Mo Doped V_2O_5 Thermochromic Thin Films for Optoelectronic Applications. *AIP Adv.* **2013**, *3* (11), 112103.
24. Liu, H. F.; Wong, S. L.; Chi, D. Z. CVD Growth of MoS_2-based Two-dimensional Materials. *Chem. Vapor Depos.* **2015**, *21* (10–11–12), 241–259.
25. Liu, H.; Jin, Y.; Yang, C. Droplet-induced Dot, Dot-in-Hole, and Hole Structures in GaGe Thin Films Grown by MOCVD on GaAs Substrates. *CrystEngComm* **2016**, *18* (24), 4499–4507.
26. Liu, H. F.; Liu, W.; Chi, D. Z. Rotation of In-plane Structural Anisotropy at the Interface of an a-Plane InN/GaN Heterostructure Grown by MOCVD on *r*-Plane Sapphire. *CrystEngComm* **2016**, *18* (11), 1871–1877.

27. Liu, H. F.; Dixit, V.; Xiang, N. Anneal-induced Interdiffusion in 1.3-μm GaInNAs/GaAs Quantum Well Structures Grown by Molecular-beam Epitaxy. *J. Appl. Phys.* **2006**, *99* (1), 013503.
28. Liu, H.; Jin, Y.; Lin, M., et al. Growth and In-plane Undulations of GaAs/Ge Superlattices on [001]-oriented Ge and GaAs Substrates: Formation of Regular 3D Island-in-Network Nanostructures. *J. Mater. Chem. C* **2018**, *6* (47), 13059–13068.
29. Muñoz-Rojas, D.; Nguyen, V. H.; Masse de la Huerta, C., et al. Spatial Atomic Layer Deposition (SALD), An Emerging Tool for Energy Materials. Application to New-generation Photovoltaic Devices and Transparent Conductive Materials. *Comptes Rendus Physique* **2017**, *18* (7), 391–400.
30. Liu, H. F.; Antwi, K. K. A.; Wang, Y. D., et al. Atomic Layer Deposition of Crystalline Bi_2O_3 Thin Films and Their Conversion into Bi_2S_3 by Thermal Vapor Sulfurization. *RSC Adv.* **2014**, *4* (102), 58724–58731.
31. Liu, H. F.; Wang, Y. D.; Lin, M., et al. Cobalt Sulfide Nanoparticles Decorated on TiO_2 Nanotubes via Thermal Vapor Sulfurization of Conformal TiO_2-coated $Co(CO_3)_{0.5}(OH)\cdot 0.11H_2O$ Core–Shell Nanowires for Energy Storage Applications. *RSC Adv.* **2015**, *5* (60), 48647–48653.
32. Liu, H.; Yang, R. B.; Yang, W., et al. Atomic Layer Deposition and Post-growth Thermal Annealing of Ultrathin MoO_3 Layers on Silicon Substrates: Formation of Surface Nanostructures. *Appl. Surf. Sci.* **2018**, *439*, 583–588.
33. Chatterjee, S.; Saha, S. K.; Pal, A. J. Formation of All-oxide Solar Cells in Atmospheric Condition based on Cu_2O Thin-films Grown through SILAR Technique. *Solar Energy Mater. Solar Cells* **2016**, *147*, 17–26.
34. Patil, U. M.; Gurav, K. V.; Joo, O. -S., et al. Synthesis of Photosensitive Nanograined TiO_2 Thin Films by SILAR Method. *J. Alloys Compd.* **2009**, *478* (1), 711–715.
35. Nwanya, A. C.; Deshmukh, P. R.; Osuji, R. U., et al. Synthesis, Characterization and Gas-sensing Properties of SILAR Deposited ZnO-CdO Nano-composite Thin Film. *Sens. Actuators B: Chem.* **2015**, *206*, 671–678.
36. Abdel Rafea, M.; Roushdy, N. Determination of the Optical Band Gap for Amorphous and Nanocrystalline Copper Oxide Thin Films Prepared by SILAR Technique. *J. Phys. D: Appl. Phys.* **2008**, *42* (1), 015413.
37. Groner, M. D.; Elam, J. W.; Fabreguette, F. H., et al. Electrical Characterization of Thin Al_2O_3 Films Grown by Atomic Layer Deposition on Silicon and Various Metal Substrates. *Thin Solid Films* **2002**, *413* (1), 186–197.
38. Groner, M. D.; Fabreguette, F. H.; Elam, J. W., et al. Low-Temperature Al_2O_3 Atomic Layer Deposition. *Chem. Mater.* **2004**, *16* (4), 639–645.
39. Liu, H.; Guo, S.; Yang, R. B., et al. Giant Blistering of Nanometer-Thick Al_2O_3/ZnO Films Grown by Atomic Layer Deposition: Mechanism and Potential Applications. *ACS Appl. Mater. Interfaces* **2017**, *9* (31), 26201–26209.
40. Liu, H.; Yang, R. B.; Guo, S., et al. Effect of Annealing on Structural and Optical Properties of ZnO/Al_2O_3 Superlattice Structures Grown by Atomic Layer Deposition at 150 °C. *J. Alloys Compd.* **2017**, *703*, 225–231.
41. Wilk, G. D.; Wallace, R. M.; Anthony, J. M. High-κ Gate Dielectrics: Current Status and Materials Properties Considerations. *J. Appl. Phys.* **2001**, *89* (10), 5243–5275.
42. Dillon, A. C.; Ott, A. W.; Way, J. D., et al. Surface Chemistry of Al_2O_3 Deposition Using $Al(CH_3)_3$ and H_2O in a Binary Reaction Sequence. *Surf. Sci.* **1995**, *322* (1), 230–242.
43. Liu, H. F.; Chua, S. J.; Hu, G. X., et al. Effects of Substrate on the Structure and Orientation of ZnO Thin Film Grown by rf-magnetron Sputtering. *J. Appl. Phys.* **2007**, *102* (8), 083529.
44. Kukli, K.; Ritala, M.; Leskelä, M., et al. Atomic Layer Epitaxy Growth of Aluminum Oxide Thin Films from a Novel $Al(CH_3)_2Cl$ Precursor and H_2O. *J. Vac. Sci. Technol. A* **1997**, *15* (4), 2214–2218.

45. Matero, R.; Rahtu, A.; Ritala, M., et al. Effect of Water Dose on the Atomic Layer Deposition Rate of Oxide Thin Films. *Thin Solid Films* **2000**, *368* (1), 1–7.
46. Kim, J. B.; Kwon, D. R.; Chakrabarti, K., et al. Improvement in Al_2O_3 Dielectric Behavior by Using Ozone as an Oxidant for the Atomic Layer Deposition Technique. *J. Appl. Phys.* **2002**, *92* (11), 6739–6742.
47. Goldstein, D. N.; McCormick, J. A.; George, S. M. Al_2O_3 Atomic Layer Deposition with Trimethylaluminum and Ozone Studied by in Situ Transmission FTIR Spectroscopy and Quadrupole Mass Spectrometry. *J. Phys. Chem. C* **2008**, *112* (49), 19530–19539.
48. von Gastrow, G.; Li, S.; Repo, P., et al. Ozone-based Batch Atomic Layer Deposited Al_2O_3 for Effective Surface Passivation. *Energy Procedia* **2013**, *38*, 890–894.
49. Elliott, S. D.; Scarel, G.; Wiemer, C., et al. Ozone-Based Atomic Layer Deposition of Alumina from TMA: Growth, Morphology, and Reaction Mechanism. *Chem. Mater.* **2006**, *18* (16), 3764–3773.
50. Woo-Seok, J.; Sung, Y.; Choon-soo, L., et al. Atomic Layer Deposition of Al_2O_3 Thin Films Using Trimethylaluminum and Isopropyl Alcohol. *Electrochem. Soc.* **2002**, *149* (6), C306–C310.
51. Cho, W.; Sung, K.; An, K. -S., et al. Atomic Layer Deposition of Al_2O_3 Thin Films Using Dimethylaluminum Isopropoxide and Water. *J. Vac. Sci. Technol. A* **2003**, *21* (4), 1366–1370.
52. Bao, Y.; Laitinen, M.; Sajavaara, T., et al. Ozone-Based Atomic Layer Deposition of Al_2O_3 from Dimethylaluminum Chloride and Its Impact on Silicon Surface Passivation. *Adv. Electron. Mater.* **2017**, *3* (6), 1600491.
53. Li, X.; Chen, Q.; Sang, L., et al. Atomic Layer Deposition Al_2O_3 Thin Films in Magnetized Radio Frequency Plasma Source. *Phys. Procedia* **2011**, *18*, 100–106.
54. Taechakumput, P.; Taylor, S.; Buiu, O., et al. Optical and Electrical Characterization of Hafnium Oxide Deposited by Liquid Injection Atomic Layer Deposition. *Microelectron. Reliabil.* **2007**, *47* (4), 825–829.
55. Avril, L.; Reymond-Laruinaz, S.; Decams, J. M., et al. TiO_2 Anatase Films Obtained by Direct Liquid Injection Atomic Layer Deposition at Low Temperature. *Appl. Surf. Sci.* **2014**, *288*, 201–207.
56. Kim, S. K.; Hoffmann-Eifert, S.; Mi, S., et al. Liquid Injection Atomic Layer Deposition of Crystalline TiO_2 Thin Films with a Smooth Morphology from $Ti(O\text{-}i\text{-}Pr)_2(DPM)_2$. *J. Electrochem. Soc.* **2009**, *156*, D296.
57. Hušeková, K.; Dobroc˘ka, E.; Rosová, A., et al. Growth of RuO_2 Thin Films by Liquid Injection Atomic Layer Deposition. *Thin Solid Films* **2010**, *518* (16), 4701–4704.
58. Jimenez-Garcia, F. N.; Segura-Giraldo, B.; Restrepo-Parra, E., et al. Synthesis of TiO_2 Thin Films by the SILAR Method and Study of the Influence of Annealing on its Structural, Morphological and Optical Properties. *Ingeniare* **2015**, *23* (4), 8.
59. Kim, H. Atomic Layer Deposition of Metal and Nitride Thin Films: Current Research Efforts and Applications for Semiconductor Device Processing. *J. Vac. Sci. Technol. B* **2003**, *21* (6), 2231–2261.
60. Utriainen, M.; Kröger-Laukkanen, M.; Johansson, L. -S., et al. Studies of Metallic Thin Film Growth in an Atomic Layer Epitaxy Reactor Using $M(acac)_2$ (M = Ni, Cu, Pt) Precursors. *Appl. Surf. Sci.* **2000**, *157* (3), 151–158.
61. Lee, Y. J.; Kang, S. -W. Atomic Layer Deposition of Aluminum Thin Films Using an Alternating Supply of Trimethylaluminum and a Hydrogen Plasma. *ECS Solid State Lett.* **2002**, *5* (10) C91-C93.
62. Aaltonen, T.; Ritala, M.; Sajavaara, T., et al. Atomic Layer Deposition of Platinum Thin Films. *Chem. Mater.* **2003**, *15* (9), 1924–1928.
63. Dendooven, J.; Ramachandran, R. K.; Devloo-Casier, K., et al. Low-Temperature Atomic Layer Deposition of Platinum Using (Methylcyclopentadienyl)trimethylplatinum and Ozone. *J. Phys. Chem. C* **2013**, *117* (40), 20557–20561.

64. Li, Z.; Rahtu, A.; Gordon, R. G. Atomic Layer Deposition of Ultrathin Copper Metal Films from a Liquid Copper(I) Amidinate Precursor. *J. Electrochem. Soc.* **2006**, *153* (11), C787–C794.
65. Blakeney, K. J.; Winter, C. H. Atomic Layer Deposition of Aluminum Metal Films Using a Thermally Stable Aluminum Hydride Reducing Agent. *Chem. Mater.* **2018**, *30* (6), 1844–1848.

CHAPTER 8

New paradigm for efficient thermoelectrics

Durga Venkata Maheswar Repaka, Ady Suwardi, Pawan Kumar
Institute of Materials Research and Engineering, Agency for Science Technology and Research, Singapore

1 Introduction

Thermoelectrics convert thermal energy into electricity and vice versa *(1,2)*. There are two types of devices, one is based on Seebeck effect called thermoelectric generators, which generate the electrical power provided the temperature gradient across the thermoelectric legs, another one is based on the Peltier effect, called thermoelectric cooler, which generates the cooling depends on the electric current direction *(3)*. Thermoelectric devices have wide variety of applications from high power spacecraft power systems to low power Internet of Things (IoT) *(4,5)*. SiGe-based radioisotope thermoelectric generators (RTG) have been used in space exploration as they possess high tensile strength, low dislocation density, low thermal expansion, and operate at high temperatures ~1300°C without degradation *(6)*. On the other hand, there are different waste heat sources, especially the low grade heat sources from industries, boilers, air compressors, and so on, which are tremendous, and recovery of low grade heat energy is more challenging and high on demand for green energy technologies *(7)*. The conversion efficiency of a thermoelectric generator is given by $\eta = T_H - T_C / T_H\left[\sqrt{1+zT_{avg}} - 1 / \sqrt{1+zT_{avg}} + {}^{T_C}\!/_{T_H}\right]$, and is proportional to Carnot efficiency, $\frac{T_H - T_C}{T_H}$; and the cooling efficiency is characterized by the coefficient of performance (COP), $COP_{max} = \left(T_C / T_H - T_C\right)\left(\sqrt{1+zT_{avg}} - \frac{T_H}{T_C} / \sqrt{1+zT_{avg}} + 1\right)$, both are proportional to the figure of merit, $zT_{avg} = S^2\sigma/\kappa, T_{avg} = PFT_{avg}/k$; power factor $(PF) = S^2\sigma$; T_H and T_C are the hot and cold side temperatures, $T = T_{avg}$ is absolute temperature, $T_{avg} = \left(T_H + T_C\right)/2$; S is Seebeck coefficient, σ is the electrical conductivity and $\kappa = \kappa_e + \kappa_l$ is the total thermal

Energy Saving Coating Materials
http://dx.doi.org/10.1016/B978-0-12-822103-7.00008-X

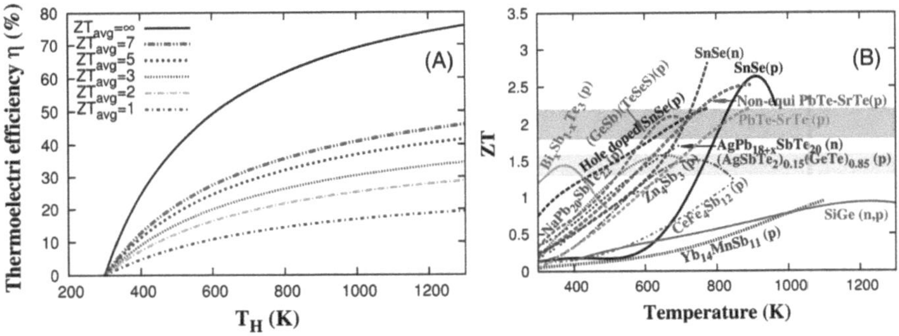

Figure 8.1 (A) The thermoelectric generator efficiency of thermoelectrics with different *zT* values and comparison with other energy-conversion technologies as a function of T_H at T_C = 300 K. The *zT* values are assumed to be temperature independent, and the heat-sink temperature is set at room temperature. (B) *zT* values as function of temperature for different thermoelectric systems *(9)*.

conductivity, κ_e and κ_l are the electronic and lattice thermal conductivities, respectively *(8)*. Fig. 8.1A shows the power generation efficiency of thermoelectric generator with different zT values as function of T_H while keeping T_C at room temperature, and Fig. 8.1B shows the state of the art of zT values in different thermoelectric systems.

Thus, in order to have high efficiency, at least $zT \sim 1$ and above near room temperature, we need high values of S and σ, and low values of κ. On the other hand, theses parameters, S, σ, and κ are strongly correlated through carrier concentration and different scattering phenomena. Hence, the zT cannot be quantified with any of these thermoelectric parameters independently. The Seebeck effect is discovered in 1821 by Thomas J. Seebeck, but very few efficient thermoelectric materials have been investigated, among all chalcogenides-based materials are best conventional thermoelectrics. The performance of thermoelectric materials can be improved by increasing the PF without affecting the thermal conductivity and/or by decreasing the total thermal conductivity without deteriorating the electrical properties. Simply, the figure of merit can be represented as, $zT = S^2\sigma / kT = S^2 ne\mu / kT$; large Seebeck can be reached with high $m_S^\star$ *and low* n, whereas for high electrical conductivity, one needs high n and large carrier mobility, $\mu = e\tau / m_C^\star$, small $m^\star$ (narrow bands). Eventually, the electronic performance of a thermoelectric materials is directly related to their intrinsic electronic band parameters, especially when the transport is contributed from multiple bands, thus need to account for the effective band parameters, such as valley band degeneracy (N_v), band effective mass $(m_b^\star)$, weighted mobility, fermi surface complexity factor $\left(N_V^\star K^\star = \left(m_S^\star / m_C^\star\right)^{3/2}\right)$, where $m_S^\star$ is Seebeck effective mass

$\left(m_S^* = N_v^{2/3} m_b^*\right)$, m_C^* is the conduction effective mass $\left(m_C^* = m_b^* / K^{*2/3}\right)$, N_V^* is effective valley band degeneracy, K^* is effective anisotropy constant, energy dependent scattering time $\left(\tau = \tau_0 \eta^r\right)$, where η is reduced fermi energy, r energy dependent scattering parameter, which can quantifies the dominant scattering phenomenon ($r = 0$ for energy independent, -0.5 for acoustic phonon, 0.5 polar optical phonon, and 1.5 for impurity scattering mechanisms) *(10)*. In order to decouple the interdependency of thermoelectric parameters and to improve the zT, nanostructured approach has been proposed by Prof. Dresselhaus and Hicks, who quantizes the electronic Density of states (DOS) to improve S^2n and evolve different scattering mechanisms between carriers and phonons are associated with quantization at interfaces, which decouple the μ and κ_{latt} *(11)*. Besides, other efforts to improve zT such as quantum confinements, the lattice thermal conductivity significantly decreases in two-dimensional superlattice structure thermoelectrics. There are several reports on nan-structuring bulk, nano inclusions at grain and grain boundaries affect the phonon and carrier scattering mechanisms, and improve the thermoelectric properties compare to their bulk materials. There are several strategies, that is, increasing the anharmonicity, resonant/rattling bonding, and presence of lone pair electrons, which develop the soft phonon modes, have been approached to decrease the thermal conductivity *(12,13)*. As increasing the temperature, the minority carriers participate in the electrical conduction, bipolar/mixed conduction state, where, zT tends to show a peak. Since the Seebeck coefficient show a peak/mxima due to minority carrier contribution as increasing temperature, $S_{max} = E_g / 2eT_{max}$, which is directly proportional to the band gap, E_g, as mentioned by Goldsmith and also increases in the electronic thermal conductivity due to bipolar conduction *(14)*. Consequently, the high performance of a thermoelectric generator (TEG), that is a peak in zT is also depend on the band gap of the materials, which decide the operating temperature range of TEG, that is, high bandgap means optimum performance will be at high temperature, and vice versa. Therefore, different systems exhibit high performance at different temperature ranges, by modulating the band gap can improve the thermoelectric performance *(15)*.

2 Progress and strategies to improve the performance of conventional thermoelectrics

The layered, narrow band gap chalcogenides based on Bi_2Te_3 system show remarkable thermoelectric properties, which is a semiconductor with trigonal unit cell. The intrinsic defects like antisite defects and density play crucial role in V_2IV_3 thermoelectrics *(16)*. The antisite defects and density are solely responsible for the p- and n-type characteristics. Bi_2Te_3 is an n-type and Sb-rich compounds of $Bi_{2-x}Sb_xTe_3$ exhibit p-type characteristics. The p- and n-type Bi_2Te_3 compounds are the benchmark materials for cooling and electric power generation applications due to their high zT values near room temperature. From few decades, there are several strategies have been applied to improve the figure of merit to reach >1at room temperature. The nanostructured bulk synthesis helps to improve the zT and reaches 1.4 by reducing the lattice thermal conductivity for p-type BiSbTe at 373 K *(17)*, and through dense dislocations formed at grain boundary and point defects, zT is reached 1.86 at 320 K *(18)* by lowering the lattice thermal conductivity. The performance of several thermoelectric materials improved by the nanostructuring bulk synthesis *(19–21)* (Fig. 8.2).

Mostly, the high performance has been found for $x = 1.5$ in $Bi_{2-x}Sb_xTe_3$ series is due to the converging valence bands of Bi_2Te_3 and Sb_2Te_3, that is, increase of band degeneracy (N_v) *(22)*. The band convergence has been observed in several thermoelectric systems, PbTe *(23)*, $Mg_2Si_{1-x}Sn_x$ *(24,25)*, which improved their performance. The zT of GeTe reaches 2.4 at 600 K due to structure change accompanied by the band convergence and decrease

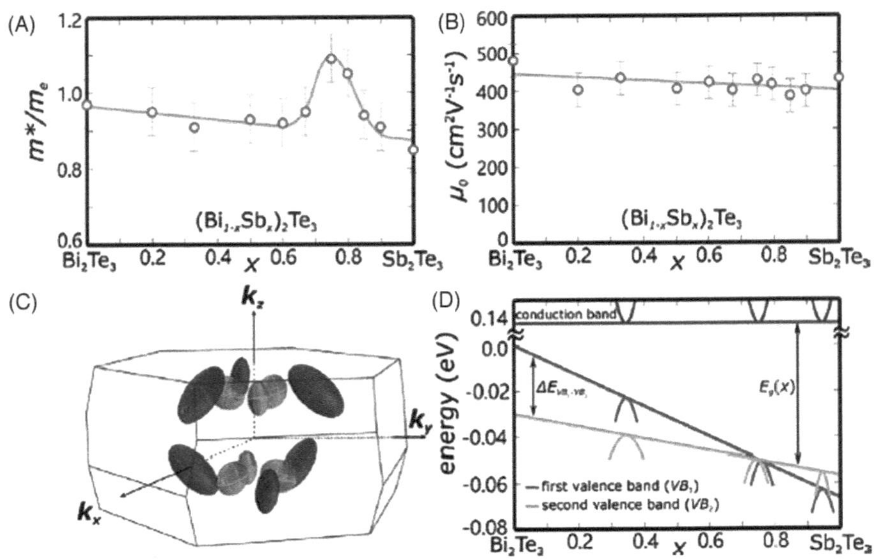

Figure 8.2 (A) Effective mass m^* (B) mobility pre factor as function of x in $Bi_{2-2x}Sb_{2x}Te_3$, respectively. (C) Brillouin zone of Bi_2Te_3 showing hole pockets for first valance band and (D) semi-empirical band structure represents the convergence of valance bands of Bi_2Te_3 and Sb_2Te_3 *(22)*.

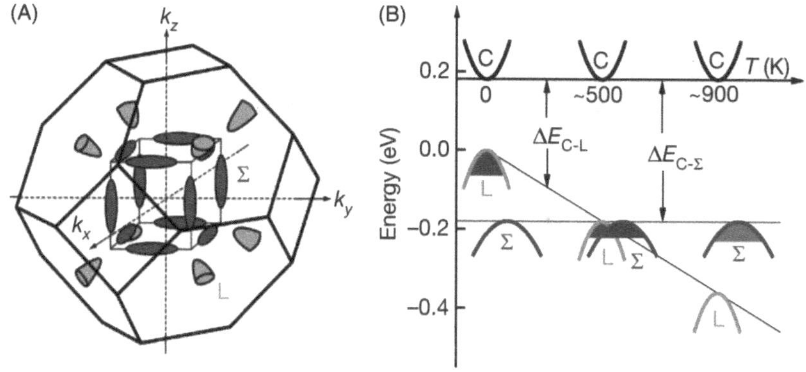

Figure 8.3 (A) Brillouin zone showing the low degeneracy hole pockets (orange) centered at the L point, and the high degeneracy hole pockets (blue) along the Σ line. The figure shows eight half-pockets at the L point so that the full number of valleys, N_v, is 4, while the valley degeneracy of the Σ band is 5 *(23)*. (B) Relative energy of the valence bands in $PbTe_{0.85}Se_{0.15}$. At 500 K, the two valence bands converge, resulting in transport contributions from both the L and Σ bands. C, conduction band; L, low degeneracy hole band; Σ, high degeneracy hole band.

in lattice thermal conductivity, the zT of PbTe system increases to 1.8 by alloying with PbSe at 850 K due to alignment of L and Σ bands, and alloy scattering, which also lowers the thermal conductivity *(26,27)* (Fig. 8.3).

The band convergence improves the band degeneracy, improves the Seebeck coefficient, because m_S^* is proportional to $N_v^{2/3}m_b^*$. One can realize that convergence of flat band (high m^*, higher Seebeck) and narrow (low m^*, high electrical conductivity) bands can improve the power factor, by increasing the Seebeck and electrical conductivity. The literature clearly represents that nano inclusions, band alignment, lowering the band offsets and large anharmonicity with complex phonon dispersion curves, promote to get best thermoelectric performance (Fig. 8.4).

3 Data driven and machine learning approach for investigating potential thermoelectrics

The physical properties of the materials from the literature and computations are available on several data base *(28–31)*. The fundamental physical characteristics of individual element and compounds are also listed and available to use. There are several tools available to mine the materials data, which can help to understand the materials and help to develop new and novel materials discovery *(32,33)*. Recently, so-called, data-driven studies and machine learning approaches on identifying potential thermoelectrics

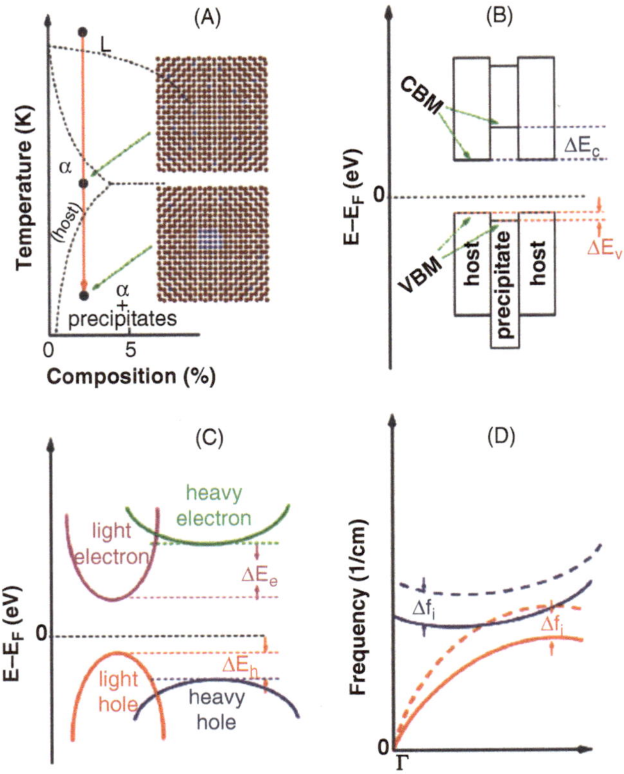

Figure 8.4 (A) Schematic representation of nanoscale precipitates to create centers for phonon scattering to reduce the lattice thermal conductivity. (B) Band alignment between host and dopant to retain a high carrier mobility. (C) Band structure engineering of alloyed host materials to minimize the energy difference between light and heavy carrier bands to enhance Seebeck coefficients and power factors. (D) Creating the anharmonicity by maximizing the frequency differences at different volumes to achieve low-lattice thermal conductivity *(9)*.

been reported *(34)*. There are very huge libraries available with materials data, a right criteria and descriptors are needed to examine and analyze the data. Even, the high-throughput Density Functional Theory (DFT) calculations and estimated thermoelectric parameter values are available in Materials Project and BoltzTrap, etc.

With prior physical understanding on the thermoelectrics, one can find the right descriptor to identify the potential thermoelectrics. Gibbs et al. use the Materials project, and proposed the fermi surface complexity factor, $N_V^* K^*$, could be a one of the right descriptors to mine the thermoelectrics materials *(35)*. As discussed earlier, concept of band convergence and band

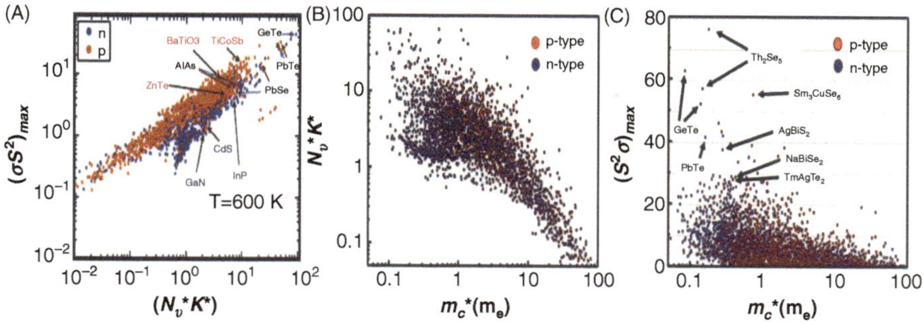

Figure 8.5 (A) Maximum power factor ($S^2\sigma$) as function of Fermi complexity factor ($N_v^*K^*$) for cubic systems *(35)*, (B) $N_v^*K^*$ vs m_c^*, and (C) S^2 σ vs m_c^* for all p- and n-type compounds.

alignment, which improve the power factor can be another key descriptor to identify the high zT materials. Recently, Ady et al. proposed the inertial effective mass as an important descriptor as well for identifying efficient TE material *(36)* (Fig. 8.5). However, the calculations for the physical properties of the materials in those databases obtained without taking the effect of carrier scattering phenomena and spin-orbit coupling, which may not be a true value to match with the experimental values. To accelerate the materials development, one need to have machine learning guided high-throughput materials synthesis and fast thermoelectric characterization tools.

In summary, tremendous progress has been made in the past decades both in terms of technological as well as the physical understandings. For instance, band convergence can be used as independent tuning knob to increase the electronic transport in thermoelectrics and not sacrificing the lattice thermal conductivity. More recently, data-driven aided studies also helped to identify descriptor like inertial effective mass, m^*. Machine learning tools need to be developed in order to do rapid characterization of thermoelectrics than conventional measurement techniques. Armed with this rich level of physical insights, we can utilize them to develop organic or hybrid organic–inorganic thermoelectrics that may enable flexible wearable applications.

4 Progress and strategies to improve the performance of hybrid thermoelectrics

Poly(3,4-ethylenedioxythiophene): polystyrene sulfonate (PEDOT:PSS) is one of the well-known conducting polymers, although the conductivity of the pristine PEDOT:PSS is small, ~0.01 S/cm *(37)*. The conductivity can be drastically increased by adding different solvents such as DMSO and EG and the reported value of conductivity is ~1000 S/cm *(38,39)*. It has shown remarkable thermoelectric properties due to improved electrical conductivity. Pipe et al. have reported zT of 0.4 at room temperature for PEDOT:PSS with different solvent treatments. Later, Bubnova et al. have also reported zT value of 0.25 at room temperature for PEDOT:Tos (tosylate) by replacing PSS with small TOS ion and by changing the oxidation states of the PEDOT *(40)*. But, the value of zT is still low compared to well-known inorganic material such as Bi_2Te_3. To improve further a new class of thermoelectric materials known as hybrid thermoelectrics have been developed. The combination of organic and inorganic has shown higher thermoelectric properties compared to their individual parts *(41,42)*. The solution-processability of these hybrid materials utilize inexpensive, scalable, and roll-to-roll techniques. Particularly, poly(3,4-ethylenedioxythiophene) (PEDOT) and Te nanowire-based hybrid materials have demonstrated potential thermoelectric properties near to room temperature *(43,44)* (Fig. 8.6).

There are several strategies that have been developed to improve the electronic and thermoelectric performance of these hybrid systems,

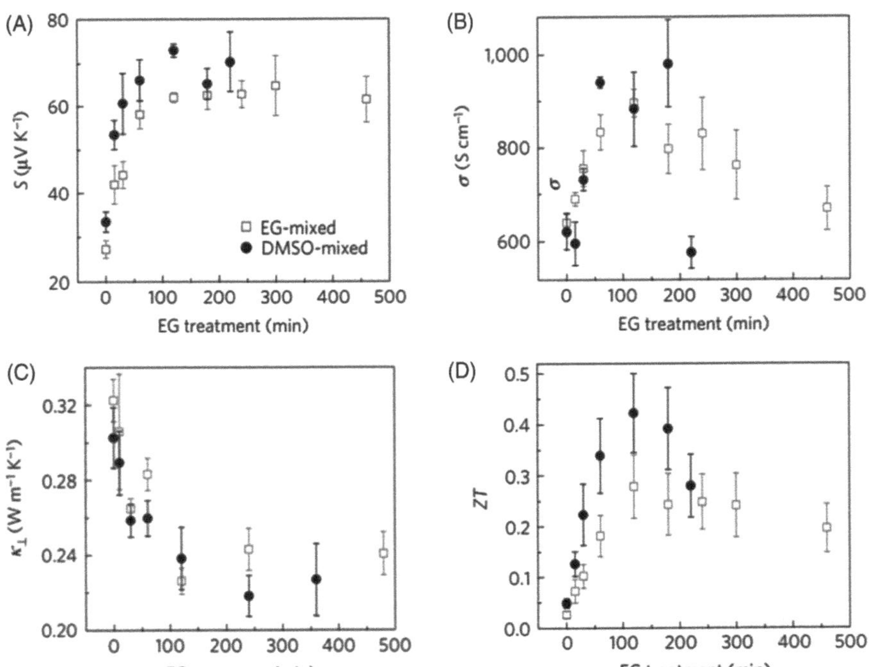

Figure 8.6 Thermoelectric properties of PEDOT:PSS at various de-doping times. Seebeck coefficients (A), electrical conductivities (B), vertical (cross-plane) thermal conductivities (C), and thermoelectric figure-of-merit (D), at $T = 297$ K in EG-mixed and DMSO-mixed PEDOT:PSS measured during the EG treatment (de-doping) process *(20)*.

interfacial transport *(45)*, structural/morphological effects *(46)*, and modifications to the energy dependence of carrier scattering *(47)*. Among these strategies, the energy-dependent scattering *(48)*is very difficult to understand in the hybrid systems *(49)*. Recently, Kumar et al. *(50)* have studied the above strategies on electrical and thermoelectric transport in PEDOT- and P3HT-based hybrid materials. To improve the performance of hybrid materials, a strong templating effect is required to boost the mobility of the carrier without disturbing the carrier concentration. Templating will improve the mobility of the free carriers in a polymer system. A large scattering parameter, indeed needed to improve the mobility as well, then to improve the electrical and thermoelectric properties in these hybrid materials. A new class of polymers with better scattering parameter with strong templating effect on inorganic surface is required for further enhancement. Wang et al. *(51)* reported $zT \sim 0.5$ for a n-type CNT/PEDOT composites treated by tetrakis(dimethylamino)ethylene (TDAE). Du et al. *(52)* reported thermoelectric properties of PEDOT:PSS/graphene quantum dot hybrid films. A 10 times enhancement in power factor was reported with adding graphene quantum dot in PEDOT:PSS. PEDOT:PSS is used as a p-type material for flexible TEG due to its high TE properties.

Recently, Kang et al. proposed a model to explain electrical and thermal transport in conducting polymers using Boltzmann transport theory *(53)*. They represented the electrical conductivity, $\sigma = \sigma_{E_0}(T) \times sF_{s-1}(\eta)$ and Seebeck coefficient $S = \frac{k_B}{e}\left[\frac{(s+1)F_s(\eta)}{sF_{s-1}(\eta)} - \eta\right]$, where, s is the energy-dependence of scattering of charge parameter which depends on the level of doping and the environment surrounding the conjugated backbone chains, η is the reduced chemical potential which determines the doping level and $\sigma_{E_0}(T)$ is the energy-independent prefactor, which is related to the alignment and mobility. According to this model, polymers with side chains, such as P3HT and PBTTT can show higher thermoelectric performance, compared to PEDOT:PSS, as they show better alignment which can give rise to large and higher s values *(54,55)*. P3HT and PBTTT possess a distinct monomer structure with additional side chains, which helps in enhancing the electronic coupling in the stacking direction by facilitating the backbones to be aligned in 2D planes *(56)*. The Coulombic interaction between carriers and impurities also reduces charge mobility, therefore decreasing electrical conductivity and the thermoelectric performance. In order to overcome, the dielectric environment of the material can be tuned, which can enhance charge carrier mobility *(57)*.

5 Conventional thermoelectric generators and wearable technologies

In order to achieve high power, the p- and n-type thermoelectric legs connected electrically in series and thermally in parallel as shown in Figure 8.7A. TEGs can generate power from few watts to microwatts, depend on the size (on the number of thermoelectric legs), temperature gradient across TEGs, and materials, etc. Bi_2Te_3-based TEGs are commercially available for near room temperature applications.

There several applications with thermoelectric generators, they have been used in the spacecraft/satellites, automobile engines, biosensors, industry areas, hot pipelines, solar heating systems, also used as refrigerators like portable wine coolers, temperature controlled car seating, as power sources in remote areas, wireless technologies, and low power IOT.

The power output by the TEG is proportional to the zT of thermoelectric materials, that is, Seebeck coefficient (S), internal resistance (R_{in}), and heat flow through the thermoelectric legs (thermal conductivity of legs), and also depends on the external factors, temperature difference, contact resistance (R_C), geometry of TEGs, dimensions of TE legs (area and length), and number of p and n pairs, heat sink, and so on. The main advantages are: TEGs have very long span of life, no moving parts, renewable resources, no harmful gases involved, no operating costs, solid-state operations, and reliable.

Recent developments of wireless sensor electronic technologies, miniaturization of electronic devices, and microelectromechanical systems (MEMS) demand power supply from few microwatts to milliwatts. Thermoelectric generators can be used in the sensors, and IOTs as the renewable

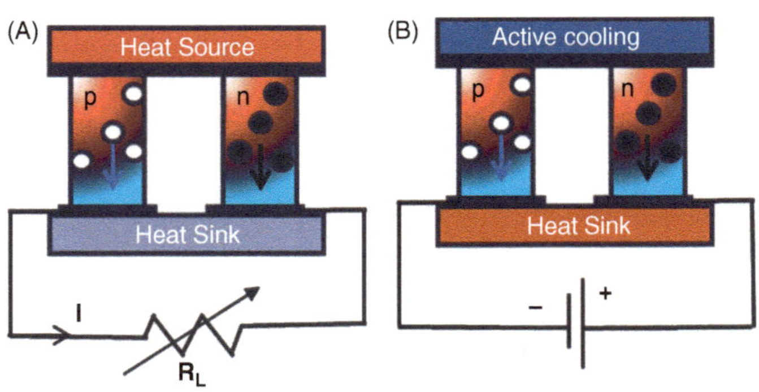

Figure 8.7 Schematic pictures of TEGs (A) for electric power generation and (B) for refrigeration.

electrical power sources, biomedical devices like health monitoring and tracking systems, e-paper, smart watches, fitness trackers, Google glasses, and alternative to portable battery-based systems. They can generate few micro-watt energy under few degrees of temperature difference, which is enough for low energy wireless sensors. The human body can be a good source of heat, generates through the metabolic functions. The human body temperature varies from 27 to 35°C and heat flow varies from 50 to 150 W/m^2. Conventionally, TEGs have a flat and rigid platform. In order to improve their applications and using on irregular heat source surfaces, flexible TEG are needed. However, there are several limitations on the thermoelectric materials growth, since all conventional inorganic thermoelectric materials are solid/brittle and require very high temperatures for preparation. Thus, for flexible TEGs, new thermoelectric materials are much needed. Especially, organic, hybrid, and composite thermoelectric materials are more suitable for wearable technologies. The thin films based TEGs can be prepared using these solution/printable process thermoelectric materials *(58)*.

6 SUMMARY

Thermoelectrics have certainly come a long way in the past few decades, both in terms of advancements in processing and characterization technologies as well as the abundance of new physics derived from careful studies. While most of these state-of-the-art progresses are done in inorganic materials, the understanding of organic materials is still at its early stage. Fortunately, some of the physical understanding derived from inorganic compounds has been found to be relevant to organic thermoelectric as well. In addition, recent efforts in machine learning and data-driven-based approach to discover new thermoelectric materials is expected to accelerate the progress and developments of both inorganic and organics thermoelectrics, which will open a commercially viable pathway towards flexible thermoelectric devices.

References

1. Tritt, Terry M.; Subramanian, M. A. Thermoelectric Materials, Phenomena, and Applications: A Bird's Eye View. *MRS Bull.* **2006**, *31* (3), 188–198.
2. Jeffrey, S. G.; Toberer, E. S. Complex Thermoelectric Materials. In Materials for Sustainable Energy: A Collection of Peer-Reviewed Research and Review Articles from Nature Publishing Group, 2011; pp 101-110.
3. Bell, Lon E. Cooling, Heating, Generating Power, and Recovering Waste Heat with Thermoelectric Systems. *Science* **2008**, *321* (5895), 1457–1461.

4. Yang, Jihui; Caillat, Thierry Thermoelectric Materials for Space and Automotive Power Generation. *MRS Bull.* **2006**, *31* (3), 224–229.
5. Gangyi, H.; Hal, E.; Lee, M. Silicon Integrated Circuit Thermoelectric Generators With a High Specific Power Generation Capacity. *Nat. Elect.* **2019**, *2* (7), 300–306.
6. Rowe, D. M. Recent Advances in Silicon-Germanium Alloy Technology and An Assessment of the Problems of Building the Modules for a Radioisotope Thermo-electric Generator. *J. Power Sourc.* **1987**, *19* (4), 247–259.
7. Ali, S. Recent Developments in Semiconductor Thermoelectric Physics and Materials. *Ann. Rev. Mater. Res.* **2011**, *41*.
8. Rowe, David, Michael, Eds. *CRC handbook of thermoelectrics*; CRC press, 2018.
9. Hao, S.; Dravid, V. P.; Kanatzidis, M. G., et al. Computational Strategies for Design and Discovery of Nanostructured Thermoelectrics. *NPJ Comput. Mater.* **2019**, *5* (1), 58.
10. Lundstrom, M. *Fundamentals of Carrier Transport*. Cambridge University Press, 2009.
11. Hicks, L. D.; Dresselhaus, M. S. Thermoelectric Figure of Merit of a One-dimensional Conductor. *Phys. Rev. B* **1993**, *47* (24), 16631.
12. Nielsen, Michele D.; Ozolins, Vidvuds; Heremans, Joseph P. Lone Pair Electrons Minimize Lattice Thermal Conductivity. *Ener. Environ. Sci.* **2013**, *6* (2), 570–578.
13. Pal, Koushik; Xia, Yi; He, Jiangang; Wolverton, Christopher Intrinsically Low Lattice Thermal Conductivity Derived from Rattler Cations in an AMM′ Q3 Family of Chalcogenides. *Chem. Mater.* **2019**, *31* (21), 8734–8741.
14. Goldsmid, H. J.; Sharp, J. W. Estimation of the Thermal Band Gap of a Semiconductor from Seebeck Measurements. *J. Elect. Mater.* **1999**, *28* (7), 869–872.
15. Pichanusakorn, P.; Bandaru, P. R. The Optimal Seebeck Coefficient for Obtaining the Maximum Power Factor in Thermoelectrics. *Appl. Phys. Lett.* **2009**, *94* (22), 223108.
16. Zhu, T.; Hu, L.; Zhao, X., et al. New Insights into Intrinsic Point Defects in V2VI3 Thermoelectric Materials. *Adv. Sci.* **2016**, *3* (7), 1600004.
17. Poudel, B.; Hao, Q.; Ma, Y., et al. High-Thermoelectric Performance of Nanostructured Bismuth Antimony Telluride Bulk Alloys. *Science* **2008**, *320* (5876), 634–638.
18. Kim, S. I.; Lee, K. H.; Mun, H. A., et al. Dense Dislocation Arrays Embedded in Grain Boundaries for High-performance Bulk Thermoelectrics. *Science* **2015**, *348* (6230), 109–114.
19. Kanatzidis, M. G. Nanostructured Thermoelectrics: The New Paradigm? *Chem. Mater.* **2009**, *22* (3), 648–659.
20. Zebarjadi, M.; Esfarjani, K.; Dresselhaus, M. S., et al. Perspectives on Thermoelectrics: From Fundamentals to Device Applications. *Energy Environ. Sci.* **2012**, *5* (1), 5147–5162.
21. Minnich, A. J.; Dresselhaus, M. S.; Ren, Z. F., et al. Bulk Nanostructured Thermoelectric Materials: Current Research and Future Prospects. *Energy Environ. Sci.* **2009**, *2* (5), 466–479.
22. Kim, H.-S.; Heinz, N. A.; Gibbs, Z. M., et al. High Thermoelectric Performance in (Bi0. 25Sb0. 75) 2Te3 due to Band Convergence and Improved by Carrier Concentration Control. *Mater. Today* **2017**, *20* (8), 452–459.
23. Zhang, Q.; Cao, F.; Liu, W., et al. Heavy Doping and Band Engineering by Potassium to Improve the Thermoelectric Figure of Merit in p-type PbTe, PbSe, and $PbTe_{1-y}Se_y$. *J. Am. Chem. Soc.* **2012**, *134* (24), 10031–10038.
24. Liu, W.; Tan, X.; Yin, K., et al. Convergence of Conduction Bands as a Means of Enhancing Thermoelectric Performance of n-type $Mg_2Si_{1-x}Sn_x$ Solid Solutions. *Phys. Rev. Lett.* **2012**, *108* (16), 166601.
25. Zaitsev, V. K.; Fedorov, M. I.; Gurieva, E. A., et al. Highly Effective $Mg_2Si_{1-x}Sn_x$ Thermoelectrics. *Phys. Rev. B* **2006**, *74* (4), 045207.
26. Pei, Y.; Wang, H.; Snyder, G. J. Band Engineering of Thermoelectric Materials. *Adv. Mater.* **2012**, *24* (46), 6125–6135.
27. Pei, Y.; Shi, X.; LaLonde, A., et al. Convergence of Electronic Bands for High Performance Bulk Thermoelectrics. *Nature* **2011**, *473* (7345), 66.

28. Jain, A.; Ong, S. P.; Hautier, G.; Chen, W.; Richards, W. D.; Dacek, S.; Cholia, Shreyas, et al. Commentary: The Materials Project: A Materials Genome Approach to Accelerating Materials Innovation. *Apl Mater.* **2013**, *1* (1), 011002.
29. Taylor, R. H.; Rose, F.; Toher, C.; Levy, O.; Yang, K.; Nardelli, M. B.; Curtarolo, S. A RESTful API for Exchanging Materials Data in the AFLOWLIB. org Consortium. *Comput. Mater. Sci.* **2014**, *93*, 178–192.
30. Kirklin, S.; Saal, J. E.; Meredig, B.; Thompson, A.; Doak, J. W.; Aykol, M.; Rühl, S.; Wolverton, C. The Open Quantum Materials Database (OQMD): Assessing the Accuracy of DFT Formation Energies. *Npj Comput. Mater.* **2015**, *1* (1), 1–15.
31. Otsuka, S.; Isao, K.; Junko, H.; Yibin, X.; Masayoshi, Y. PoLyInfo: Polymer Database for Polymeric Materials Design. In *2011 International Conference on Emerging Intelligent Data and Web Technologies*, 2011, pp. 22-29. IEEE.
32. Ward, L.; Dunn, A.; Faghaninia, A.; Zimmermann, N. ER.; Bajaj, S.; Wang, Q.; Montoya, J., et al. Matminer: An Open Source Toolkit for Materials Data Mining. *Comput. Mater. Sci.* **2018**, *152*, 60–69.
33. Ong, S. P.; Richards, W. D.; Jain, A.; Hautier, G.; Kocher, M.; Cholia, S.; Gunter, D.; Chevrier, V. L.; Persson, K. A.; Ceder, G. Python Materials Genomics (Pymatgen): A Robust, Open-Source Python Library for Materials Analysis. *Comput. Mater. Sci.* **2013**, *68*, 314–319.
34. Recatala-Gomez, J.; Suwardi, A.; Nandhakumar, I.; Abutaha, A.; Hippalgaonkar, K. Toward Accelerated Thermoelectric Materials and Process Discovery. *ACS Applied Energy Materials* **2020**, .
35. Gibbs, Z. M.; Ricci, .; Li, G., et al. Effective Mass and Fermi Surface Complexity Factor from ab initio Band Structure Calculations. *NPJ Comput. Mater.* **2017**, *3* (1), 8.
36. Suwardi, A.; Bash, D.; Ng, H. K., et al. Inertial Effective Mass as an Effective Descriptor for Thermoelectrics via Data-driven Evaluation. *J. Mater. Chem. A* **2019**, 7 (41), 23762–23769.
37. Crispin, X.; Jakobsson, F. L. E.; Crispin, A., et al. The Origin of the High Conductivity of Poly(3,4-ethylenedioxythiophene)-Poly (styrenesulfonate) (PEDOT-PSS) Plastic Electrodes. *Chem. Mater.* **2006**, *18* (18), 4354–4360.
38. Kim, G. H.; Shao, L.; Zhang, K., et al. Engineered Doping of Organic Semiconductors for Enhanced Thermoelectric Efficiency. *Nat. Mater.* **2013**, *12* (8), 719.
39. Bubnova, O.; Khan, Z. U.; Wang, H., et al. Semi-Metallic Polymers. *Nat. Mater.* **2014**, *13* (2), 190.
40. Bubnova, O.; Khan, Z. U.; Malti, A., et al. Optimization of the Thermoelectric Figure of Merit in the Conducting Polymer Poly(3,4-ethylenedioxythiophene). *Nat. Mater.* **2011**, *10* (6), 429.
41. Wang, H.; Hsu, J.-H.; Yi, S.-I., et al. Thermally Driven Large n-type Voltage Responses from Hybrids of Carbon Nanotubes and Poly(3,4-ethylenedioxythiophene) with Tetrakis(dimethylamino) Ethylene. *Adv. Mater.* **2015**, *27* (43), 6855–6861.
42. Cho, E. S.; Coates, N. E.; Forster, J. D., et al. Engineering Synergy: Energy and Mass Transport in Hybrid Nanomaterials. *Adv. Mater.* **2015**, *27* (38), 5744–5752.
43. See S K.C.; Feser, J. P.; Chen, C. E., et al. Water-processable Polymer−Nanocrystal Hybrids for Thermoelectrics. *Nano Lett.* **2010**, *10* (11), 4664–4667.
44. Zaia, E. W.; Sahu, A.; Zhou, P., et al. Carrier Scattering at Alloy Nanointerfaces Enhances Power Factor in PEDOT:PSS Hybrid Thermoelectrics. *Nano Lett.* **2016**, *16* (5), 3352–3359.
45. Coates, N. E.; Yee, S. K.; McCulloch, B., et al. Effect of Interfacial Properties on Polymer–Nanocrystal Thermoelectric Transport. *Adv. Mater.* **2013**, *25* (11), 1629–1633.
46. Yao, Q.; Chen, L.; Zhang, W. Enhanced Thermoelectric Performance of Aingle-walled Carbon Nanotubes/Polyaniline Hybrid Nanocomposites. *ACS Nano* **2010**, *4* (4), 2445–2451.
47. Choi, J.; Lee, J. Y.; Lee, S.-S., et al. High-Performance Thermoelectric Paper Based on Double Carrier-Filtering Processes at Nanowire Heterojunctions. *Adv. Energy Mater.* **2016**, *6* (9), 1502181.

48. Kang, S. D.; Snyder, G. J. Charge-transport Model for Conducting Polymers. *Nat. Mater.* **2017**, *16* (2), 252.
49. Liang, Z.; Boland, M. J.; Butrouna, K., et al. Increased Power Factors of Organic–Inorganic Nanocomposite Thermoelectric Materials and the Role of Energy Filtering. *J. Mater. Chem. A* **2017**, *5* (30), 15891–15900.
50. Kumar, P.; Zaia, E. W.; Yildirim, E., et al. Polymer Morphology and Interfacial Charge Transfer Dominate Over Energy-Dependent Scattering in Organic-Inorganic Thermoelectrics. *Nat. Commun.* **2018**, *9* (1), 5347.
51. Wang, H.; Hsu, J.-H.; Yi, S.-I., et al. Thermally Driven Large n-type Voltage Responses from Hybrids of Carbon Nanotubes and Poly(3,4-ethylenedioxythiophene) with Tetrakis(dimethylamino) ethylene. *Adv. Mater.* **2015**, *27* (43), 6855–6861.
52. Du, F.-P.; Cao, N.-N.; Zhang, Y.-F., et al. PEDOT:PSS/Graphene Quantum Dots Films with Enhanced Thermoelectric Properties via Strong Interfacial Interaction and Phase Separation. *Scientific Rep.* **2018**, *8* (1), 6441.
53. Kang, S. D.; Snyder, G. J. Charge-Transport Model for Conducting Polymers. *Nat. Mater.* **2017**, *16* (2), 252–257.
54. Bubnova, O.; Khan, Z. U.; Wang, H.; Braun, S.; Evans, D. R.; Fabretto, M.; Hojati-Talemi, P., et al. Semi-Metallic Polymers. *Nat. Mater.* **2014**, *13* (2), 190–194.
55. Glaudell, A. M.; Cochran, J. E.; Patel, S. N.; Chabinyc, M. L. Impact of the Doping Method on Conductivity and Thermopower in Semiconducting Polythiophenes. *Adv. Ener. Mater.* **2015**, *5* (4), 1401072.
56. Sirringhaus, H.; Brown, P. J.; Friend, R. H.; Nielsen, M. M.; Bechgaard, K.; Langeveld-Voss, B. M. W.; Spiering, A. J. H., et al. Two-Dimensional Charge Transport in Self-Organized, High-Mobility Conjugated Polymers. *Nature* **1999**, *401* (6754), 685–688.
57. Torabi, S.; Jahani, F.; Severen, I. V.; Kanimozhi, C.; Patil, S.; Havenith, R. W. A.; Chiechi, R. C., et al. Strategy for Enhancing the Dielectric Constant of Organic Semiconductors Without Sacrificing Charge Carrier Mobility and Solubility. *Adv. Func. Mater.* **2015**, *25* (1), 150–157.
58. Wan, C.; Gu, X.; Dang, F.; Itoh, T.; Wang, Y.; Sasaki, H.; Kondo, M., et al. Flexible n-Type Thermoelectric Materials by Organic Intercalation of Layered Transition Metal Dichalcogenide TiS2. *Nat. Mater.* **2015**, *14* (6), 622–627.

CHAPTER 9

Design of thermochromic materials and coatings for cool building applications

Ashwini Nawade, Kunchanapalli Ramya, Sabyasachi Mukhopadhyay
Department of Physics, SRM University, AP– Amaravati, Guntur, Andhra Pradesh, India

1 Introduction

Conservation of energy in building is a serious issue for achieving sustainable environment. Significant energy loss or gain by windows makes it difficult to save energy in building, whereas without windows, it is easier to save energy in building but they are recommended for the natural source of light and have biological significance for development of human being. Hence, windows are significant part of the building architecture for energy conservation *(1)*. Windows are known as one of the most energy inefficient components of buildings in cold environments. Preventing heat losses by improving the window's thermal performance will result in reduced electricity costs and less greenhouse gas emissions. While controlling transmitted infrared (IR) radiation, an ideal window should be capable of sufficient transmission of visible light. The most significant parameters influencing the heat transfer through windows include outdoor conditions, shading, building orientation, type and area of window, glass properties, and glazing characteristics *(2)*. Improving glazing characteristics of windows such as thermal transmittance and solar transmittances is one of the most important criterion to be considered in building windows standards *(3)*. Fig. 9.1 describes how a perfect thermal window should act.

Glazing windows play essential role in energy conservation as well as the architecture of buildings *(4)*. Normal windows make it difficult for conservation of energy since it transmits major part of solar radiation, to enhance the energy conservation properties of windows extensive research is done over the years. Initially, static glazing windows were developed, which conserves energy but have a lot of challenges such as high maintenance and high cost that needs to be overcome with advance configuration *(5)*.

Energy Saving Coating Materials
http://dx.doi.org/10.1016/B978-0-12-822103-7.00009-1

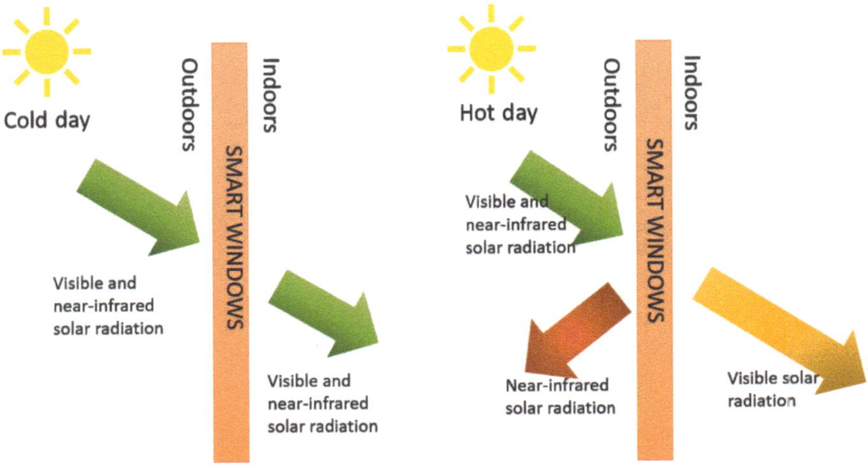

Figure 9.1 ***An ideal smart window in cold climate allows the visible and near-infrared solar radiation whereas in hot climate, it is obstructing the near-infrared solar radiation allowing only the visible solar radiation.***

1.1 Static window glazing

Static glazing refers to glazing that possesses single fixed thermal and optical properties, which means it can have either a particular thermal transmittance value (U-Value) or visible transmittance value or solar heat gain coefficient. These properties affect the thermal energy performance and day lighting level of the building. Glazed windows incorporate coatings (e.g., thin transparent silver films) to retain or reject solar heat gain but are completely static and cannot respond to changing weather or seasons. To enhance the thermal energy performance, various orientation and sizes are used, which mainly include the double glazing and triple glazing of windows *(6)*. The double-glazing windows are a combination of two glass windows with space between them whereas triple glazing windows as the name suggests consist of three glass planes with a specific space between them. The spacing between the glass planes leads to a reduction in the loss of heat from window hence leading to the energy conservation of building *(7)*. From the previous studies, researchers have found that the double-glazing windows have higher consumption of thermal energy than the triple glazing windows; hence, the triple glazing windows are a better option in reference to conservation of thermal energy. There are many different materials or methods used for preparing a static glazing window. One of the most prominent and promising materials is low emissivity (low-E) coatings. The low-E coatings are spectrally selective films that are aimed to let the visible light pass through and block the infrared (IR) and ultraviolet (UV) wavelengths, which generally create excess heat, here higher amount of IR radiations is reflected by the low emissive glazing for a certain visible transmittance. There are two types of low-E coatings, the tin oxide-based hard coating and the silver-based soft coating with higher IR reflectance and lower transmittance than the other one. The insulated low-E coatings should save in lighting, cooling, and energy in both summer and winter climates, but these coatings have a major drawback of the high cost and maintenance, making it financially unfeasible.

1.2 Dynamic window glazing

Dynamic window glazing is one of the promising active research fields nowadays in designing the buildings. The optical switching devices for dynamic glazing include chromogenic technologies, thermotropic, liquid crystals (LCs), and suspended particles *(8)*. Chromogenic technologies comprise photochromic, gasochromic, electrochromic, and thermochromic technologies *(9,10)*. Photochromic materials used for glazing, change their transparency in response to the visible light intensity falling on it. It changes into a darkened state leading to obstruction of the IR radiation, whereas thermochromic materials portray both visible and IR transmission changes with the temperature and heat surrounding it. Photochromic materials are mostly used in eyeglasses that change from clear in the dim indoor light to dark in the bright outdoors. Electrochromic materials get active by supply of particular external voltage causing change in their optical properties such as transmission of visible light and obstruction of IR radiations. They modulate its transmittance in visible and near IR when a low direct current (DC) potential is applied. The most common electrodes used in electrochromic materials are ITO (tin oxide doped with indium) or FTO (tin oxide doped with fluorine). Self-powered electrochromic windows are developed using semitransparent photovoltaic cell, which provide the required activating electricity. Gasochromic devices work on the similar mechanism as that of electrochromic materials but instead of DC voltage, a hydrogen gas is applied to switch between the two states that is, clear and dark state. They provide a better transmittance modulation as compared to others but control over the gas exchange is difficult (see Fig. 9.2).

Liquid crystal layers are one of the methods used in dynamic glazing windows, which are made up of two transparent conducting layers on a plastic film, which is pressed between two glass layers. When the power is switched between the two electrically conducting layers, the production

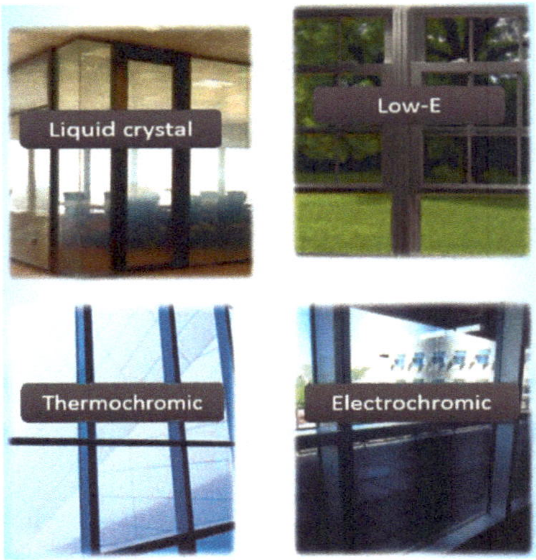

Figure 9.2 ***Types of dynamic glazing windows.*** Source: *Internet.*

of electric field causes the crystal molecules to be aligned in a particular direction, which leads change in transmittance. These are mostly used in wrist watches and have major application in interior design of any building. This material has major drawbacks due to the high-power demand in transparent mode, long-term UV stability, and higher cost. Suspended particle devices (SPD) are similar to the LC devices since both consist of liquid like active layer. Here, mostly poly-halide suspended in organic fluid or gel is sandwiched between two sheets of glass coated with transparent conductive films *(11)*. When electric field is applied to the device in the dark reflective state, it will align resulting in clear transmissivity state because of the random pattern of the light absorbing particles.

Smart windows and smart coating are available commercially worldwide. The suppliers claim blockage of 90%–95% of ultraviolet-infrared radiation, which can save up to 30% of electricity consumed by building. These smart windows or/and coating are available in the form of thin film or glass or paints (sprayed by aerosol technique on window shield or buildings). These windows can be used in large infrastructures, such as in hospitals, banks, and officers, where electrical energy efficiency is a large constraint. They can be costlier as compared to the normal glass but are definitely useful for the energy conservation. Some of the leading companies proving smart windows are Saint Gobain (France), View Inc. (US), Corning Inc. (US), Gentex Corporation (US), AGC Inc. (Japan), Polytronix Inc. (US), Vision Systems (France), Star Shield (India), Royal tint (India) (see Fig. 9.3).

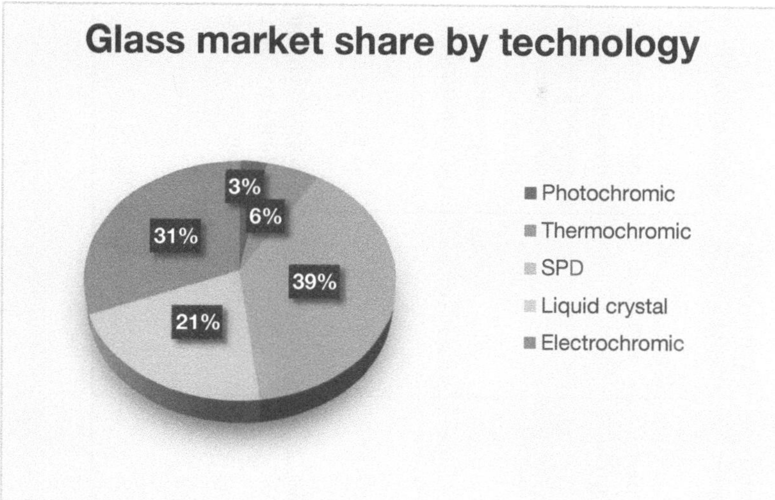

Figure 9.3 ***Smart window market share by different technology.*** Source: *Market Research Report, June 2019, USA.*

In these windows, the technologies used are mostly suspended particle deposition or electrochromic or polymer-dispersed liquid crystals or LC as stated in Fig. 9.3. The commercialization of thermochromic windows has not yet been covered on large scale. To understand the working advantages of thermochromic material, and its application in smart window, here we have reviewed application-specific different methodologies to develop these materials, their characteristics, which might be useful for readers working on these topics (Table 9.1).

2 Thermochromic materials

A variety of compounds change color on heating and revert to its original color on cooling. This reversible change in the color of a compound with temperature is defined as thermochromism. Thermochromism is defined operationally as an easily noticeable reversible color change with the change of temperature, limited by the boiling point of each liquid compound as the upper temperature or melting point for solids. The mechanism responsible for thermochromism varies with materials structure at its molecular level *(12, 13)*. Thermochromism can occur even with the optical absorption at visible spectral range in such a way that no color change can be visible. In thermochromic material, the temperature at which change in the phase and/ or color occurs is defined as transition temperature also known as critical

Table 9.1 Summarized comparison between various smart windows.

Smart window	Activation	Advantages	Disadvantages	Color	Thermal performance	Optical performance	Application
Thermochromic	Heat	High energy saving, no activation energy required	Manufacturing difficulties for large sizes, no outside visibility	Colored	Reflecting heat gain	Reflecting IR radiation	Thermochromic paints, window coating, duracell battery state indicators, thermochromic paper used for thermal printers, thermochromic polymers
Photochromic	Light	No activation electricity, No foggy effect, automatic	Manufacturing difficulties for large sizes, limited application, cannot reduce heat gain. Darker in winter than the summer	Bleach		UV protector, darken and lighten at sunlight and dusk	Sunglasses, supramolecular chemistry, data storage

Electrochromic	Voltage/ current	Energy saving	Activation electricity needed	Colored	Control the amount of heat	Reversibly changing color when a burst of charge is applied, control the amount of light	Electrochromic devices such as windows and smart glass, automobile industry
Suspended particles	Current	Instantaneous control of light, wide range of transmittance, outside visibility	Activation electricity needed, limited in size	Bleach	Energy saving	UV protector, reduction of infrared light	Windows, polarized camera
Liquid crystal	Voltage	Control privacy	No energy saving, no outside visibility, activation electricity is needed	Bleach	High heat gain	Transmit incident light	Electrooptical devices, hyperspectral imaging

temperature of the respective structural/phase change. These materials are majorly classified into two categories according to its material composition:

2.1. Organic thermochromic materials
2.2. Inorganic thermochromic materials

2.1 Organic thermochromic materials

Organic thermochromic materials show sharp color changes in provided conditions and are easy to control the operational temperature. These systems have a wide range of applications in the field of fibers, optics, photostorage instruments, and optical sensors. Thermochromic behavior of organic materials results in molecular rearrangement, stereoisomerism, and change in crystal structure.

Several intermediate phases called mesophases are formed when organic materials pass from crystalline solid state to isotropic liquid state. This phase transition can be under the influence of temperature or solvent. One of the most important types of LC for thermochromic system is cholesteric crystal, with helical structure as shown below. When temperature varies, the pitch length of the helical arrangement varies and results in alteration of the wavelength of the reflected light, which changes color spectrum (Fig. 9.4).

Thermochromic materials can also be result of equilibrium between stereoisomers, where the stereoisomers refer to same molecular formulae but different molecular structure. The phenomenon of thermochromism due to variation in stereoisomers is known as stereoisomerism. The switching of the stereoisomers of these compounds occurs when the temperature increases resulting in color change of that compound. Stereoisomerism is mostly observed when the temperature of that compound crosses its melting point. The stereoisomerism is mostly associated with ethylene. The first compound in this class used as thermochromic was Bianthrone, which undergo color transmission from lemon yellow to green, it was observed by Mayer in 1909 *(14)*. In 1928, Schonberg suggested that dixanthylene changes from colorless in liquid air and deep blue green in solution at 281 degrees *(15)*. In both cases, the reason behind this color change was observed to be the bi-radical formation due to stereoisomerism (Table 9.2).

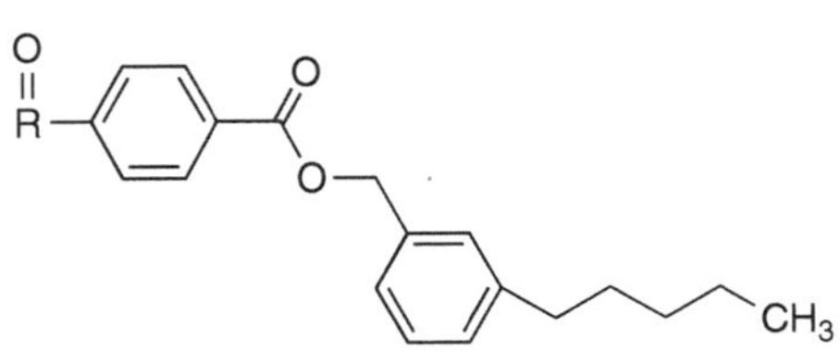

Figure 9.4 ***Chemical structure of a chiral molecules that form cholesteric mesophase.***

Table 9.2 A list of different organic thermochromic materials.

Name	Structure	Color changes depicting thermochromism
10-m-Nitrobenzylidene anthrone	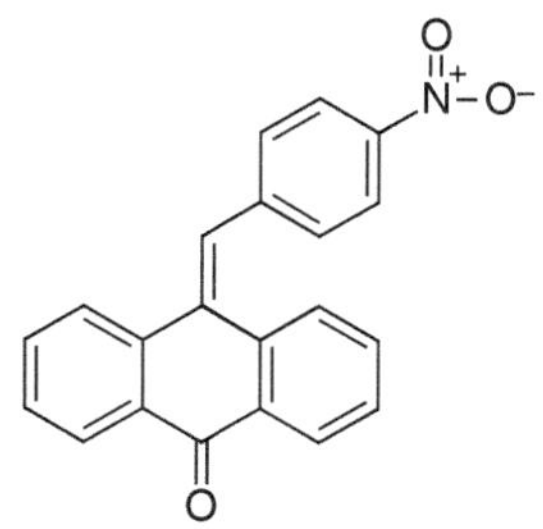	Yellow prisms
N-methylacridan	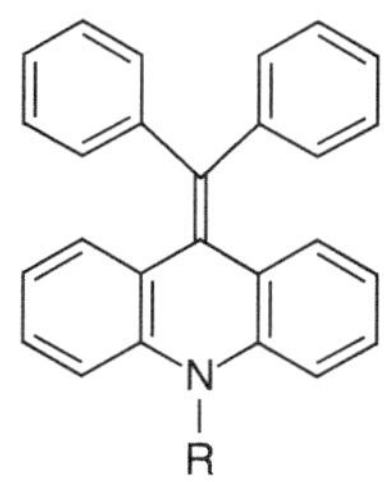	Deep red melt
Dibiphenyleneethane	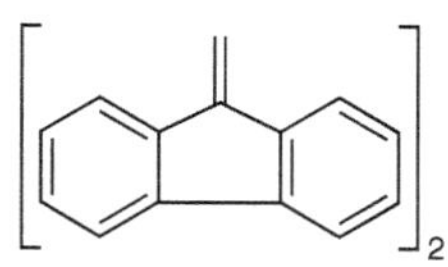	Intense red
10-(9-Fluorenylidene)-thioxanthene	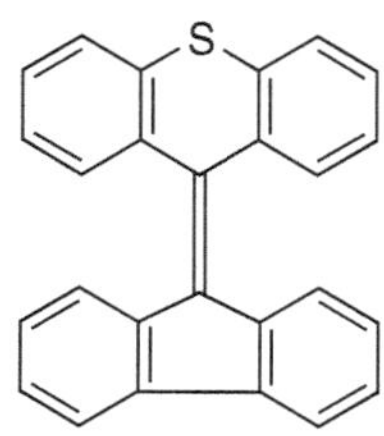	Yellow solid give green melt
2-Hydroxy-methyl-fluorenylidene-propiophenone	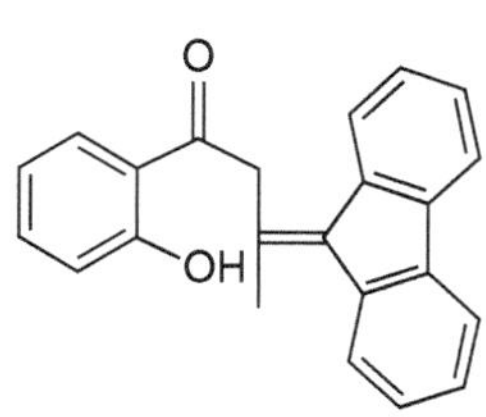	Colorless solid yellow melt
Di-(2-phenyl)-1,3-benzoxaz-4-ylidene	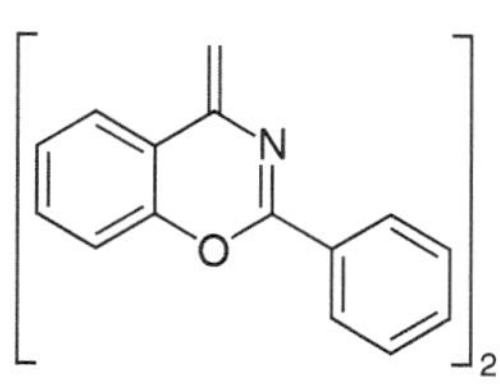	Yellow solid deep red melt

(Continued)

Table 9.2 A list of different organic thermochromic materials. (*Cont.*)

Name	Structure	Color changes depicting thermochromism
Diphenylchromylene	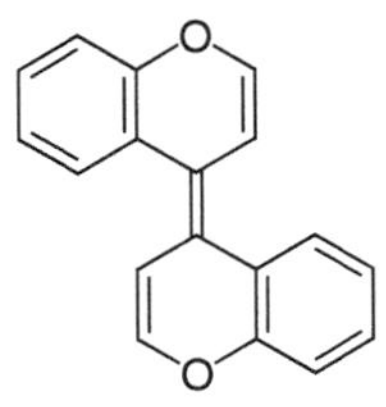	Yellow solid

Thermochromism in organic compounds can occur due to one more mechanism that is, molecular rearrangement. Molecular rearrangement in organic compounds originates from the increase in conjugation of molecules, which result in formation of new chromophores. This rearrangement is mostly affected by the change temperature. The molecular rearrangement of Spiro lactone dyes can be illustrated using crystal violet lactone as an example (Fig. 9.5).

Many polymeric and molecular systems exhibit thermochromism, which can be activated by processes like thermally driven reversible reactions or temperature-dependent structural rearrangements. Thermochromism can be observed in various polymers such as thermoplastics, duro-plastics, coatings, or gel. Thermochromic effect in polymers is by the interaction of polymer with incorporated additives. Thermochromic polymer systems are

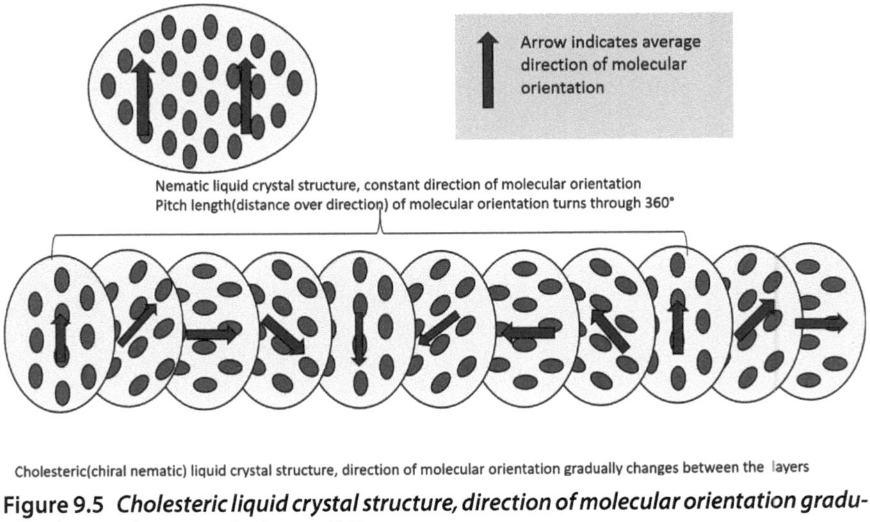

Figure 9.5 ***Cholesteric liquid crystal structure, direction of molecular orientation gradually changes between the layers*** *(16).*

obtained by doping the polymers matrix with thermochromic additives *(17)*. Leuco dye-developer complexes became the most important systems to achieve thermochromic properties for different polymer materials by endowing a separate phase of the thermochromic system in a nonthermochromic polymer matrix.

Conjugated polymers consist of a polymer backbone with conjugated π-electron system. Polyacetylenes, polydiacetylenes, polythiophenes, and poly (phenylene vinylidenes) belong to this class of polymers. These ranges of polymers absorb the radiation in visible range and showcase color change. The length of the conjugation π-bond in the system in the polymer backbone results in the shift of the absorption wavelength and thus shows thermochromism. This shifting of the planarity leads to the color change. These conjugated polymers can be used as pigments or binders in paints.

2.2 Inorganic thermochromic materials

Inorganic thermochromism develops via various mechanisms such as change in ligand geometry, band-gap energy (e.g., semiconductor), and the solvation sphere of a dissolved compound, metal coordination, reflectance properties, the arrangement and distribution of defects in the crystalline solids, and majorly due to phase transition. In 1871, Houston found first time by heating various solids on copper strips over Bunsen burner and reported that copper ferrocyanide, antimony sulfide, iron oxide, copper sub iodide, mercury sulfide, lead chromate, lead oxide, and potassium dichromate change its color from red to black on heating *(18–20)*. These reactions were found to be reversible. Zinc oxide and tin oxide also show color change on heating but are irreversible. The most promising thermochromic material in inorganic class is vanadium oxide (VO_2) (see Fig. 9.6).

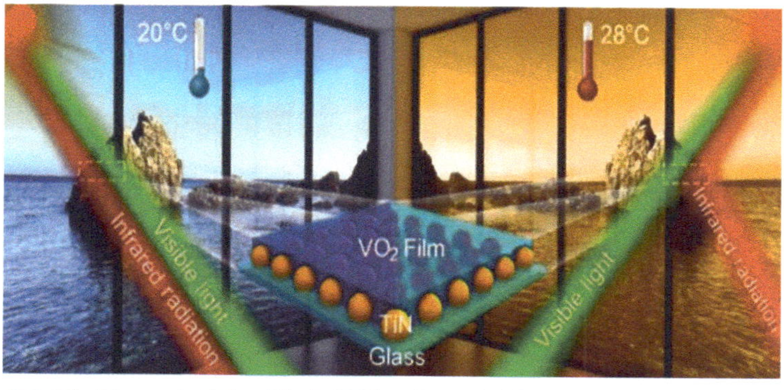

Figure 9.6 ***Working principle of hybrid Titanium nitride (TiN)/Vanadium oxide (VO2)-based thermochromic windows at different temperature environments*** *(21)*.

Vanadium oxide is found in four polymorphic forms: monoclinic VO_2(M) and rutile VO_2(R) and two metastable forms: VO_2(A) and VO_2(B) by monoclinic to rutile temperature 68°C. VO_2 acts as semiconductors and transmits infrared light below 68°C, at 68°C, there is semiconductor to metal phase transition of VO_2 that is, monoclinic to rutile resulting in the change of optical properties of IR radiation *(22–24)*. This feature has made vanadium oxide, a prime candidate for thermochromic coating. Despite being a prime candidate, it has its own challenges while using it as a smart coating. Some of them are listed below:

1. The transition temperature of VO_2 as compared to the surrounding temperature is very high, making it difficult to use for room temperature application.
2. The luminous transmittance is less than 40% and a notable magnitude of modulation of solar energy is too low for practical purpose.
3. In the near-mid-IR region of solar spectrum, VO_2 displays higher thermochromic reflectance modulation, whereas the modulation of solar energy in another region is weaker.

To improve the thermochromic properties of VO_2, especially increasing the luminous transmittance and modulation of solar energy, whereas decreasing the transition temperature to room temperature, various methods are implemented such as change in the material synthesis, using of favorable dopants and use of multilayered coating *(25)*.

2.2.1 Synthesis of VO_2 films

The variation in synthesis condition alters the product morphology and structure, which influences the phase transition characteristics and functional properties of VO_2 films. Two techniques are majorly used to accomplish this goal that is, gas-based deposition and solution-based deposition.

(A) Gas-based deposition

Gas deposition method is one of the popular methods, which consist of chemical vapor deposition (CVD) and physical vapor deposition (PVD).

i. Chemical vapor deposition method

VO_2 film was first time deposited by using CVD technique in 1967. It is a common industrial technique for high-performance and high-quality coatings. The most popular CVD methods are atmospheric pressure chemical vapor deposition (APCVD) and aerosol-assisted chemical vapor deposition (AACVD) *(26–30)*. APCVD was conducted for the fabrication of titanium dioxide–vanadium dioxide (TiO_2–VO_2) films; these films were developed on glass plate by the reaction of titanium tetrachloride ($TiCl_4$), vanadium tetrachloride (VCl_4), ethyl acetate, and water (H_2O) at 550°C *(27)*. The produced VO_2 films with different phases and composition had multifunctional properties, which in turn reduce the critical or transition temperature value by almost 18°C. Thin films prepared by this method demonstrated critical transition at room temperature, which makes TiO_2–VO_2 films more applicable for smart window applications *(31)*.

Similarly, AACVD was used to prepare nanocomposite VO_2 films by using TiO_2 or cerium (IV) oxide (CeO_2) from vanadyl acetylacetonate and an aerosol of nanoparticles at temperature around 500° to 600°C. The resultant film exhibited large change in IR reflectivity along with other multiple functionalities upon exceeding the critical temperature *(32)*.

The latest development of CVD is the electric field-assisted chemical vapor deposition (EACVD). Electric field affects crystallographic orientation, reduces particle size, and decreases the critical temperature. The experiment was carried out in a cold wall reactor where the electric field was generated by applying potential difference between top plate and the substrate of the reactor and VO_2 films were deposited at 530°C from vanadyl acetylacetonate in the presence of ethanol under the presence of electric field. The increase in the electric field led to an increase in the particle size, hence increasing the surface area. It also influences the microstructure and growth of the VO_2 films with predefined orientation. These specific oriented films have variety of application such as for super-capacitor electrodes and chemical sensors *(31,32)*.

A low-temperature CVD method has been developed to synthesize patterned VO_2/ZnO nanotetrapods on indium tin oxide glass substrate by using zinc oxide (ZnO) nanotetrapods as templates. At low temperature, the field emission performance was notably not good as compared to that with higher temperature because of the high resistance of VO_2 *(33)*. Further research is been carried out to overcome this issue over the synthesis temperature. We can include SEM images of VO_2 by different methods of CVD from recent progress paper.

ii. Physical vapor deposition method

To fulfil the requirement of lower critical phase transition temperature, environment friendliness, ease of developing multilayer thin films, and compatibility of the material to a wide range of substance ultrathin films prepared by physical vapor deposition technology are used *(34)*. It has four steps: evaporation, transportation, reaction, and deposition. Targets or the materials, which are to be deposited, are usually solid-state metals, which are bombarded by high energy source under reduced pressure. One of the most used PVD methods is the reactive sputtering method used to fabricate VO_2 films. VO_2 films also prepared utilizing other sputtering methods: DC, radio frequency, and magnetron sputtering. The advantages of sputtering process are high efficiency of deposition, good scalability to larger substrate, and high homogeneity of products. Very recently developed technique, radio frequency-inverted cylindrical magnetron sputtering have been found to be most cost-effective for the production of uniform nanocomposite of large area, exhibiting plasmonic properties *(35)*.

High power impulse magnetron sputtering is helpful to achieve transmission modulation in the near IR region for films deposited at substrate temperature of 300°C. This method has brought down the critical phase transition temperature to 50°C *(36)*.

There has been further development in the sputtering techniques; one of which is a two-step process that combines sputtering and post annealing or oxidation processes. High-quality VO_2 films were fabricated using DC magnetron sputtering process and then the film was oxidized at high temperature in the presence of air, which made the film to deliver promising phase transition near room temperature. By using DC magnetron sputtering followed by a postannealing process, VO_2 films were fabricated with the transition temperature of 45°C. Nanostructured VO_2 films exhibited an excellent phase transition property with critical temperature value of 35°C. Cosputtering method was developed for the preparation of doped VO_2 films to further decrease the critical temperature; it was all the more advantageous as the deposition could be done without the substrate temperature, presence of oxygen, using single mixed metal target, and annealing in air, and it did not change the preferred orientation growth but improved the surface morphology of the VO_2 films *(36)*.

Pulsed laser deposition (PLD) is suitable for oxide film growth. PLD was first used to deposit VO_2 films containing gold nanoparticles. The switching behavior of such VO_2 films exhibited a resistivity change of 4 orders of magnitude and the decrease in the transition temperature *(37)*. Pure vanadium metal was used as target; the target was rotated with the frequency of a range of 1–10 hertz for the purpose of uniformly sputtering the target. VO_2 films were deposited using different energy density at the target and different oxygen pressure in the chamber *(38)*. Electron beam evaporation method is used for high-quality VO_2 films, which were fabricated using the same method as vanadium target in pure oxygen atmosphere.

Nanoporous VO_2 films were synthesized using oblique angle deposition technique; these are also called as sculptured thin films. This technique is used to adjust the microstructure of thin films. With the increase in deposition angle, the degree of orientation gets increased resulting in anisotropic structure. Using this technique, solar energy reflection/transmission modulation properties of the thin film was improved and the critical temperature of VO_2 showed a large reduction to 54.5°C *(39)*. This type of sculptured thin films is more suitable for smart window application compared to normal VO_2 films.

(B) Solution-based deposition

The equipment used in gas-based deposition techniques is usually complicated and expensive. Thin films developed with solution-based deposition method is more effective, inexpensive, high yielding, and flexible as well as useful for smart window coating application *(40)*.

i. Sol–gel technique

The basic principle of sol–gel process is to create an oxide network by a polymerization reaction of chemical precursors dissolved in liquid medium. The most commonly used precursor is the tetravalent alkoxide. For the sol–gel derived VO_2 films, the precursor used was vanadium tetra butoxide *(41)*.

Inorganic vanadium sol–gel was obtained by mixing water with molten vanadium pentoxide powder. It requires low annealing temperature of 400–600°C, and the VO_2 film prepared by following the method has steady phase transition properties below 200°C. At higher annealing temperature (>600°C), critical phase transition temperature also moves to >300°C; that could be originated because of formation of larger grain size of VO_2 films when annealed at >600°C *(42)*. The value of critical temperature was reduced by controlling the grain size of VO_2 films by annealing temperature of substrate *(41)*. VO_2 films showed better growth orientation, smaller transition width, and larger grain size by prolonging annealing time at lower annealing temperature. These films have a synergetic optical performance with a good visible transmittance and excellent near-infrared modulation abilities.

ii. Using peroxide precursor

Vanadium pentoxide ($V_2O_5 \cdot n(H_2O)$) precursors were prepared by dissolving V_2O_5 powder in 30% hydrogen peroxide solution. Precursor gels such as $V_2O_5 \cdot 2(H_2O)$ and $V_2O_5 \cdot 1.8(H_2O)$ were prepared using sol–gel process at annealing temperature of 750°C. These films depicted better growth orientation, smaller transition width, and larger grain size and the critical temperature value for phase transition of 62.5°C along with comparable solar energy modulation and luminous transmittance values to those of films fabricated by using other methods *(41)*. Silicon dioxide (SiO_2)-coated

VO_2 films fabricated by sol–gel process increase the visible transmittance by 14.6%, whereas the infrared reflectance power reduces with the lowering of critical phase transition temperature *(24)*.

iii. Hydrothermal method

Hydrothermal synthesis involves water-soluble precursors to develop desired thin film at low operational temperature (150–200°C). Film structure and morphology are controlled by synthesis parameters such as reaction time, temperature, pH, concentration, pressure, and reducing agent. Hydrothermal process is useful to control thin film structure without high temperature solid state reaction.

Hydrothermal processing of VO_2 has focus on the formation of pure phase VO_2(M) with controllable morphology and enhanced thermochromic properties. VO_2(M) can also be obtained by annealing VO_2(B) under vacuum for 1 hour. Phase pure VO_2(B) can also be obtained by rapid hydrothermal process with a combination of low temperature that is, 180°C for two hours *(43)*. By using V_2O_5 as the precursor, Ban et al. Synthesized VO_2(M) and VO_2(A) nanostructures with desirable structure and morphology at high yield. The decrease in VO_2 particles size or grain size improves its' optical transmission properties at visible range, along with increase in critical phase transition temperature. Tungsten doping to VO_2 nanoparticles also reduces critical phase transition temperature to ambient temperature. Magnesium-doped VO_2 nanoparticles prepared by hydrothermal synthesis displayed excellent visible transmittance of around 54.2% and modulation of solar energy of 10.6% *(44)*. Hydrothermal method has been used to prepare nanocomposites such as molybdenum-doped VO_2(M)/TiO_2 composite. The method is beneficial for morphology control, particle size reduction, and infrared-range light modulation. Enhancing TiO_2 particles concentration size of the nanocomposite was reduced, which makes the nanocomposite films operational at room temperature following the phase change and optical property modulations. This method further utilized to synthesis nanofibers such as VO_2(M)/SiO_2, which display excellent thermochromic properties *(24)* (Fig. 9.7).

iv. Polymer-assisted depositions

Polymer-assisted deposition (PAD) is an alternative method for sol–gel process, which employs coordination between metal alloys and polymers for homogenous distribution for the formation of metal oxide films *(46)*. This method is especially important for the fabrication of nanostructured VO_2 films. The film is prepared by film coating or dip coating method by preparing the solution with soluble polymers and doping agent.

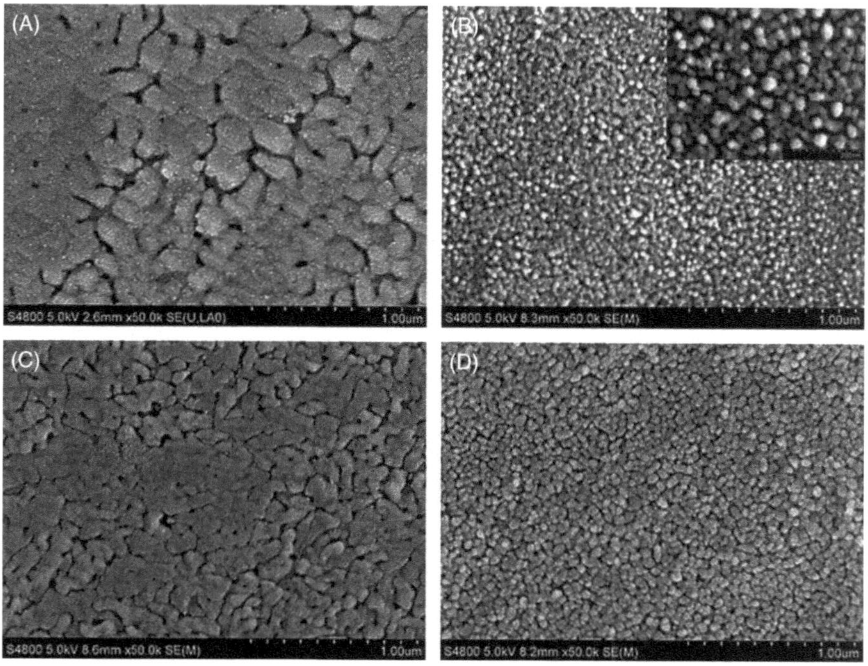

Figure 9.7 SEM images of samples obtained at different synthesis conditions, (A) pure VO_2 film, (B) VO_2 film with CTAB, (C) Mo and W codoped VO_2 film, and (D) Mo and W codoped VO_2 film with CTAB. *Reproduced with permission from Reference* (45).

Polyvinylpyrrolidone (PVP) is the most commonly used polymer in the PAD technique as the optical properties of PVP added films in the visible range and the measured IR range were noteworthy. In the preparation of composite, another polymer F-127 is used instead of PVP, which can boost luminous transmittance and modulation of solar energy *(47)*.

v. Electrochemical approach

This method is utilized to deposit thin films with complex geometrical structures. The technique is routinely utilized in device fabrication on different semiconducting or nearly insulating substrates such as silicon or gallium arsenic (Si/GaAs) *(48)*. Highly oriented VO_2 films were deposited on Si substrate by electrochemical approach, which involves three steps: electrodeposition, storage, and vacuum thermal annealing. Generally, deposition is carried out with a cathodic potential of 0.75 V for 30 min for about 150 millimolar per liter solution of $VOSO_{4x,}$ H_2O, with the pH value kept constant at 2.8 throughout the process. VO_2 films show a [011] preferred orientation. This method produces VO_2 layers of only few nanometers in thickness and has to be stored for 45 days prior to annealing in vacuum (Fig. 9.8).

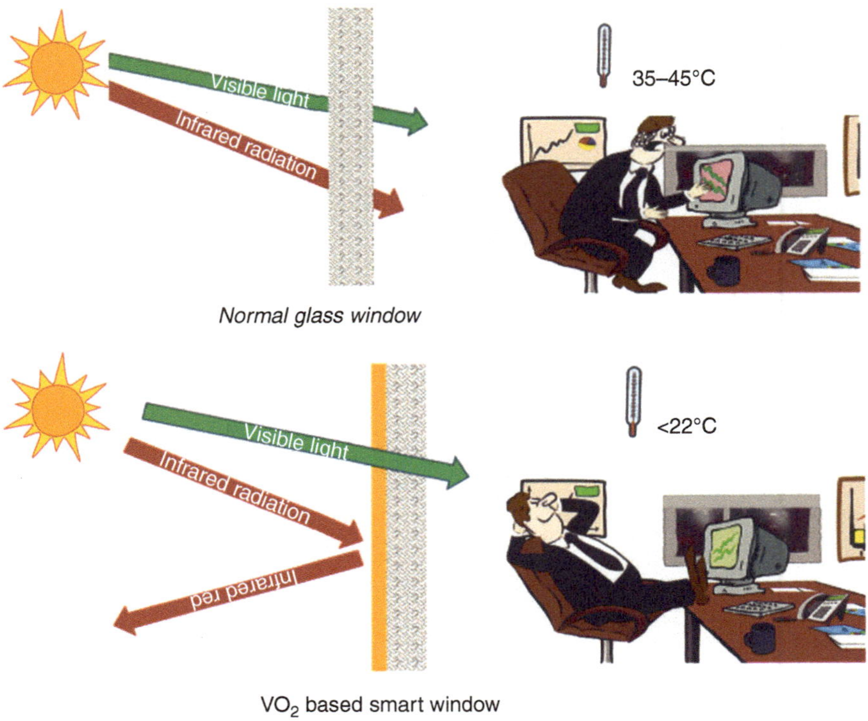

Figure 9.8 ***Schematic illustration of comparison of normal glass window and thermochromic W doped VO2-based smart window shows reversible phase transition resulting turn down of temperature to 22°C*** *(49).*

3 Strategies in improving thermochromic properties

Thermochromic materials are mainly utilized to serve the purpose of maintaining the room temperature by inhibiting the solar radiation especially the infrared radiations and allowing the visible transmittance to occur. In order to achieve infrared light rejection and visible light transmission, these materials should attain the critical temperature where they will undergo phase transition. Hence, the critical temperature of the material should not be very different from room temperature but most of the readily available materials have the critical temperature above common room temperature (18–50°C). As example, most common thermochromic coating VO_2 has a critical temperature of 68°C. To overcome these issues, some of the strategies which have already been implemented are discussed below:

3.1 Elemental doping

Elemental doping of VO_2 films is a good technique to lower its critical phase transition temperature. Critical phase transition temperature gets reduced by doping VO_2 thin films with high valence cations such as Niobium ion (Nb^{5+}), Molybdenum ion (Mo^{6+}), Tungsten ion (W^{6+}), Tantalum ion (Ta^{5+}), and the value of critical phase transition temperature gets enhanced by doping with low valence cations such as Titanium ion (Ti^{2+}), Gallium ion (Ga^{3+}), and Chromium ion (Cr^{3+}). The elemental doping technique contributes to improve thermochromic properties in two ways that is, the reduction of critical phase transition temperature and improvement of optical transmission properties at visible range with enhanced reflective properties at infrared regime *(50–52)*. The metallic Tungsten (W) is one of the most effective ions to reduce the critical phase transition temperature of VO_2 thin films. Strong interaction between doped W atoms and VO_2 lattices drives the symmetric W atom to nearby asymmetric monoclinic VO_2 lattice and induced monoclinic to rutile VO_2 transition by lowering the thermal energy barrier for this phase transition *(53,54)*. The thermochromic and electric properties of metal ion-doped VO_2 thin films prepared be sol–gel process were investigated by Cho et al. It was reported that the addition of dopant Tungsten trioxide (WO_3) leads to the change in average grain size in VO_2 films with the reduction of critical temperature by 15.5°C per mole *(55)*. Barron et al. studied electrical and optical properties of $V_{1-x}M_xO_2$ where M is metal impurity (M can be Tantalum (Ta), Niobium (Nb), or Tungsten(W)) *(56)*. Batatista et al. studied physical properties of Nb-doped VO_2 thin films (varying Nb doping concentration) and reported significant modifications on critical phase transition temperature, optical properties, and surface morphology of doped VO_2 thin films compare to undoped VO_2 thin films *(57)*. Critical phase transition temperature shows drastic reduction from 59°C to 34°C with increase of Nb content in VO_2 thin films. But the defect formation in VO_2 crystal lattice, with the addition of Nb atoms, depresses the modulation of solar radiation energy and luminous transmittance across the film *(52)*. Further studies found that W has showed a significant effect on the reduction of critical phase transition temperature of VO_2 thin films. Changes in critical temperature of $V_{1-x}W_xO_2$ film were observed due to the orbital structure stress. With increasing tungsten doping concentration in VO_2 thin film, there is a reduction in the strength of V–V interaction, which initiate monoclinic to rutile VO_2 transition even at room temperature. Codoping of two different metal ions is another mechanism for lowering the value of critical phase transition temperature without any change of thin film optical properties *(45)*. Mo–W ions were utilized as codopants for VO_2 films, the codoped film showed critical phase transition temperature value of 35°C with faster phase transition.

Elemental doping technique also aids to enhance the visible transparency and reduction of diminish yellow color of VO_2 thin films. The frequently used dopants for enhancing visible transmittance are Fluorine (F), Magnesium (Mg), and Titanium (Ti). Binions et al. fabricated VO_2 thin films incorporating F via AAVCD process and observed that the enhancement in visible transmittance to 48.7% and modulation of solar energy to 10.7% while the thermochromic properties were retained and the value of critical phase transition temperature was lowered by 10–12°C compared to the undoped one *(58)*. Besides these dopants, alkaline earth metals such as Ca, Sr, and Ba are also been used as dopants for VO_2 thin films. It was reported that these dopants also lower critical phase transition temperature of thin films and improve the luminous transmittance and solar modulation energy, which are more crucial for smart window applications *(59)*.

3.2 Design of multilayers

VO_2 thin films-coated glass or polyethylene terephthalate (PET) substrate is most desirable for smart window applications as scientists achieved its critical phase transition temperature around 46–60°C *(54)*. But the stability of VO_2 films is one of the drawbacks for its application in smart window modules. To stabilize and enhance the stability of VO_2 films, composite thin films were developed where a VO_2 layer is sandwiched with the other functional layers to functional multilayer thin film with enhanced and stable thermochromic properties *(60)*. These multilayered films act smart by being transparent to visible light at lower temperatures and opaque to IR radiation at high temperatures. Most of these multilayer thin films consist of oxide buffer layers, antireflection layers, and Nobel metal layers. The oxide layers such as ZnO, TiO_2, Stannic oxide (SnO_2), and CeO_2 have the structural properties, which are similar to VO_2 films. Oxide layers serve as both protecting and buffer layers for VO_2 thin films and improve the stability of its thermochromic properties for outdoor applications. VO_2 film deposited on ZnO showed high visible transmittance, originating from good lattice match and high degree of preferred VO_2 orientation on ZnO coated substrates. TiO_2 buffer layer was utilized by many research groups and industries to enhance the oxidization durability of solution-deposited VO_2 thin films. Al_2O_3 buffer layer improved the crystallinity, texture of VO_2 thin films on glass, or PET substrates and observed the sharper phase change *(61)*. These oxide buffer layers improve the optical properties of VO_2 films. The single layer of VO_2 film reflects only the radiation having wavelength of >3 μm. An aluminum-doped zinc oxide buffer layer on VO_2 thin film improves optical properties by enhancing the values of luminous transmittance and modulation of solar energy and reducing critical phase transition temperature near to room temperature *(62)*. Dielectric antireflection layers on VO_2 films were used mainly to improve stability and optical transmission of VO_2 films on glass following refractive index matching method. The Si–Al antireflection coating on VO_2 thin film enhanced the visible transmittance by increasing the value of modulation of solar energy and reflecting the IR radiation. Fluorinated Si–Al antireflection coating enhanced the luminous transmittance by 14% without changing the value of modulation of solar energy. Recently, multifunctional $TiO_2/VO_2/TiO_2$ layers were developed on glass substrate to enhance the optical and thermochromic properties of VO_2 film. The multilayer $TiO_2/VO_2/TiO_2$ films showed optical properties with modulation of solar energy up to 10.2% and luminance transmittance up to 30%. For the formation of monoclinic phase of VO_2, the bottom rutile TiO_2 played crucial role, which acts as antireflection coating. By selecting the proper thickness of TiO_2 layers, the values of luminous transmittance and modulation of solar energy were enhanced. Noble metal layers were used on VO_2 thin films to reduce thermal radiation by reflecting infrared in sunlight *(63,64)*. Metals such as Au and Ag with VO_2 films were used to modulate optical properties such as modulation of solar energy and visible transmittance. Platinum/vanadium oxide (Pt/VO_2) composite films were grown on silica substrate prepared by PAD process, to enhance IR reflection from VO_2 thin films. In addition, the antireflection coating on Pt coated VO_2 film boosted the visible transmittance *(65)*. In Gold/Vanadium oxide (Au/VO_2) composites, the value of critical phase transition temperature reduced to 45°C due to inbreeded Au nanoparticles in VO_2 thin film during fabrication process. The Au/VO_2 thin films will reduce the critical phase transition temperature to 40°C *(66)*. In most of the above discussion, different noble metal layers or their nanoparticles were utilized in optimizing the emissivity of VO_2 thin films grown on glass or PET substrate.

3.2.1 Effects of deposition substrate

Both physical and chemical natures of the substrate on which VO_2 thin films was deposited play an essential role in switching the thermochromic performance of VO_2 thin films *(67)*. Depending on the VO_2 thin film preparation method, types of substrate include as conducting substrate that is, transparent electrodes on glass, insulating substrate, substrate with different crystal orientation and sodium (Na) ion diffusion barrier levels. When VO_2 layer was deposited on soda lime glass, Na ion or Na particles diffuse into VO_2

layers and affect its thermochromic properties. The optical properties such as modulation of solar energy and visible transmittance were explained by the diffusion of Na ions. The introduction of silicon nitrite improves the thermochromic properties of Na-diffused VO_2 layers. Because of the high IR reflectivity conducting substrate/transparent conductive oxides are useful for low-E heat shielding coatings. FTO- and ITO-coated glass or PET are most commonly used conducting substrate for VO_2 thin film deposition. The similarity in the crystalline structure of FTO and VO_2, crystal quality of VO_2 thin films developed on FTO-coated substrate enhanced its visible transmittance by reducing interface reflection. Gao et al. used VO_2/FTO-glass structure, which depict promising optical properties by using the FTO substrate and the deposition temperature was also lowered to 390°C *(63)*. A 65 nm thick VO_2/FTO film exhibited low-E of 0.27 and 0.19 after and before the phase transition. Incorporating TiO_2 antireflective coating to form TiO_2/VO_2/FTO, the visible transmittance was enhanced up to 80%. The bottom ITO layer acts as barrier between films and glass or PET substrate producing higher quality of the VO_2 films with smaller grain size and reduced critical phase transition temperature of thin VO_2 film. Advantage of this configuration is phase transition of VO_2 film was triggered externally by applying an electrical power through conducting ITO and FTO layer. It reduced the critical phase transition temperature further lower as desired by local environmental conditions. Due to the ITO and FTO films, an active and efficient control of thermochromic switching of smart windows was realized and same time TiO_2/VO_2/FTO serve as thermochromic low emittance energy efficient layers *(68,69)*.

As reported, metal substrate such as aluminum (Al) also enhances the positive emittance switching performance of VO_2 films. Benkahoul et al. studied the properties of VO_2 thin films deposited on Al-coated substrates. Compared to the Si substrate, VO_2 thin film on Al-coated substrate has enhanced emissive properties of VO_2 thin films. These studies depict the emissive properties of VO_2 thin films dependent on infrared optical properties of buffer metal thin film on glass or PET substrate.

Thermal expansion of substrate also has impact on the value of critical phase transition temperature of VO_2 thin films and was investigated by Sakai et al. The substrates with larger expansion coefficient cause an enhanced out-of-plane lattice spacing in both substrate and films by lowering the critical temperature of films *(70)*. VO_2 films were grown on glass or PET substrate with various buffer layers such as SiO_2, calcium fluoride (CaF_2), and potassium iodate (KI_2O_3). The substrate with a larger thermal expansion coefficient develops out-of-plane lattice spacing in both the thin films, thus lowering the critical phase transition temperature of VO_2. Sapphire is commonly used insulating substrate, which is transparent to terahertz waves, thus suitable for epitaxial growth of VO_2 films. Fan et al. investigated on the properties of VO_2 films grown on *c*-, *r*-, and *m*- plane sapphire substrate under different growth conditions. The properties of VO_2 thin films were greatly affected by the orientation of the sapphire substrate and concluded that VO_2 films grown on *r*-plane sapphire substrate exhibited lower critical phase transition temperature and terahertz intensity modulation wavelength. VO_2 films deposited on soda lime glass showed different transmittance due to sodium (Na) ion diffusion. Silicon nitride (SiN_x) layer also been used as diffusion barrier of VO_2 films to improve thermochromic properties, Si_3N_4-buffered VO_2 films exhibited long-term stability and lowered the value of critical phase transition temperature of VO_2 films to 30°C *(71)*. Deposition of VO_2 thin films on above-mentioned substrate with different buffer layer modifies the structural and optical properties with enhanced crystallinity and lower phase transition critical temperature.

3.2.2 Effects of VO_2 microstructures

The thermochromic property of VO_2 thin films varies with the grain size of the deposited thin films on different substrates. Different research group studies thin films prepared with VO_2 nanoparticle to reveal these effects. It was reported that critical phase transition temperature of nanostructured VO_2 film varies with its grain size; VO_2 nanoparticle of 8 nm depicts a phase transition temperature of 35°C, which is much lower than the microstructured VO_2 thin film (above 80°C) *(58)*. Gao's group prepared VO_2/SiO_2 nanofiber coating and measured their optical properties *(25)*. VO_2/SiO_2 composite nanofiber thin film exhibited a high transmittance even at 25°C, showcased the IR reflectance about 30% and visible transmittance as 50%. Small radius VO_2/SiO_2 composite nanofiber makes it important to have a practical application for smart window coating applications *(25)*. VO_2/TiO_2 composite thin film exhibited remarkable modulation ability for solar infrared light and enhanced luminous transmittance at visible range; that changes the intrinsic color of VO_2 thin film from yellowish to light blue, favorable for practical window application *(72)*.

Thus, nanotechnology (nanostructured VO_2 films coated with SiO_2 or TiO_2) plays important role in developing high-performance thermochromic thin film on glass or PET substrates, maintaining balance between luminous transmittance and modulation of solar energy. Industry grade performance with high luminous transmittance of about 76.5% and modulation of solar energy of about 14% obtained from nano porous VO_2 films when coated on different substrates *(73)*.

4 Theoretical predictions for phase transition mechanism and energy modeling

Besides experimental work on developing thermochromic materials for practical applications, theoretical work has been carried out by research groups to progress the development process by understanding two concepts that are phase transition mechanism and energy modeling *(2)*.

4.1 Phase transition mechanism

From last few decades, substantial research has been conducted to understand the various factors that determine the phase transition of different thermochromic materials. The possible altering factor can be stain induced transition, photon-induced transition, or an electric field induced. The research also focuses on the development of a new methodology to control transition at 30–50°C regime *(74,75)*. Zu et al. proposed composite modeling by combining Band theory—the modified Becke–Jonhnson exchanged potential to understand structural electronic magnetic features of metal-insulator phase transitions for thermochromic materials *(76)*. Xie et al. investigate the role of structural transformation applying X-ray absorption fine structure spectroscopy to monitor the variation in atomic and electronic kinetics during phase transition *(77)*. Bai et al. investigated mechanical pressure-induced phase transition and metallization using numerical modeling techniques, and found that the pressure-induced phase transition is driven by the structural transition in thermochromic materials *(78)*. Lu et al. concluded that the metal insulator transition in VO_2 thin film is driven only by electronic level transition *(79)*. Recently, Gao et al. mapped reduced dimension potential energy surface using first principle calculations and concluded that a staged electron corelation-driven Mott phase transition and structure-driven Peierls phase transition were involved in the phase transition of VO_2 *(80)*. Miller et al. reported a general mechanism for photo-induced phase transition *(81)*. Park et al. measured solid-state triple point for metal-insulator transition and found that the triple point is related to the phase transition temperature *(82)*. Aetukuri et al. found that the metal-insulator transition was controlled by the modification of orbital occupancy in VO_2 structure *(83)*. Wall et al. reported ultrafast structural transition process in VO_2 thin film utilizing optical probes *(84)*. Photo-induced structural phase transition in VO_2 thin film depicts that the phase transition threshold and photoexcitation completely change the lattice potential of VO_2 structure on ultrafast time scale.

4.2 Energy modeling of thermochromic materials

Energy saving performance of windows energy consumption index was evaluated as the ratio of energy consumption of a given window to the corresponding value of perfect window. Various software tools are commercially developed and utilized to numerically estimate energy efficiency of thermochromic windows. This software is utilized to simulate energy saving effect and optical performance for different smart window configurations *(85)*. Some of the commonly used software tools are as follows: the window simulation tool, Winsel *(86)*, Simulation tool SOLENE, self-developed simulation software based on ASHRAE tables, DOE building energy simulation program, TRNSYS building energy simulation program, and simulation package integrated energy used simulation. Saeli et al. modeled thermochromic materials-based smart windows by energy plus software and demonstrated energy saving purpose for both in lighting and ventilation *(87)*. Xu et al. compared the cooling energy consumption of white glass and four low emissive glasses by using TRNSYS 16 and proved that the best performance was obtained by double glass when VO_2 thin films were coated on inside surface *(88)*. Ye et al. evaluated energy consumption of different windows by Building Energy programme and concluded that controlling the emissivity of the window is more beneficial than regulating solar transmissivity *(89)*.

5 Conclusion

In this chapter, we discussed the importance of glass windows in energy saving application for any building architecture. Smart windows such as thermochromic windows are being developed rapidly because of their great energy saving potential in building application. The smartness of window was discussed for providing general information on smartness on energy performance of thermochromic windows. There are many new methods and materials coming up to be used as thermochromic materials. Above, we have discussed mostly about the two major classes of these materials on the basis of their composition along with the mechanism taking place during the phase transition in accordance to the transition temperature. The two major classes discussed are organic and inorganic thermochromic materials. The most frequently used thermochromic material is VO_2, which is basically inorganic material, which undergoes phase transition at 68°C making it an excellent material for infrared reflectance, solar energy modulation along with high visible transmittance. Even though it has excellent properties for energy saving, it has a major drawback that is the requirement of high transition temperature as compared to the surrounding room temperature, which can be reduced with different techniques such as doping, method of preparation of films, and deposition of those films, which are extensively discussed in the above chapter. We also discussed the recent advances on how to improve the thermochromic performance VO_2 films.

Along with the experimental research on the thermochromic properties, there have been a lot of work done over the theoretical approach to improve the performance of these materials. Research is been conducted to understand the various factors to determine the transition of the phase of a particular material, whether the transition is stain induced, photon induced, or electric field induced as well as development methods to control transition. Various software tools can be used for energy modeling of thermochromic windows for evaluating their energy saving effect and optical performance, different smart window configuration have been simulated.

Application of thermochromic materials in the form of coating on windows or in paints has been proven to be beneficial for energy saving application for buildings.

References

1. Long, L.; Ye, H. How to be Smart and Energy Efficient: A General Discussion on Thermochromic Windows. *Sci. Rep.* **2014**, *4*, 6427.
2. Ye, H.; Meng, X.; Xu, B. Theoretical Discussions of Perfect Window, Ideal Near Infrared Solar Spectrum Regulating Window and Current Thermochromic Window. *Energy Build.* **2012**, *49*, 164–172.
3. Baetens, R.; Jelle, B. P.; Gustavsen, A. Properties, Requirements and Possibilities of Smart Windows for Dynamic Daylight and Solar Energy Control in Buildings: A State-of-the-art Review. *Sol. Energy Mater. Sol. Cells* **2010**, *94* (2), 87–105.
4. Warwick, M. E. A.; Ridley, I.; Binions, R. The Effect of Transition Gradient in Thermochromic Glazing Systems. *Energy Build.* **2014**, 77, 80–90.
5. Hee, W. J., et al. The Role of Window Glazing on Daylighting and Energy Saving in Buildings. *Renew. Sustain. Energy Rev.* **2015**, *42*, 323–343.
6. Tahmasebi, M. M.; Banihashemi, S.; Hassanabadi, M. S. Assessment of the Variation Impacts of Window on Energy Consumption and Carbon Footprint. In 2011 Int. Conf. Green Build. Sustain. Cities, vol. 21, Jan. 2011; pp. 820–828.
7. Hassouneh, K.; Alshboul, A.; Al-Salaymeh, A. Influence of Windows on the Energy Balance of Apartment Buildings in Amman. In Glob. Conf. Renew. Energy Effic. Desert Reg. GCREEDER 2009, vol. 51, no. 8, Aug. 2010; pp. 1583–1591.

8. Jaber, S.; Ajib, S. Thermal and Economic Windows Design for Different Climate Zones. *Energy Build.* **2011**, *43* (11), 3208–3215.
9. Thummavichai, K.; Xia, Y.; Zhu, Y. Recent progress in Chromogenic Research of Tungsten Oxides Towards Energy-related Applications. *Prog. Mater. Sci.* **2017**, *88*, 281–324.
10. Juknevicius, L. The 10th International Conference On Modern Building Materials, Structures and Techniques. In Vilnius Gediminas Technical University, 01-Dec-2009.
11. Lampert, C. M. Chromogenic Smart Materials. *Mater. Today* **2004**, 7 (3), 28–35.
12. Dalapati, G. K., et al. Transparent Heat Regulating (THR) Materials and Coatings for Energy Saving Window Applications: Impact of Materials Design, Micro-structural, and Interface Quality on the THR Performance. *Prog. Mater. Sci.* **2018**, *95*, 42–131.
13. Miodownik, M. The Time For Thermochromics. *Mater. Today* **2008**, *11* (11), 6.
14. Day, J. H. Thermochromism. *Chem. Rev.* **1963**, *63* (1), 65–80.
15. Pfundt, G.; Schenck, G. O. Chapter 11 – Cycloadditions to o-Quinones, 1,2-Diketones, and Some of Their Derivatives. *Organic Chemistry*, vol. 8, Hamer, J., Ed.; Elsevier, 1967; pp. 345–417.
16. Chowdhury, M. A.; Joshi, M.; Butola, B. S. Photochromic and Thermochromic Colorants in Textile Applications. *J. Eng. Fibers Fabr.* **2014**, *9* (1), 107–123 155892501400900.
17. Seeboth, A.; Lötzsch, D.; Ruhmann, R. First Example of a Non-toxic Thermochromic Polymer Material – Based on a Novel Mechanism. *J. Mater. Chem. C* **2013**, *1* (16), 2811–2816.
18. Cui, Y., et al. Thermochromic VO_2 for Energy-Efficient Smart Windows. *Joule* **2018**, *2* (9), 1707–1746.
19. Gao, Y., et al. Nanoceramic VO_2 Thermochromic Smart Glass: A Review on Progress in Solution Processing. *Nano Energy* **2012**, *1* (2), 221–246.
20. Babulanam, S. M.; Eriksson, T. S.; Niklasson, G. A., et al. Thermochromic VO_2 Films for Energy-Efficient Windows. *Sol. Energy Mater.* **1987**, *16* (5), 347–363.
21. Hao, Q., et al. VO_2/TiN Plasmonic Thermochromic Smart Coatings for Room-Temperature Applications. *Adv. Mater.* **2018**, *30* (10), 1705421.
22. Morin, F. J. Oxides Which Show a Metal-to-Insulator Transition at the Neel Temperature. *Phys. Rev. Lett.* **1959**, *3* (1), 34–36.
23. Cocker, T. L., et al. Phase Diagram of the Ultrafast Photo-Induced Insulator-Metal Transition in Vanadium Dioxide. *Phys. Rev. B* **2012**, *85* (15), 155120.
24. Li, D.; Huang, W.; Song, L., et al. Thermal Stability of VO_2 Thin Films Deposited by Sol–Gel Method. *J. Sol.-Gel. Sci. Technol.* **2015**, *75* (1), 189–197.
25. Li, S. -Y.; Niklasson, G. A.; Granqvist, C. G. A Thermochromic Low-Emittance Coating: Calculations for Nanocomposites of In_2O_3:Sn and VO_2. *Appl. Phys. Lett.* **2011**, *99* (13), 131907.
26. Sahana, M. B.; Dharmaprakash, M. S.; Shivashankar, S. A. Microstructure and Properties of VO_2 Thin Films Deposited by MOCVD from Vanadyl Acetylacetonate. *J. Mater. Chem.* **2002**, *12* (2), 333–338.
27. Vernardou, D.; Louloudakis, D.; Spanakis, E., et al. Thermochromic Amorphous VO_2 Coatings Grown by APCVD Using a Single-Precursor. *Sol. Energy Mater. Sol. Cells* **2014**, *128*, 36–40.
28. Feng, Q. J. et al., Influence of Sb Doping on the Structural, Optical and Electrical Properties of p-ZnO Thin Films Prepared on n-GaN/Al2O3 Substrates by a Simple CVD Method. In Spec. Top. Issue Wide-Bandgap Semicond. Mater. vol. 29, Jan. 2015, pp. 188–192.
29. Soo Choi, D., et al. Flexible Electrochromic Films Based on CVD-Graphene Electrodes. *Nanotechnology* **2014**, *25* (39), 395702.
30. Liu, C.; Wang, J. -H.; Weng, J. Growth of Micro- and Nanocrystalline Dual Layer Composite Diamond Films by Microwave Plasma CVD: Influence of CO_2 Concentration on Growth of Nano-Layer. *J. Cryst. Growth* **2015**, *410*, 30–34.
31. Warwick, M. E. A.; Ridley, I.; Binions, R. Thermochromic Vanadium Dioxide Thin Films from Electric Field Assisted Aerosol Assisted Chemical Vapour Deposition. In 19th Eur.

Conf. Chem. Vap. Depos. EuroCVD19 Varna Bulg. 1st–6th Sept. 2013, vol. 230, Sep. 2013, pp. 163–167.
32. Warwick, M.; Binions, R. On the Effects of Electric Fields in Aerosol Assisted Chemical Vapour Deposition Reactions of Vanadyl Acetylacetonate Solutions in Ethanol. *J. Nanosci. Nanotechnol.* **2011**, *11*, 8126–8131.
33. Yin, H.; Yu, K.; Song, C., et al. Low-Temperature CVD Synthesis of Patterned Core–Shell VO_2@ZnO Nanotetrapods and Enhanced Temperature-Dependent Field-Emission Properties. *Nanoscale* **2014**, *6* (20), 11820–11827.
34. Joanna Rymarczyk; Elżbieta Czerwosz; Mirosław Kozłowski, et al. The Influence of Technological PVD Process Parameters on the Topography, Crystal and Molecular Structure of Nanocomposite Films Containing Palladium Nanograins. *Pol. J. Chem. Technol.* **2014**, *16* (3), 18–24.
35. Kana Kana, J. B., et al. Thermochromic Nanocrystalline Au–VO_2 Composite Thin Films Prepared by Radiofrequency Inverted Cylindrical Magnetron Sputtering. *Thin Solid Films* **2010**, *518* (6), 1641–1647.
36. Jiang, M.; Cao, X.; Bao, S., et al. Regulation of the Phase Transition Temperature of VO_2 Thin Films Deposited by Reactive Magnetron Sputtering Without Doping. *Thin Solid Films* **2014**, *562*, 314–318.
37. Borek, M.; Qian, F.; Nagabushnam, V., et al. Pulsed Laser Deposition of Oriented VO_2 Thin Films on R-cut Sapphire Substrates. *Appl. Phys. Lett.* **1993**, *63* (24), 3288–3290.
38. Novodvorsky, O. A.; Parshina, L. S.; Khramova, O. D., et al. Influence of the Conditions of Pulsed Laser Deposition on the Structural, Electrical, and Optical Properties of VO_2 Thin Films. *Semiconductors* **2015**, *49* (5), 563–569.
39. Leroy, J.; Bessaudou, A.; Cosset, F.; et al. Structural, Electrical and Optical Properties of Thermochromic VO2 Thin Films Obtained by Reactive Electron Beam Evaporation. In Proc. EMRS 2011 Spring Meet. Symp. Synth. Process. Charact. Nanoscale Multi Funct. Oxide Films III, vol. 520, no. 14, May 2012, pp. 4823–4825.
40. Zhao, L. et al., A Low Cost Preparation of VO2 Thin Films with Improved Thermochromic Properties from a Solution-Based Process. In Int. Conf. NanoSEA NANOstructures SElf Assem. 2012, vol. 543, Sep. 2013, pp. 157–161.
41. Wang, N.; Magdassi, S.; Mandler, D., et al. Simple Sol–Gel Process and One-Step Annealing of Vanadium Dioxide Thin Films: Synthesis and Thermochromic Properties. *Thin Solid Films* **2013**, *534*, 594–598.
42. Li, S. -Y.; Niklasson, G. A.; Granqvist, C. G. Nanothermochromics: Calculations for VO_2 Nanoparticles in Dielectric Hosts Show Much Improved Luminous Transmittance and Solar Energy Transmittance Modulation. *J. Appl. Phys.* **2010**, *108* (6), 063525.
43. Popuri, S. R.; Miclau, M.; Artemenko, A., et al. Rapid Hydrothermal Synthesis of VO_2 (B) and Its Conversion to Thermochromic VO_2 (M1). *Inorg. Chem.* **2013**, *52* (9), 4780–4785.
44. Zhou, J., et al. Mg-doped VO_2 Nanoparticles: Hydrothermal Synthesis, Enhanced Visible Transmittance and Decreased Metal–Insulator Transition Temperature. *Phys. Chem. Chem. Phys.* **2013**, *15* (20), 7505–7511.
45. Xu, Y.; Huang, W.; Shi, Q., et al. Synthesis and Properties of Mo and W Ions Co-Doped Porous Nano-Structured VO_2 Films by Sol–Gel Process. *J. Sol.-Gel. Sci. Technol.* **2012**, *64* (2), 493–499.
46. Jia, Q. X., et al. Polymer-Assisted Deposition of Metal-Oxide Films. *Nat. Mater.* **2004**, *3* (8), 529–532.
47. Zhang, Z.; Gao, Y.; Kang, L., et al. Effects of a TiO_2 Buffer Layer on Solution-Deposited VO_2 Films: Enhanced Oxidization Durability. *J. Phys. Chem. C* **2010**, *114* (50), 22214–22220.
48. Minch, R.; Es-Souni, M. Nanostructured VO_2 Thin Films via Cathodic Deposition. *Cryst. Eng. Commun.* **2013**, *15* (34), 6645–6647.
49. Majid, S. Vanadium Dioxide (VO2) Based Smart Thermochromic Windows, 2019.
50. Goodenough, J. B. The Two Components of the Crystallographic Transition in VO_2. *J. Solid State Chem.* **1971**, *3* (4), 490–500.

51. Béteille, F.; Livage, J. Optical Switching in VO_2 Thin Films. *J. Sol.-Gel. Sci. Technol.* **1998**, *13* (1), 915–921.
52. Barron, S. C.; Gorham, J. M.; Patel, M. P., et al. High-Throughput Measurements of Thermochromic Behavior in $V_{1-x}Nb_xO_2$ Combinatorial Thin Film Libraries. *ACS Comb. Sci.* **2014**, *16* (10), 526–534.
53. Tan, X., et al. Unraveling Metal-Insulator Transition Mechanism of VO_2 Triggered by Tungsten Doping. *Sci. Rep.* **2012**, *2*, 466.
54. Wang, N.; Goh, Q. S.; Lee, P. L., et al. One-Step Hydrothermal Synthesis of Rare Earth/W-Codoped VO_2 Nanoparticles: Reduced Phase Transition Temperature and Improved Thermochromic Properties. *J. Alloys Compd.* **2017**, *711*, 222–228.
55. Cho, J.-H. et al., Thermochromic Characteristics of WO3-Doped Vanadium Dioxide Thin Films Prepared By Sol–Gel Method. In 7th Asian Meet. Electroceramics AMEC-7 Conjunction 7th Asian Meet. Ferroelectr. AMF-7, vol. 38, Jan. 2012, pp. S589–S593.
56. Barron, S.; Gorham, J. (Invited) Thermochromic Phase Transitions in VO2-Based Thin Films for Energy-Saving Applications, vol. 61, 2014.
57. Batista, C.; Carneiro, J. O.; Ribeiro, R., et al. Reactive Pulsed-DC Sputtered Nb-Doped VO(2) Coatings for Smart Thermochromic Windows with Active Solar Control. *J. Nanosci. Nanotechnol.* **2011**, *11*, 9042–9045.
58. Wang, S.; Liu, M.; Kong, L., et al. Recent Progress in VO_2 Smart Coatings: Strategies to Improve the Thermochromic Properties. *Prog. Mater. Sci.* **2016**, *81*, 1–54.
59. Dietrich, M. K.; Kramm, B. G.; Becker, M., et al. Influence of Doping with Alkaline Earth Metals on the Optical Properties of Thermochromic VO_2. *J. Appl. Phys.* **2015**, *117* (18), 185301.
60. Kim, H., et al. Enhanced Optical Response of Hybridized VO_2/Graphene Films. *Nanoscale* **2013**, *5* (7), 2632–2636.
61. Xiong, Y., et al. Tuning the Phase Transitions of VO_2 Thin Films on Silicon Substrates Using Ultrathin Al_2O_3 as Buffer Layers. *J. Phys. Appl. Phys.* **2014**, *47* (45), 455304.
62. Chu, X.; Tao, H.; Liu, Y. et al., VO2/AZO Double-Layer Films with Thermochromism and Low-Emissivity for Smart Window Applications. In XIII Int. Conf. Phys. Non-Cryst. Solids, vol. 383, Jan. 2014; pp. 121–125.
63. Zhang, Z., et al. Solution-Based Fabrication of Vanadium Dioxide on F:SnO_2 Substrates with Largely Enhanced Thermochromism And Low-Emissivity for Energy-Saving Applications. *Energy Environ. Sci.* **2011**, *4*, 4290–4297.
64. Zheng, J.; Bao, S.; Jin, P. TiO2(R)/VO2(M)/TiO2(A) Multilayer Film as Smart Window: Combination of Energy-Saving, Antifogging and Self-Cleaning Functions. *Nano Energy* **2015**, *11*, 136–145.
65. Kang, L.; Gao, Y.; Chen, Z., et al. Pt/VO_2 Double-Layered Films Combining Thermochromic Properties with Low Emissivity. *Sol. Energy Mater. Sol. Cells* **2010**, *94* (12), 2078–2084.
66. Zhu, Y.; Na, J.; He, F., et al. The Characteristics of Au:VO_2 Nanocomposite Thin Film for Photo-Electricity Applications. *Phys. E Low-Dimens. Syst. Nanostruct.* **2013**, *52*, 112–115.
67. Binions, R.; Hyett, G.; Piccirillo, C., et al. Doped and Un-Doped Vanadium Dioxide Thin Films Prepared by Atmospheric Pressure Chemical Vapour Deposition from Vanadyl Acetylacetonate and Tungsten Hexachloride: The Effects of Thickness and Crystallographic Orientation on Thermochromic Properties. *J. Mater. Chem.* **2007**, *17*.
68. Peng, S.; Cheng, F.; Liang, J., et al. Facile Solution-Controlled Growth of CuInS2 Thin Films on FTO and TiO_2/FTO Glass Substrates for Photovoltaic Application. *J. Alloys Compd.* **2009**, *481* (1), 786–791.
69. Hussain, S. Q., et al. Influence of High Work Function ITO:Zr Films for the Barrier Height Modification in a-Si:H/c-Si Heterojunction Solar Cells. *Sol. Energy Mater. Sol. Cells* **2014**, *122*, 130–135.
70. Sakai, J.; Zaghrioui, M.; Matsushima, M., et al. Impact of Thermal Expansion of Substrates on Phase Transition Temperature of VO_2 Films. *J. Appl. Phys.* **2014**, *116* (12), 123510.

71. Koo, H.; You, H.; Ko, K. -E., et al. Thermochromic Properties of VO_2 Thin Film on SiNx Buffered Glass Substrate. *Appl. Surf. Sci.* **2013**, *277*, 237–241.
72. Lu, Q.; Liu, C.; Wang, N., et al. Periodic Micro-Patterned VO_2 Thermochromic Films by Mesh Printing. *J. Mater. Chem. C* **2016**, *4* (36), 8385–8391.
73. Li, S.-Y.; Niklasson, G. A.; Granqvist, C. G. Thermochromic Fenestration with VO2-Based Materials: Three Challenges and How They Can Be Met. In 7th Int. Symp. Transparent Oxide Thin Films Electron. Opt. TOEO-7, vol. 520, no. 10, Mar. 2012, pp. 3823–3828.
74. Qazilbash, M. M., et al. Mott Transition in VO_2 Revealed by Infrared Spectroscopy and Nano-Imaging. *Science* **2007**, *318* (5857), 1750.
75. Rice, T. M.; Launois, H.; Pouget, J. P. Comment on VO_2: Peierls or Mott-Hubbard? A View from Band Theory. *Phys. Rev. Lett.* **1994**, *73* (22), 3042.
76. Zhu, Z.; Schwingenschlögl, U. Comprehensive Picture of VO_2 from Band Theory. *Phys. Rev. B* **2012**, *86* (7), 075149.
77. Yao, T., et al. Understanding the Nature of the Kinetic Process in a VO2 Metal-Insulator Transition. *Phys. Rev. Lett.* **2010**, *105* (22).
78. Bai, L., et al. Pressure-Induced Phase Transitions and Metallization in VO2. *Phys. Rev. B* **2015**, *91* (10), 104110.
79. Kittiwatanakul, S.; Wolf, S. A.; Lu, J. Large Epitaxial Bi-Axial Strain Induces a Mott-Like Phase Transition in VO_2. *Appl. Phys. Lett.* **2014**, *105* (7), 073112.
80. Chen, S.; Liu, J.; Luo, H., et al. Calculation Evidence of Staged Mott and Peierls Transitions in VO_2 Revealed by Mapping Reduced-Dimension Potential Energy Surface. *J. Phys. Chem. Lett.* **2015**, *6* (18), 3650–3656.
81. Hada, M., et al. Hot Electron Injection Driven Phase Transitions. *Phys. Rev. B* **2012**, *86* (13), 134101.
82. Park, J. H., et al. Measurement of a Solid-State Triple Point at the Metal–Insulator Transition in VO_2. *Nature* **2013**, *500*, 431.
83. Aetukuri, N. B., et al. Control of the Metal–Insulator Transition in Vanadium Dioxide by Modifying Orbital Occupancy. *Nat. Phys.* **2013**, *9*, 661.
84. Wall, S., et al. Ultrafast Changes in Lattice Symmetry Probed by Coherent Phonons. *Nat. Commun.* **2012**, *3* (1), 721.
85. Crawley, D. B.; Hand, J. W.; Kummert, M., et al. Contrasting the Capabilities of Building Energy Performance Simulation Programs. *Part Spec. Build. Perform. Simul* **2008**, *43* (4), 661–673.
86. Jonsson, A.; Roos, A. Evaluation of Control Strategies for Different Smart Window Combinations Using Computer Simulations. *Sol. Energy* **2010**, *84* (1), 1–9.
87. Saeli, M.; Piccirillo, C.; Parkin, I. P., et al. Energy Modelling Studies of Thermochromic Glazing. *Energy Build.* **2010**, *42* (10), 1666–1673.
88. Xu, X.; Wu, X.; Zhao, C., et al. Simulation and Improvement of Energy Consumption on Intelligent Glasses in Typical Cities of China. *Sci. China Technol. Sci.* **2012**, *55*, 1999–2005.
89. Ye, H.; Meng, X.; Xu, B. Theoretical Discussions of Perfect Window, Ideal Near Infrared Solar Spectrum Regulating Window and Current Thermochromic Window. *Energy Build.* **2012**, *49*, 164–172.

CHAPTER 10

Recent developments in smart window engineering: from antibacterial activity to self-cleaning behavior

Busi Kumar Babu[a], Siddhartha Ghosh[b], Sabyasachi Chakrabortty[a]
[a]Department of Chemistry, SRM University, AP – Amaravati, Andhra Pradesh, India
[b]Department of Physics, SRM University, AP – Amaravati, Andhra Pradesh, India

1 Introduction

Smart windows are attracting researcher's interest as these are dynamically switchable with respect to their functions compared to current static windows. In that respect, the choice of materials recognizing their prospective potential is highly important. One such example is demonstrated as an effort to improve the performance of a rechargeable battery provided for the next generation energy-efficient windows that switch from transparent to opaque spontaneously upon exposure to increasing power of sunlight, or by the application of a small voltage *(1)*. Similarly, smart windows can effectively modulate the transmittance of solar irradiation depending on weather conditions, thus improving energy efficiency and indoor human comfort. They are intended to control the visible light transmission and regulate the buildings heat transformation of near-infrared (NIR) solar radiation between clear state to blocking state. However, examples on antibacterial coating for preparing smart window to reduce harmful microorganism are rare. Therefore, much research effort is needed to identify proper hybrid materials to address this issue. Smart or switchable windows are widely used for thermal management, energy harvesting, and privacy applications by changing light transmittance when light, heat, or voltage applied. As depicted in Fig. 10.1, smart window coatings are energy efficient to store the energy, and the most important thing to limit the NIR light between the indoor and outdoor environments *(2)*. Also, these energy-efficient smart coatings can successfully destroy harmful pollutants dispersed in the environment because of owing good photocatalytic and hydrophilic properties,

Energy Saving Coating Materials
http://dx.doi.org/10.1016/B978-0-12-822103-7.00010-8

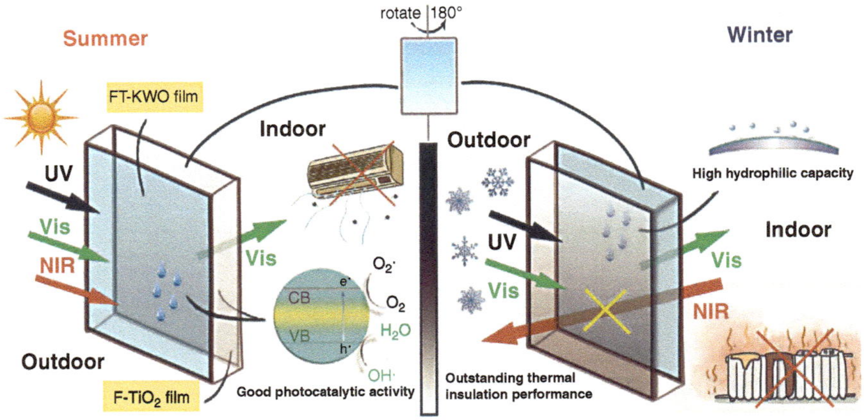

Figure 10.1 ***Distinguishing the working principle of smart window based on weather condition*** **(2).**

which is clearly described in photocatalysis of TiO_2 NPs on smart window below section.

Recently, nanoparticles (NPs) are extensively used in various applications ranging from renewable energy to healthcare materials and in particular, as antibacterial agents where the inhibition of bacterial activities is demonstrated. Those areas include textile industry, medicine/hospitals, food preparation processing, water purification units, air conditioning, and ventilation systems. Inorganic nanoparticles, such as metal oxide, are extensively used for water purification as other organic compounds are proven to be toxic for human. Also, those surface-modified inorganic nanostructured materials exhibit good antimicrobial activity without affecting the other functions.

In general, smart window is an engineered glass where light transmission properties and/or energy of incident photons are reformed when voltage, light, or heat are applied. Based on those at present, three different kinds of materials are known to hybrid in smart windows and they are chromic materials, liquid crystals, and suspended colloidal particles. Also, the chromic materials may be divided into four categories, that is, electrochromic, gasochromic, photochromic, and thermochromic devices. Simultaneously, recent improvements in applied nanostructured materials facilitate better opportunities for next-generation smart window technology. This is mainly due to their unique structure property relationship along with enhanced coloration efficiency, faster switching kinetics, and longer lifetime of the working nanomaterials. Materials like TiO_2 NPs exhibit good photocatalytic properties and therefore, they are widely used in commercial scale as self-cleaning surfaces over a wide gamma of window application. Thus, functional coatings on smart windows that are able to control outside climate from inside the room in a more energy efficient way by reducing both heating and cooling loads are highly desirable.

In general, static window technologies are simply not capable of maximizing both energy efficiency, and visual and thermal comfort of building's occupants. However, smart window can change the transmittance of sunlight to provide variable solar heating and daylight according to changing the weather and surrounding environment. Earlier simulation studies show that smart windows can decrease a building energy needs by up to 40% relative to static windows. There are three main categories of smart windows, which can further be generalized by their operating principles: (1) electrochromic windows, which change the transmittance under applied voltages, (2) photochromic windows where the working principle is based on the incidence of light, and (3) thermochromic windows where we get response with respect to the applied temperature.

This chapter mainly discusses about fundamental principles and highlights the recent advancements in electrochromic, photochromic, thermochromic materials, which are typically NP based and their devices are used in smart windows. In recent year, various novel methods are developed for nanomaterial synthesis that facilitate precise control over the nanocrystal size, shape, composition, and surface functionalization and those advancements provide better chances to open new avenues to get unique applications of NPs in chromogenic devices. Smart window materials made from chemically synthesized nanoscale building blocks with their low-cost solution deposition and low temperature processing techniques like roll-to-roll printing and lamination are envisioned as to increase the production yield and reduce the manufacturing costs *(3,4)*. In addition, a special focus will be given about the antibacterial coatings of nanoparticles on smart windows where the application on diminishing microorganisms is discussed. In addition, we will highlight on the antibacterial properties of Ag and Cu NPs and their consequences.

In the below section, we would like to discuss about the nanoparticle coatings on electrochromic smart windows of different materials.

2 Electrochromic smart window

Electrochromic materials, where upon providing sufficient electrochemical potential changes or bleaching of color occurs, are extensively used in windows as depicted in Fig. 10.2. At present, electrochromic windows

Figure 10.2 ***Switching sequence of an electrochromic laminated glass (Xu et al., 2018)*** **(8).**

are able to both control the transmittance of sunlight and regulate building's heat transformation from near-infrared (NIR) solar radiation, which accounts for nearly half of the solar energy that usually propagates through smart windows, but it does not contribute effectively to energy savings in aircraft and building. To modify the rapid color change against electrochemical potential, stimuli are mainly used *(5,6)*. The important characteristics of electrochromic constituents are electrolytes, transparent conductors, and storage materials must be highly translucent to visible light and must also have good refractive index matching at their interfaces to decrease reflections. All the electrochromic components must be stable within the switching voltage range to be chromogenic for their application. In practical applications, the usage of solid state electrolytes such as polymer ion gels or ion conducting ceramics is much advantageous compared to liquid electrolytes because they can leak, catch fire, or break the glass owing to gravity, especially for large windows. The active materials used in an electrochromic device must exhibit a combination of properties that have high coloration efficiency, high optical contrast, long-term redox, and photo stability and fast switching kinetics.

Most of the previous reports on electrochromic devices continue to focus on the enhancement of the smart windows transmission performance in the visible region and the controllable transmission range. Current studies of smart windows are extended to near-infrared light region *(9)*. For example, Milliron and her group members found that charged NbO_x-indium tin oxide nanocrystals could tune visible and near infrared light transmission by varying the nanoparticle percentage. Other metal-oxide semiconductors such as TiO_2, WO_3, and NbO_x are also capable similar performances as they dynamically modulate the localized surface plasmon resonance via capacitive charging processes. This allows the metal-oxide layer to support a tunable localized surface plasmon response that interacts with the NIR light *(10)*. Also, amorphous WO_3, as one of the capacitive metal oxides, has been widely used in smart windows and devices because of its outstanding tunability in the visible range, fast switching speed, and excellent cycling life time *(11)*.

Gold (Au) nanostructures, when compared with other noble metals, exhibit strong, stable, and tunable localized surface plasmon resonance (LSPR) absorption in both visible and NIR bands that is dependent on the size and shape of the structures. In particular, strongly absorbed NIR radiation (nonradiative plasmon decay) can be transformed into heat because of the related electron–phonon and phonon–phonon interaction processes. To date, plasmons have been applied in various regions such as photothermal therapy (PTA) and biological sensors and used as contrast agents for optical detection *(9, 12)*. However, in smart windows, the photothermal conversion is successfully done by Jingwen Xu, using a 3D honeycomb-like WO_3 film acting as the electrochromic substrate and gold nanostructures (nanoparticles acting as anchors and nanorods acting as the NIR surface plasmon resonance amplifiers). Jingwen Xu and his group members demonstrated two mechanisms for photothermal conversion by using electrochromic–photothermal materials: (1) Cation interaction process-enriched NIR-localized surface plasmon resonance (LSPR) annihilation and (2) The strong dependence of LSPR performance on the states of the WO_3 film. The as-prepared electrochromic film can also effectively reduce the numbers of attacking microorganisms, thus promising use for sterile window for advanced applications *(8)*, which will be explained later in this chapter. Next, we will discuss about photochromic material and their usage.

3 Photochromic smart window

A photochromic material is one that changes from transparent state to a colored state when it is exposed to certain wavelengths of light, and it returns to transparency when the light is dimmed or blocked. Photochromic coatings are generally made by embedding active materials into host matrices, such as polymers or ceramics, and they have already found applications in translational lenses and after-market window tints for buildings and automobiles. A good photochromic material should possess a high photocoloration difference, fast and reversible switching, and satisfactory photochemical stability *(13)*.

Many electrochromic transition metal oxides, including WO_3, MoO_3, VO_5, Nb_2O_5, TiO_2, ZnO, can also exhibit photochromism via proton-coupled electron transfer, which leads to the formation of reduced and colored metal oxides, such as a deep blue color for reduced WO_3 *(14–16)*. Bleaching occurs when the colored oxides are exposed to air or electrochemically oxidized. Some researchers have used this effect to store or regenerate charge for WO_3-based electrochromic devices. The Mayer and Gamelin groups have also observed similar effects for TiO_2 and WO_3 nanocrystals after exposure to UV light. In addition, a photo-electrochromic cell can be fabricated by combining a photovoltaic film (e.g., dye-sensitized TiO_2) and an electrochromic film (e.g., WO_3). Under illumination, the photovoltaic film converts absorbed photons into electrons and holes. The former are injected into the electrochromic film and induce coloration, whereas latter oxidize the redox shuttle in the electrolyte. Although this system can be potentially self-powdered, the use of dye sensitized electrodes lowers the transmittance of the clear state owing to the dyes absorption, which is not ideal for window applications. In addition, while comparing with electrochromic material, photochromic materials have not received much attention for use in smart widows, because they heavily rely on light intensity and cannot be user controlled effectively. In the next part, we will discuss about thermochromic material, which are used in preparation of smart window.

4 Thermochromic smart window

Thermochromic smart windows work on the principle of temperature alteration between transparent state and the darkened state. Generally, at low temperatures, clear state predominates and we observe darkened state at high temperatures. Mostly, thermochromic materials achieve coloration by undergoing a chemical reaction by heating at critical temperatures. At present, researchers' aim is to control the critical temperatures through various techniques such as doping, nanostructuring, or other processing, with the goal of getting critical temperature to normal room temperature. Also, they aim to enhance the optical distinction between bright and dark states by increasing the modulation of solar energy for thermochromic materials. Similar to photochromic properties, thermochromic smart windows cannot be user controlled *(17–19)*. Thermochromic smart windows can allow the visible light and modulate the transmittance of solar radiation in NIR region as described in Fig. 10.3.

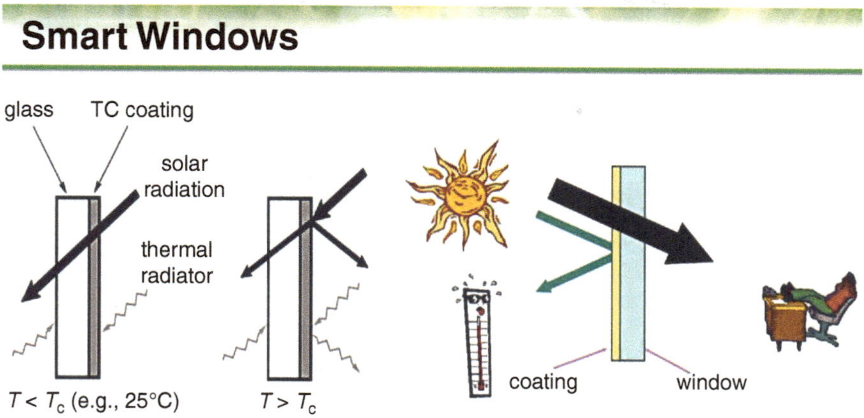

Figure 10.3 ***A typical diagram describing the working of thermochromic smart windows*** **(20).**

Many of the transition metal oxides of vanadium, titanium, iron, molybdenum, and niobium have good thermochromic properties for usage in smart windows. In particular, inorganic solids vanadium dioxide (VO_2) is most well-studied material and undergoes an insulator to metal transition at 68°C, where it changes from low-temperature semiconducting monoclinic phase to the high temperature metallic tetragonal rutile phase. Lowering the transition temperature is normally achieved by doping VO_2 with supervalent metals like Nb^{5+} and W^{6+}. Such dopants increase the vanadium dioxide lattice parameter and likely inject electrons into the vanadium d-band; both effects decrease the insulator to metal transition temperature, and 2% tungsten-doped VO_2 has a transition temperature near 25°C. Thus, it would be very advantageous as there exist two phases, where the insulating phase is transparent and the metallic phase strongly reflects infrared light. Also, different methods are present to enhance the visible transmittance of VO_2, such as doping with Mg^{2+} to increase the bandgap or using high refractive index materials to construct numerous layers (e.g., $TiO_2/VO_2/TiO_2$) that transmit more visible light *(21)*. On the other hand, heat reflecting properties of the metallic nanoparticles will be discussed in the next section.

5 Heat reflecting coating on smart window

Compared to current normal glasses, these engineered glasses are capable of heat reflectors, heat transmitters etc., because of doping different materials (e.g., Au, Ag, Cu, TiO_2, VO_2, etc.) on the smart windows. Current normal windows pass the visible light and but not able to control the

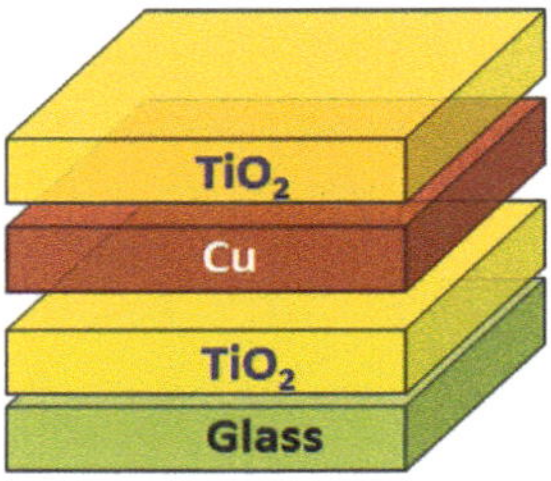

Figure 10.4 ***Schematic illustration of transparent heat reflector (THR) using copper layer sandwiched between the dielectric (TiO_2) materials*** **(22).**

near-infrared light (heat), which makes discomfort for the human beings to survive inside the room. Pertaining to this problem, Dalapati et al. demonstrated a transparent heat reflecting (THR) coating on suitable substrate *(22)*. As shown in Fig. 10.4, THR coating contains a multilayered thin film of copper layer sandwiched between titanium dioxide dielectric materials (TiO_2/Cu/TiO_2). This is, in particular, very useful techniques for cost saving window application as copper is low cost material compared to other metal such as silver. In this paper, sputter-deposited TiO_2/Cu/TiO_2 thin film was prepared and annealed at 500°C, which shows 85% heat reflectance at NIR region and 90% transmittance in the visible region at a wavelength of 1200 nm *(22–25)*. Therefore, it is able to perform reflectance of near-infrared light and the transmittance of visible spectrum both continuously and these kinds of prominent coating techniques are extremely important for energy saving building industry. Similarly, in the next section, photocatalysis properties will be reviewed.

6 Photocatalysis of TiO_2 nanoparticles on smart window

Photocatalysis is the process where light and a catalyst are required to accelerate a chemical transformation. The photocatalytic activity depends on the ability of the catalyst to create electron–hole pairs upon light irradiation, which therefore create free radicals (e.g., hydroxyl radicals) or other reactive species that are able to undergo secondary reactions. In that respect, semiconductor materials are very potent as they produce exciton (electron–hole pair) under suitable incident photon. When a semiconductor catalyst is exposed to light source with a higher energy than its bandgap energy, the light source generates excited states in the catalyst that are able to perform redox reactions, thereby creating molecular reactions.

One of the most studied materials upon photocatalytic and superhydrophilic properties is nanostructured semiconductor titanium dioxide

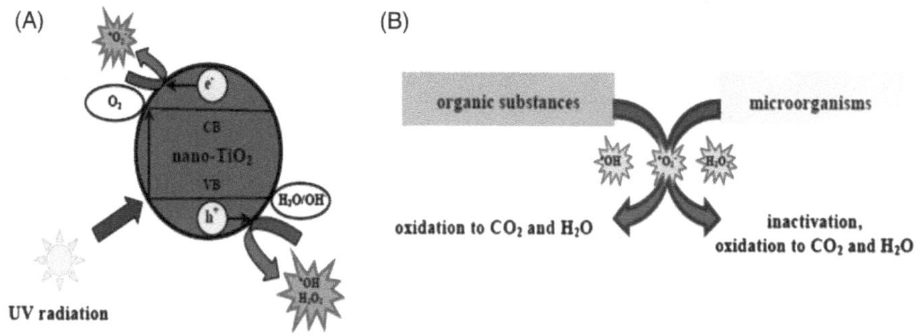

Figure 10.5 ***Generation of photocatalytic active species (reactive oxygen species) at the surface of TiO_2 nanoparticles*** **(26).**

(TiO_2). Also, TiO_2 can be effectively used in windows coating material as it possesses antibacterial and self-cleaning applications. When nanostructured TiO_2 is irradiated with UV light, it is able to absorb incident photons and creates an electron–hole pair on the surface, as described in Fig. 10.5. Later, the electron–hole pair reacts with environment moisture (i.e., H_2O) and oxygen to generate free radicals. These free radicals are very unstable and energetic species, which can participate in various chemical transformations. Especially, it can destroy the pollutants on the surface of a smart window. The hydroxyl radical is a dominant oxidizing agent, which may attack the organic bodies present in the atmosphere or near the surface of TiO_2 and it is capable of degrade toxic and bio-resistant compounds into harmless species (e.g., CO_2, H_2O, etc.) *(26)*.

Nanostructured TiO_2 blended in ceramics, cement, concrete, tiles, glass, and tents can act as self-cleaning material. Adsorbed organic materials like oil can be decomposed by the photocatalytic property of TiO_2, while the organic contaminants and dust can be washed away by rain water because of super-hydrophilic property of TiO_2. Thus, both cleaning and superior hydrophobic properties of the surface are preserved with the help of sunlight and rain.

Such super-hydrophilic and self-cleaning products will never get dirty. Professor Fujishima at University of Tokyo found a phenomenon on the photocatalyst. When the surface of photocatalytic film is exposed to light, the contact angle of photocatalyst with water is reduced gradually. After sufficient exposure to light, the surface gradually reaches super-hydrophilicity. It means the TiO_2 surface will never drive back water, so the rain water does not exist as the shape of a droplet but the water drop spreads flatly like thin sheet on the surface of the photocatalyst as described in Fig. 10.6. Also, by

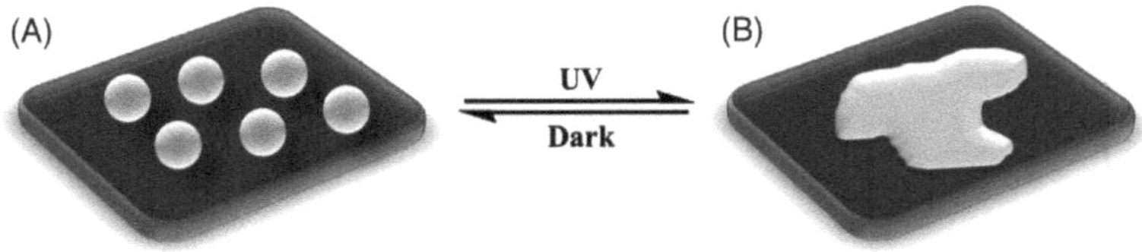

Figure 10.6 Comparing ordinary float glass (A) with anti-fogging glass along with super-hydrophilicity effect, (B) the water forms of a thin sheet, with no fogging effect [inspired from 27]. This phenomenon was observed upon UV light irradiation.

using these intelligent products, we can save our time and cost of cleaning maintenance for buildings. Now, smart windows are also working on the basis of photocatalysis to perform multiple applications in commercial scale. Next, we will elaborate on antibacterial material used in smart window *(28–30)*.

7 A special focus on antibacterial activity

7.1 Antibacterial coatings in various applications

Antibacterial material took huge interest of researchers, strongly stimulated by the increasing demand of identifying alternatives from the traditional administrative to the antibiotics. In spite of rigorous and considerable research achievements by material scientists and biologists, the difficulties of infections related to medical devices and implants remain an open challenge *(31)*. Microorganisms have various existences, in the form of virus, bacteria, algae, fungi, and several other living infectants in the last million years ago, and have unique effect in almost every human life. Most of the human beings are getting diseases related to infections with different kind of microorganisms. Especially, in hospital acquired infections, where a large number of patients who go to hospital with a disease and they are returning with something more serious problem because of these nosocomial infections *(32)*. According to the World Health Organization (WHO) survey, nearly 2 million people are suffering with this nosocomial or hospital acquired infections, among them 80,000 deaths were recorded per year due to this. In that aspect, antibacterial coatings on exposed surface become handy with respect to completely restricting the family of bacteria accumulated in the closed atmosphere. Some of the important antibacterial surfaces in different working environments are healthcare systems, food packagings, paints, ceramics, industrial pipes, smart windows, smart glasses, transportation systems, and they are schematically shown in Fig. 10.7.

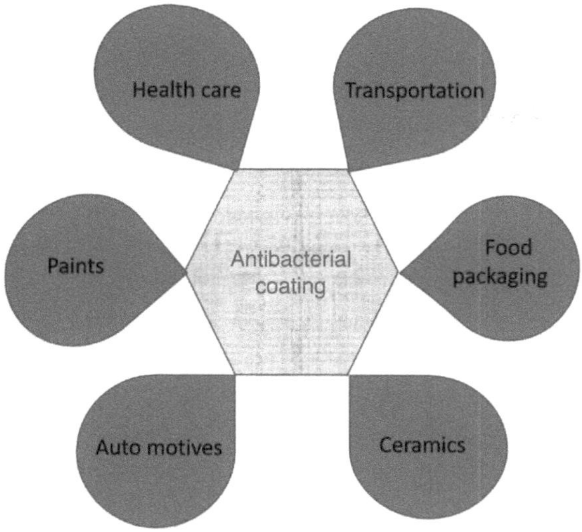

Figure 10.7 ***Illustrating antibacterial coatings in different areas.***

Nanoparticles are envisioned as the suitable materials to act as good carriers and fit for conducting antimicrobial operations in antibiotics. These nanoparticles can effectively abolish bacteria by their inherent antibacterial properties for even implantable devices such as heart valves, dental implants, and catheter. Also, they provide medicinal materials to inhibit bacterial infection and promote wound healing, in antibiotic drug-controlled delivery systems to treat disease, in bacterial detection systems to produce microbial diagnostics, and in antibacterial vaccines to control bacterial infections *(31–33)*. In antibiotic resistance mechanisms, nanoparticle-based materials became more exciting and great attention has been focused because nanoparticles are directly in contact with the bacteria cell wall without need to enter the cell. Thus, the nanoparticles would be less prone to promoting resistance in bacteria than antibiotics. The major processes underlying the antibacterial effects of nanoparticles are as follows: (1) interruption of the microbial cell membrane; (2) production of reactive oxygen species (ROS); (3) diffusion of the bacterial cell membrane; (4) induction of intracellular antibacterial effects, including interactions with DNA and proteins *(34)*. Additionally, exploration of the sterile strategies of nanoparticles is essential for the improvement of more effective antimicrobial materials.

The production of antibacterial surfaces for biomedical devices has been accomplished with different antibacterial agents containing organic molecules and inorganic metal ions. However, metal nanoparticles such as copper, silver, gold, titanium, magnesium, and zinc are acting as excellent bactericidal activators and these were used in number of in vivo and in vitro applications. Here, we will discuss about some of the extensively used materials as antibacterial agents for the antimicrobial activity.

Functionalization of suitable surfaces by applying antibacterial materials onto a working substrate as surface coatings were most extensively used methods for bactericidal activity. To achieve the bactericidal activity in bio-medical applications, there are two different strategies were adopted: (1) by applying the chemical or antibacterial agent on the bio-medical device surface, which can target the surrounding bacteria; (2) antibiotic molecules can be covalently immobilized (immovable) on the device surface that kills the attached bacteria on the surface. The safety of patients from the bacterial infections (especially in hospitals) is the utmost important issue, which needs proper solution. Along with the cleanliness of hospitals and in medical facilities, antimicrobial electric smart glass might be the optimum solution to limit the amount of bacteria that can stick on glass. Simultaneously, those developments in health care products such as medical devices, textiles, and paints with antimicrobial properties can significantly reduce various types of health acquired infections.

In Table 10.1, we have highlighted the information of numerous antibacterial materials and their characteristics and applications, with different mechanisms to perform bactericidal activity in different areas from reference *(32)*.

7.2 Antibacterial paints

In antibacterial paints, it is expected that the additives will provide effective protection against harmful bacteria, mould and fungi with the capabilities to minimize bad odors and material degradation. They are highly essential in hospitals, bathrooms, kitchens, and other locations where the bacterial population are known to be the highest *(35)*. Also, we know that with time we are experiencing rapid increase of pollutants and bacterial family in the atmosphere and researcher have started exploring the usefulness of nanoparticles to limit those. Silver nanoparticles are acting as one of the promising materials for the antibacterial activity and it is widely used as nanoparticles in painting industry.

Among many antibacterial paints available, recently, Kumar et al. illustrated the use of Ag nanoparticles based on vegetable oil, for their antimicrobial properties *(36)*. In water-based paints, nanostructured silver vanadate (Ag_3VO_4) has been used as an antibacterial agent. Among others,

Table 10.1 Proposed antibacterial mechanism for antibacterial materials their characteristics and potential applications.

Antibacterial material and their mechanism	Characteristics	Prospective applications
Copper, silver ion with phosphate counter ion— slow release metal ion with sterilization	High chemical activity provides long-term usage, which enhances slow release of antibacterial material.	Widely used in medical applications, stainless steel, water treatment. Prevent bask in liquid coatings and fabrics. But these materials tarnish easily and are expansive, which limits their applications
Hydroxyapatite, Ag-carrying phosphate— slow release metal ion and photocatalytic sterilization	Phosphoric acid double salt provides a strong adsorption function, large specific surface area, nontoxic, stable chemical properties; good combination of efficiency and lasting slow release performance.	
ZnO, TiO_2 materials— Slow release metal ion sterilization, photo-catalytic sterilization, and reactive oxygen species generation	Stable chemical properties under UV irradiation and antimicrobial properties, good pH stability, nontoxic, abundant raw material sources, low cost, etc.	Used in fiber, plastic, ceramic, coating, biomedical, and other fields

metal-oxide nanoparticles like ZnO and TiO_2 have been developed and mixed with paints. These materials are very advantageous because of their pigment nature, chemical resistance, and high thermal stability, and thus perfectly working as bactericidal additives. MoO_3 nanoplates demonstrated by Krishnamurthy et al. have shown interesting properties, such as photocatalysis, super capacitors, gas sensors, photochromic coatings, and as antibacterial additives in paints and subsequently MoO_3 gained considerable interest among the researchers to explore more *(37)*. Recently, Natalio et al. demonstrated the vanadium pentoxide (V_2O_5) nanorods merged in paints are capable of exhibiting worthy antibacterial and antifouling properties

(38). At present, these are the main contributions of researchers regarding antibacterial paints.

7.3 Silver

Although silver ion or metallic silver was used from long time, silver nanoparticles have gained much attention in the last three decades. Those nanoparticles can be utilized in medicine for burn treatment, water treatment, dental materials, coating stainless steel materials, sunscreen lotions, textile fabrics, etc. In particular, they possess very low toxicity to human cells, low volatility, and high thermal stability and therefore these can effectively work in numerous formulations of device-based applications. Silver nanoparticles have been integrated in variety of medical devices like vascular, urinary and peritoneal catheters, prosthetic heart valve sewing rings, vascular grafts, surgical sutures, and fracture fixation *(39)*. They attracted the researcher's interest because they are capable of having extreme antibiotic properties in commercial scale.

The antimicrobial properties of silver nanoparticle are correlated directly to the quantity of silver released. Although, silver is inert in its metallic state, upon reaction with the moisture in the skin and fluid of wound, it is ionized. The ionized silver is extremely reactive, as it can bind to tissue proteins and carries structural changes in the bacterial cell wall and nuclear membrane to cell distortion and ultimately death. Also, they can attach to bacterial DNA and RNA by modifying and preventing bacterial reproduction. Silver has been found to be very active against various kinds of pathogens known to occur at implant sites, including *Pseudomonas aeruginosa, Staphylococcus aureus, Escherichia coli*, and *Staphylococcus epidermidis.* Feng et al. proposed the antibacterial effect of the silver ion on *E. coli* and *S. aureus*, and recommended a mechanism that silver ion (Ag^+) reacts with the bacteria and disturbs the function of bacterial cell tissues and vital metabolic proteins and enzymes by sticking to DNA and thiol groups in proteins *(40)*. Earlier studies have shown that the silver ion is critical for antibacterial activity because of its positive charge, which allows the electrostatic attraction between the negative charge of the bacterial cell membrane and positively charged nanoparticles. Silver zeolite is fabricated by combining alkaline earth metal with a porous crystalline material of hydrated aluminosilicate that can bind up to 40% silver ions within its structure, which is partially replaced by silver ion using ion exchange method. In Japan, ceramics are prepared by coating with silver zeolite to apply antibacterial activity to their product *(41)*. These ceramics are used for food preservation, disinfection of medical products, and decontamination of materials. Silver nanoparticles are used instead of elemental silver or complex silver compounds to inhibit biofilm formation on surfaces both for biomedical and for more general use of applications.

7.4 Copper

As an alternative to Ag, Cu metals have been used for centuries as bactericidal agents. In comparison with silver metal, copper metal took wide attention of researchers against microorganisms because some of the comparative studies show that copper has similar mode of action to silver, and the main highlight of copper metal is a low-cost material than silver *(42)*. Copper and its alloys (copper–nickel–zinc, cupronickel, brass, bronze, etc.) have intrinsic properties to kill a wide range of bacteria, and these copper alloy surfaces are working as natural bactericidal materials. Cu salt has also been studied with respect to antiseptics and antifungal agents in the treatment of tinea pedis (athlete's foot) and also used effectively by coating them on different surfaces to show antimicrobial functions in medical devices and water treatment filters.Yoon et al. demonstrated even greater antibacterial activity with copper nanoparticle against *E. coli*, and micro-organism forming *Bacillus subtilis* compared with silver particles *(43)*. Similar to silver, copper metal actively reacts with proteins by combining with –SH groups of microbial enzymes. Similarly, copper oxide can also serve as antimicrobial agent *(44)*. However, extensive research is still needed for copper to prove better efficacy than silver.

7.5 Conclusion

This chapter describes a diverse set of basic principles and recent developments of smart window techniques. Compared to the current normal windows, smart windows can dynamically control the transmittance of solar irradiation for energy efficiency and human comfort and provide better healthcare. We mainly focus on the use of nanoparticles in those applications. For the last few decades, electrochromic, thermochromic, and photochromic materials have been investigating with primarily architectural applications in mind. On the other hand, silver and copper are known to contribute as antibacterial agents in various fields such as health care/medicine, food packaging, paints, transportation, and ceramics. We believe blending of these materials with existing techniques can help to improve their respective property in manyfold. We have given a special focus on antibacterial coating as human acquired infections became dangerous issue in hospitals and smart glasses can provide appropriate protection from bacteria, fungi, virus, etc. Indeed, focused research is required by the researchers to take care of human health. These developments collectively suggest that continuous invention of better material is essential to generate very active research attention from both the standpoints of impactful applications and fundamental studies.

8 A special focus on wettability studies

8.1 Fundamental interaction at solid–liquid interface

Understanding fundamental interactions at solid/liquid interface is of key importance today because of its potential application ranging from everyday life to advanced technological application. In modern industrial urban lifestyle, we are surrounded by too many hard-to-reach surfaces from building facade to window surfaces to bathroom tiles to industrial instruments to solar cell panels. In the current circumstances, we would like to keep these surfaces clean using minimum amount of water and energy—both of which are becoming less and less available with a combination of increasing human population and their energy intensive modern lifestyle. Thus, creating self-cleaning *(45–47)* and self-lubricant *(45–48)* surfaces are of utmost necessity at this juncture of human civilization for careful management of water and energy resources.

The fundamental quantification of solid/liquid interfacial interaction is carried away through measurement of contact angle (CA). For the case of liquid water, the interacting angle can be written as water contact angle (WCA). Wetting of any surface is essentially characterized by the contact angle *θ (theta)* of liquids over the solid surfaces (Fig. 10.8). Primary equation that quantifies contact angle of a liquid droplet on a flat surface was given by Thomas Young, the eminent 19th-century scholar and scientist. WCA *θ*

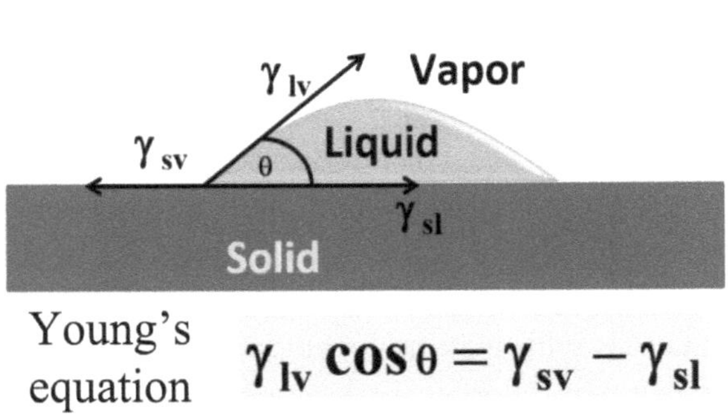

Figure 10.8 ***Formation of static contact angle under influence of three interfacial surface tension forces.***

is related to three interfacial surface tension forces (per unit length) by the Young's equation as described below *(53)*

$$Cos\ \theta_Y = \frac{\gamma_{SV} - \gamma_{SL}}{\gamma_{LV}} \tag{10.1}$$

Here, γ_{SV}, γ_{SL} and γ_{LV}—are the interfacial surface tension forces working at the solid–vapor, solid–liquid interface, liquid–vapor interfaces, respectively.

Young's equation essentially defines the equilibrium shape of the liquid droplet on a solid surface, which is governed by the action of three interfacial surface tension forces pulling at the three-phase contact line formed where the solid/liquid, liquid/gas, and solid/gas interfaces meet. For normal liquid droplets of where the role of gravity is negligible (which is the case for most of the experimental observations where the droplet size varies from millimeters to micrometers), the drop will have a shape of a spherical cap (Fig. 10.8).

Based on the values of WCA, surfaces are conventionally classified as either hydrophobic ($\theta_Y > 90°$) or hydrophilic ($\theta_Y < 90°$). There is further classification wherein surfaces with $\theta_Y > 150°$ are called super-hydrophobic and those with $\theta_Y < 10°$ are termed as super-hydrophilic.

Young's equation is a very good mathematical model for explaining the solid/liquid interaction only at the static situation. In dynamic scenarios when the size (or volume) of the liquid droplet is changing over time or when the liquid droplet is rolling over a nonhorizontal solid surface, we get two different WCA values known as advancing CA (θ_A) and receding contact angle (θ_R). Value of static contact angle θ_Y always remains within the value of advancing and receding CA. It can be said that advancing (receding) CA is the maximum (minimum) CA possible on any given surface *(10)*. Further understanding on the wetting behavior of solid surfaces reveals that advancing CA is a measure of cohesion force at the sold/liquid interfaces, whereas receding CA is a measure of adhesion force at similar interface *(45,46,54,55)*. Advancing and receding CAs can be measured on horizontal solid surface when the liquid droplet is increasing (θ_A) or decreasing (θ_R) in size (Fig. 10.9A). Another measurement scenario can emerge where the droplet size is kept constant but the measuring solid surface is "tilted"—in this scenario, the liquid droplet loses the "spherical cap" equilibrium position due to the effect of gravity (Fig. 10.9B) *(54)*. In this scenario, advancing CA (θ_A) is measured at the downhill of solid/liquid interface where receding CA (θ_R) is observed at the uphill position of the solid/liquid interface (Fig. 10.9B). The important concept of contact

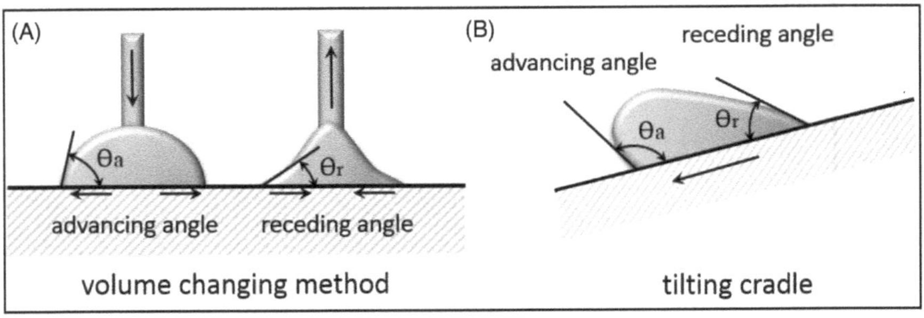

Figure 10.9 ***Concept of advancing and receding CA and their experimental realization under (A) Volume change method and (B) tilting cradle method.***

angle hysteresis (CAH) is theorized from the difference of advancing and receding CAs (CAH = $\theta_A - \theta_R$). Two different types of CAH are generally observed—"petal effect" with high CA, large CAH, and strong adhesion to water and "lotus leaf-effect" with high CA, small CAH hysteresis, and low adhesion to water *(55)*. A simple quantitative assessment of CAH can be done using pinning force (per unit length of drop perimeter) measurement at the solid–liquid interface *(1)*:

$$F_{pinning} = \gamma_{LG}\left(\cos\theta_A - \cos\theta_R\right) \tag{10.2}$$

CAH is the most important parameter for droplet motion on a tilted surface—this is the force that needs to be overcome by external forces like wind and/or gravity. If the CAH value is too high (along with a low static WCA value $\theta_Y \leq 90°$) and the external driving is not big enough, the liquid droplet will stick or is smeared out across the solid surface instead of smoothly rolling on. However, in case of sufficiently high static WCA ($\theta_Y \geq 170°$), viscous droplets indeed roll off the surfaces—they do not slide *(45,46)*.

8.2 Effect of surface chemistry and roughness on contact angle (CA)

It is widely known that static WCA (θ_Y) on a solid surface is essentially controlled by two parameters:

- *Surface chemistry*—Chemical nature of solid/liquid interface decides the surface energy of the solid surface. If the interfacial interaction energy between the surface and water is high, then water droplet would like to spread onto it and the surface would be characterized as a "hydrophilic surface." In the extreme end of high energy interaction where the sum of the surface tensions of the liquid/solid and liquid/gas interfaces

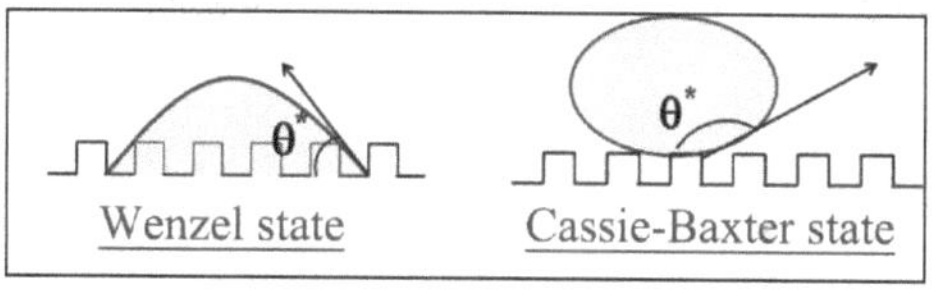

Figure 10.10 ***Schematic of Wenzel and Cassie-Baxter type of surface roughness.***

equals the surface tension of the solid/gas interface, Cosθ becomes 1, or $\theta = 0$. This condition can be termed as *complete wetting*—the certain type of super-hydrophilic behavior where drop will flatten out to form a film. On the other hand, if the interaction energy at the surface (of liquid and solid) is low, then water droplet would like to avoid/roll-off from that surface—this kind of surfaces would be hydrophobic surfaces. Usually, surface polarity acts as an important parameter that decides the affinity of a solid surface for water—generally higher (lower) surface polarity leads to increased hydrophilicity (hydrophobicity).

- *Surface roughness*—Wettability is characterized not only by the chemical composition but also by the roughness of the surface. Young's equation works very well for the case of an ideal solid surface—defined as a smooth, rigid, chemically homogeneous, insoluble, and a nonreactive surface. Hence, wetting on rough surfaces cannot be described accurately by Young's equation. For most of the cases where liquid droplet can maintain interaction with all points of rough solid surface, we can formulate wetting model that correlates surface roughness with the equilibrium contact angle based on Wenzel model (Fig. 10.10) *(12)*. The Wenzel contact angle (θ_{WZ}) of the rough surface can be related to true contact angle (or Young's contact angle—θ_Y) by the relation *(56)*:

$$\cos\theta_{WZ} = r\ \cos\theta_Y \tag{10.3}$$

Here, r is a quantification of surface roughness defined as roughness factor of the surface—it can be defined as the ratio of actual surface area over projected surface area of any solid surface. If the surface roughness increases, so will be the value of r. For rough surfaces, r is always > 1.00, which easily shows θ_{WZ} always $> \theta_Y$ for $\theta > 90°$ and $\theta_{WZ} < \theta_Y$ for $\theta < 90°$. Wenzel equation thus states that (as long as the liquid droplet keeps contact with the solid surface at all points) all chemically hydrophobic (hydrophilic) surfaces become progressively more and more hydrophobic (hydrophilic) as the surface roughness increases.

A different situation emerges when high surface roughness causes entrapment of air bubbles in the microstructures of the surface and liquid sits on top of both rough solid surface and air bubble (Fig. 10.10). Water molecules could not displace the air bubbles trapped between the liquid droplet and solid surface due to the complicated hierarchical nature of surface roughness. This particular situation is explained by the works of Cassie-Baxter *(57)* (CB) and the apparent CB WCA (θ_{CB}) is related to true contact angle (or Young's CA – θ_Y) by the relation:

$$\cos\theta_{CB} = -1 + \varphi_S(1 + \cos\theta_Y) \tag{10.4}$$

Here, φ_S is the percentage of area under direct contact with liquid droplet—a fractional number ranging between 0 and 1. For perfectly smooth surface, $\varphi_S = 1$ and $\theta_{CB} = \theta_Y$. For highly rough surface, interfacial corrugation becomes extremely large, which leads to $\varphi_S \rightarrow 0$ and $\theta_{CB} = 180°$—complete "nonwetting" happens. CB equation clearly demonstrates that with increasing amount of surface roughness, the value of φ_S will always decrease and the value of $\theta_{CB} \rightarrow 180°$ irrespective of the value of θ_Y. In case of Cassie-Baxter type surface roughness, it is clear that any surface can before hydrophobic or even super-hydrophobic by introduction of more and more surface roughness irrespective of the chemical nature of wettability of the smooth surface (i.e., hydrophilic or hydrophobic) *(57)*.

This effect, however, gets highly enhanced if the starting surface is already hydrophobic i.e., a low energy surface. It becomes energetically very costly for the liquid droplet to follow the surface undulations of a very rough surface and the Gibbs free energy minima essentially attained at a position above all the corrugations. Literally, the drop hangs like a "Fakir would be on his carpet"—a famous phenomenon seen on various biological systems capable of nearly "complete nonwetting"—most widely seen on lotus leaf *(45)* (Fig. 10.11). This super-hydrophobicity of lotus leaf can be attributed to a series of regular array of micro- and nanostructural pillars covered by hydrophobic wax crystals on their surface (Fig. 10.11).

8.3 Wettability studies on metal-oxide surfaces

Understanding the basic phenomena of wetting at the molecular level in metal oxides is a very important direction in solid/liquid interface research, and numerous studies have been undertaken to observe and study reversible wettability or enhancement of hydrophilicity/hydrophobicity at water/oxide (e.g., ZnO and TiO_2) interfaces via ultra violet (UV) and visible light illumination *(58,59)*, electro-wetting *(60)*, structural modifications *(61)*, thermal treatment *(62)*, and laser irradiation *(63)*. In-depth studies have been particularly done to understand the fundamental solid/liquid interaction

Figure 10.11 ***Super-hydrophobicity on biological surfaces—schematics of water droplet floating on lotus leaf.*** *Reprinted from YouTube video screen shot of 'Superhydrophobic surfaces/lotus effect by W.Th.' (https://www.youtube.com/watch?v=HF4blivJQ6o). Creative Commons Attribution license (reuse allowed)*

mechanism, which leads to an enhancement in hydrophilicity/hydrophobicity in transition metal oxides (TMO) for e.g., ZnO and TiO_2 and feasible explanations can come in the form of removal of organic contaminants *(64)*, a drastic change in crystal structure *(65)*, significant enhancement/deterioration of surface smoothness *(66)*, oxygen vacancies, and Ti^{3+} or Zn^{+} defect sites *(67)*. Reversible changes in water contact angle (WCA) of transition metal-oxide (TMO) surfaces have generated significant interest in recent years *(63,68–71)* where these changes in WCA can be achieved by the alteration of electronic structure in TMO surface layers due to formation of cationic defects. These types of alterations have been experimentally shown by emergence of photo-induced hydrophilicity (induced by UV irradiation) in various TMO surfaces like TiO_2, SnO_2, ZnO, WO_3, and V_2O_5 *(28–32)*.

Further studies were undertaken to explore fundamental interaction at the solid/liquid interfaces in molecular level where dissociation of water molecules into H and OH (at vacancies sites of TiO_2) has been imaged through noncontact AFM and STM *(77–80)*. However, a highly cited recent study also shows the importance of surface electronic structure where they have reported absence of any photo-induced hydrophilicity on $SrTiO_3$ surfaces unlike TiO_2—this finding was truly astonishing as the surface electronic structure of $SrTiO_3$ should completely resemble that of TiO_2, which conclude that wetting properties of TMO surfaces depend not only on stoichiometry or surface roughness *(81)*.

Along with the study (of WCA) of transition metal-oxide (TMO) surfaces, analyzing the wettability behavior of rare-earth oxide (REO) surfaces *(82–84)* also got significant attention recently. An important study involving a series of REO surfaces ranging from ceria to lutetia reported (for the first time) the hydrophobic nature of the entire REO ceramics surfaces *(83)*. Authors attributed this hydrophobic behavior to unfilled 4f orbitals of (mostly) trivalent 4F cations of REO systems with a general formula of Re_2O_3. However, most experiments were done on bulk polycrystalline pellet samples with very high surface roughness (~10 μm) and no mention of surface crystallinity and/or elemental composition at surface. Another subsequent study led by Zenkins et al. also showed very high WCA values of a series of metal oxides and metal nitrides (of low electronegativity) including REOs and rare-earth nitrides (RENs) grown on Si (1 1 1) substrate. Authors in this report have also attributed these high WCA values to their low electronegativity *(41)*. However, all these studies did not discuss anything about hydrocarbon contamination at the surface as well as smoothness of surface *(83,85,86)*. Preston et al. showed for the first time the immense importance of hydro-carbon contamination of the metal oxide by making REO surfaces super-hydrophobic by treating the surface with Argon plasma etching *(42)*. Azimi et al. subsequently demonstrated the role of surface roughness by fabricating a super-hydrophobic surface (WCA = 160°) from hydrophobic ceria pellet (WCA = 102°) using laser texturing *(40)*, whereas Lundy et al. experimentally found an unique transition from hydrophilic to hydrophobic nature of the CeO_2 pellet surfaces with increasing atmospheric exposure *(87)*.

Although highly illuminating, all these above studies were done on metal oxide bulk (primarily pellets with surface roughness in the range of >100 nm). A major problem in this direction of research is their experimental reproducibility—many research groups were/are not being able to reproduce other group's data or their own data at later times. The major challenge that the research community understood was the high surface roughness of the bulk/pellet surfaces that can subsequently generate various WCA values and lead to completely erroneous understanding of the underlying interfacial mechanism. To address this fundamental technical issue, few research groups have focused on bringing the high quality epitaxial thin-film growth techniques [like atomic layer deposition (ALD), chemical vapour deposition (CVD), pulsed laser deposition (PLD)] to

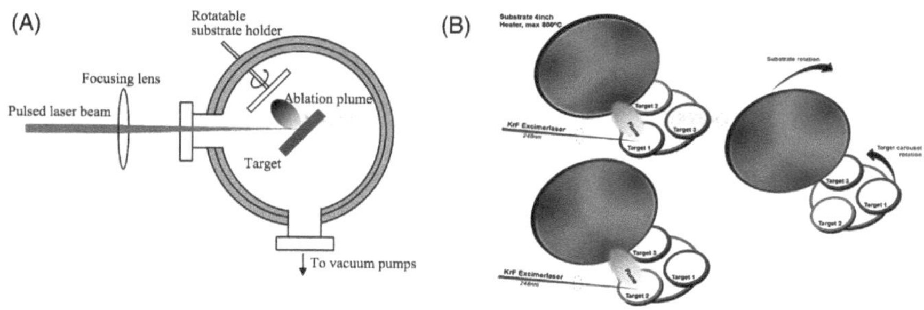

Figure 10.12 (A) Basic schematic of a PLD set-up; (B) Illustration of basic arrangement of substrate and targets during continuous composition spread deposition. After LASER hits the first target and deposits desired amount of material A, target carousel rotates and places target B at the deposition position. At the same time, substrate is rotated accordingly. As seen clearly in the image, desired rotation angle is 360/*n* (where *n* is the total number of metal-oxides or elements used for solid solution). Deposition takes place in the same way with the remaining targets. This procedure is repeated by number of times for desired thickness of the film. *Reprinted with permission from* (7). *Copyright (2017) American Chemical Society.*

grow ultra-smooth, phase pure, single crystalline thin films of metal oxides (Fig. 10.12).

All these state-of-the art thin-film growth techniques are ubiquitously used in semiconductor industry for their highest standards in maintaining perfect elemental composition in thin films with very low surface roughness (<10 nm). Among all these techniques, PLD proved to be more advantageous for its "clean" bottom-up growth process where high-power laser pulses are used (typically ~1–2 J/cm^2) to melt, evaporate, and ionize material from the surface of a target. This "ablation" event produces a transient, highly luminous plasma plume that expands rapidly away from the target surface. The short pulse duration reduces thermal heating of the target while minimizing unintentional thermal evaporation from target, hereby maintaining the stoichiometry of the target material in the plume. The ablated material is collected on an appropriately placed substrate upon which it condenses and the thin film grows. As discussed above, PLD does not use any chemical precursor (unlike CVD or ALD), thereby eliminating any chance of unwanted hydrocarbon contamination of the metal-oxide surfaces. Hydrocarbon contamination can lead to serious misrepresentation of the true nature of the wettability of any metal-oxide surface. Another advantage of using PLD technique is its capability of making continuous composition spectra (CCS) [using

PLD] i.e., an in-plane solid solution of different oxides on a 4-inch Si wafer and studying the WCA of different parts of the large sample. In this way, a high-throughput screening of different material libraries can be achieved for the search of earth abundant super-hydrophobic and/or super-hydrophilic materials. Very few systematic studies were done on wettability study of metal oxides, few recent studies came on WCA measurement on CeO_2 thin films grown by electrochemical methods *(88,89)* where surfaces roughness values were either found to be very high (in μm range) *(44)* or not reported at all without any systematic surface characterization *(89)*. Only few articles reported systematic analysis of solid/liquid interaction on REO thin-film surfaces grown by Atomic Layer Deposition (ALD) *(90)*, room-temperature sputtering *(91)*, and pulsed laser deposition (PLD) *(52)*. The first study investigated WCA of REO thin films of thickness greater than 50 nm (grown by ALD) of different thickness on Si (100), which showed hydrophobic behavior and a dependence of WCA on film thickness indicating an effect of substrate) *(90)*. Although systematic, ALD process involves organic precursors, which makes thin-film surfaces (grown by ALD process) extremely prone to very high degree of hydrocarbon contamination compared to thin-film surfaces grown by PLD or sputtering. Authors have carried out XPS analysis to address the issue of surface hydro-carbon contamination; however, the XPS analysis failed to comment on the hydro-carbon contamination at the pristine thin-film surface as the XPS data were collected after sputtering the surface with Argon for 20s *(90)*. In the next article *(91)*, authors have reported systematic and thorough analysis of REO surfaces involving five different REO thin-film samples (CeO_2& R_2O_3 where R = Tb, Gd, Ho & Er) grown by sputtering technique at room temperature *(91)*. Authors in this study discussed the role of hydrocarbon contamination in great detail by exposing the REO surfaces into atmospheric hydrocarbon as well as various specific hydrocarbons like polycyclic aromatic hydrocarbons (PAHs) *(91)*. Authors have also shown temporal evolution of WCA values of various REO surfaces like CeO_2, Ho_2O_3, and Gd_2O_3. Despite their careful and comprehensive analysis, long-term solubilization experiments revealed that most of these films get dissolved into deionized water at RT (within 24 h), which poses significant concern about the quality of the thin films prepared. Latest report regarding systematic study of wettability on REO thin films came from Prof. T. Venkatesan group (National University of Singapore) where Prakash et al. prepared entire series of PLD grown REO thin-film surfaces (from CeO_2 to Lu_2O_3) to study the intrinsic WCA of these samples *(52)* (Fig. 10.13).

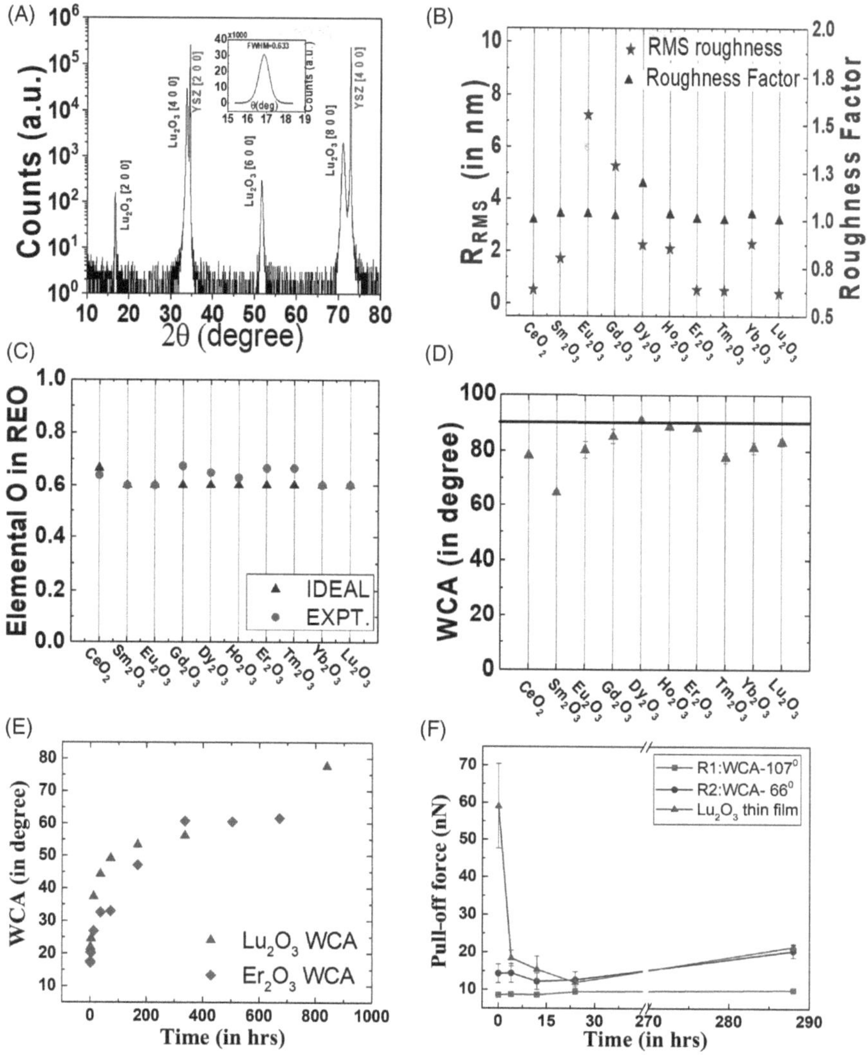

Figure 10.13 (A) Showing the θ–2θ scan (along with rocking curve of the Lu_2O_3 [4 0 0] peak showing a FWHM of 0.633 in the inset) of Lu_2O_3 (100) grown on YSZ (1 0 0). (B) AFM studies showing roughness factor [*Y*-axis RHS] (as well as RMS roughness [*Y*-axis LHS]. (C) Composition analysis of all the Re_2O_3 thin films grown by PLD using RBS. The *y*-axis is showing elemental O in REO system, that is, O:(RE + O). (D) WCA measurements done on Re_2O_3 thin films grown by PLD after a lapse of >120 h (5 days) showing technically hydrophilic behavior in all cases. (E) Temporal variation of the WCA of thin films of Lu_2O_3 and Er_2O_3. (F) f-d curve measurement done on Lu_2O_3 thin-film samples and reference samples. *Reproduced from [8] with permission from The Royal Society of Chemistry.*

In their systematic study, these thin-film samples were prepared with high surface reproducibility by ensuring that (1) all REO thin films are single-crystalline and phase-pure, (2) all REO thin-film surfaces are free from any hydro-carbon contamination, (3) all the thin films are extremely robust under atmospheric exposure, and (4) the surfaces of these thin films are ultra-smooth with roughness factor close to 1.00 (R_{RMS}< 2nm mostly), implying that they were atomically flat. They have found that the REO thin films are intrinsically hydrophilic if not super-hydrophilic (WCA < 10°) just after removal from the growth chamber! *(82)* Both the traditional sessile drop method and a novel f–d curve approach have been used to measure WCA. Although freshly prepared samples showed this highly surprising super-hydrophilic behavior with WCA values as low as 16° for Er_2O_3 and 21° for Lu_2O_3 *(82)*. However, all the WCA values do evolve with time due to atmospheric exposure and finally reach their eventual saturation values near 90°. XPS analysis and Soft Ion Flow Tube Mass Spectrometry (SIFT-MS) was used to further investigate the nature of hydrocarbon contamination on the freshly prepared as well as environmentally exposed REO thin film samples as a function of the exposure time (counted after removal from the deposition chamber). A clear correlation between the carbon coverage of the surface and increase in WCA was observed for all the rare earth films indicating the extrinsic nature of the surface wetting properties of these films and having very little to do with the electronic configuration of the rare earth atoms. As clearly seen in Fig. 10.13F, a novel technique (called *f–d curve approach*) is used here apart from traditional Sessile drop method to measure and study the static as well dynamic WCA values, which require some explanation. *f–d curve method* is a novel force spectroscopy technique, which demonstrates that the maximum pull-off force (adhesion) is an exponential function of macroscale CA and surface energy. The technique fundamentally relies on the widely accepted fact that very thin layer of water (one or at the maximum few atomic unit cells) is condensed between the AFM tip and the surface of any material in ambient conditions *(92)*. Please see Fig. 10.14 schematic for better understanding on various stages of AFM tip interaction with metal-oxide surface. Authors have further shown that this *pull-off* force correlates well with the polar component of the surface energy while finding no such relation with the dispersive component, which shows that measured force arises from electrostatic and dipole–dipole interaction (hydrogen bonding) on the surface and London forces do not play a significant role *(48)*.

Another exciting new finding from the same research group reports the effect of oxygen vacancy on wettability of 3d transition metal oxide ($SrTiO_3$) and 4f rare-earth oxide (Lu_2O_3) thin-film surfaces. These metal oxides were grown using pulsed laser deposition and then subsequently subjected to annealing to induce Oxygen vacancy in them. All the metal-oxide thin-film sample surfaces were extensively characterized before and

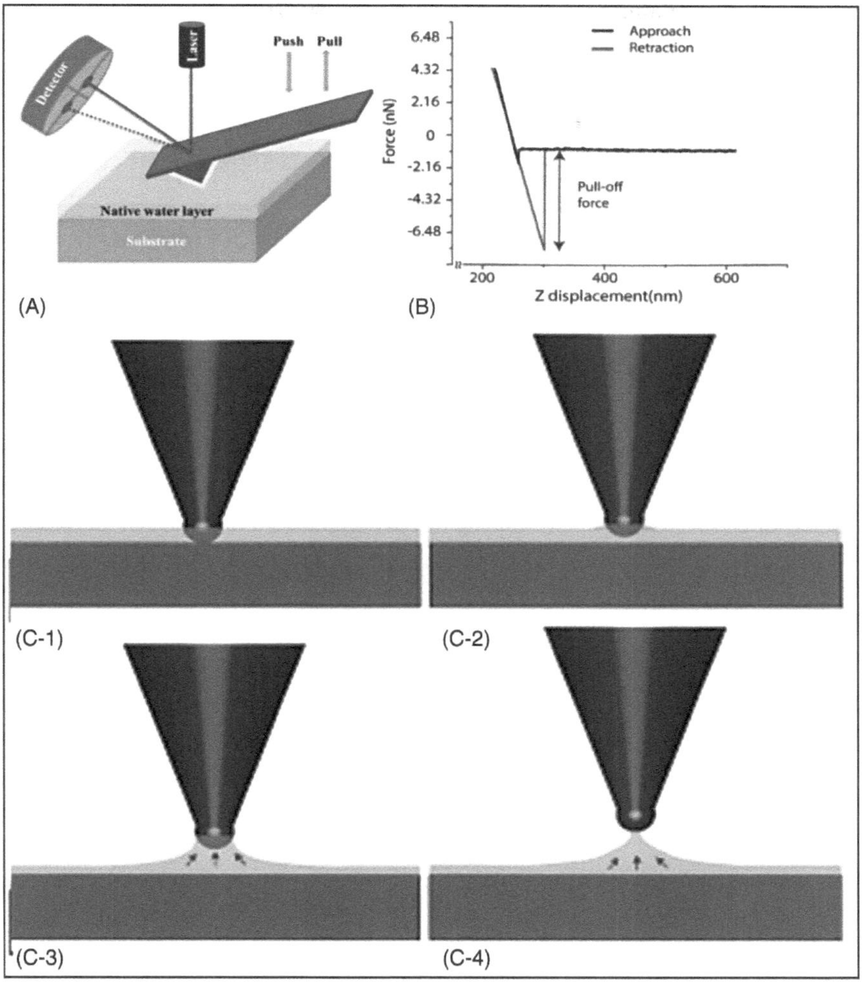

Figure 10.14 Force spectroscopy (FS) measurements, (A) schematic of measurement (B) a sample force–displacement curve obtained on lubricated hard disk media. Illustration depicting the various phases of tip interaction with the surface in the case of retraction (going away) from the sample. (C1) Tip leaves the solid surface, (C2) tip travels through the film of adsorbed water, (C3) tip extends the neck of the meniscus formed by the water, (C4) ultimately, the tip breaks free from the film of water. *Reproduced from (48) with permission from The Royal Society of Chemistry.*

after annealing (using XRD & AFM) to rule out any trivial unwanted effect originating from alteration of surface crystalline structure and/or surface roughness due to annealing. For 3d $SrTiO_3$ (STO) with oxygen vacancies, authors reported an increase in hydroxylation with consequent increase of wettability, which is in line with the previous reports, whereas an interesting opposite trend was seen in the case of rare earth oxide Lu_2O_3 thin film. Photoluminescence (PL) measurement was used to quantify oxygen vacancy in case of ternary metal oxide $SrTiO_3$, whereas RBS & XPS techniques precisely measured amount of oxygen vacancy in Lu_2O_3 thin-film samples. Finally, DFT calculations were used for theoretical support that concluded that difference between the two materials can be linked to the different nature of the oxygen vacancy in both of them ($SrTiO_3$ & Lu_2O_3). In case of $SrTiO_3$, formation of O-vacancy reduces the valence of Ti ions from +4 to +3 with a concentration of two Ti^{3+} per Oxygen vacancy site. When water is adsorbed, the two extra electrons remain in the Ti 3d states and the concentration of Ti^{3+} is maintained. This phenomenon essentially led to splitting up of water molecule into H^+ and OH^- bonds and these OH^- bonds then eventually get adsorbed at the O-vacancy site in $SrTiO_3$. Whenever OH^- bonds are formed on any metal oxide (or any other) surface, it leads to further hydroxylation and subsequent enhanced wettability (or decreased WCA values). However, for Lu_2O_3, the Lu f states are not accessible for reduction and the two electrons left in the material and when a neutral oxygen vacancy is formed, they are localized at the vacancy site. As a consequence, when water is introduced, whole water molecule stays at a metal site (near the vacancy where it can benefit from electrostatic interaction with the trapped electrons), which does not create hydroxyls groups and thus does not enhance surface wetting in agreement with the contact angle measurements (Fig. 10.15).

8.4 Potential novel applications

A novel field of wettability research in metal-oxide thin film is their potential application in the research field of microfluidics. Microfluidics is the field of research that deals with manipulation of extremely small amount of liquid (10^{-9} to 10^{-18} liters) using channels with dimensions of about 10–100 μm. Microfluidic technologies are highly valued for their potential applications in chemical and biological analysis of very small amount of samples due to their high resolution of detection and sensitivity, lower cost, short analysis time, and small instrument foot-print *(93,94)*. Growing patterned surfaces (using PLD) with hydrophilic and hydrophobic channels and arrays can create the possibility of *channel-free surface-tension confined microfluidics* or *liquid microchip*. Playing with these patterned surfaces allows novel dynamic features to be brought in, such as morphological transitions from parallel liquid-filled stripes to configurations in which liquid bulges emerge that reach out from one channel across a hydrophilic region, and merge with another

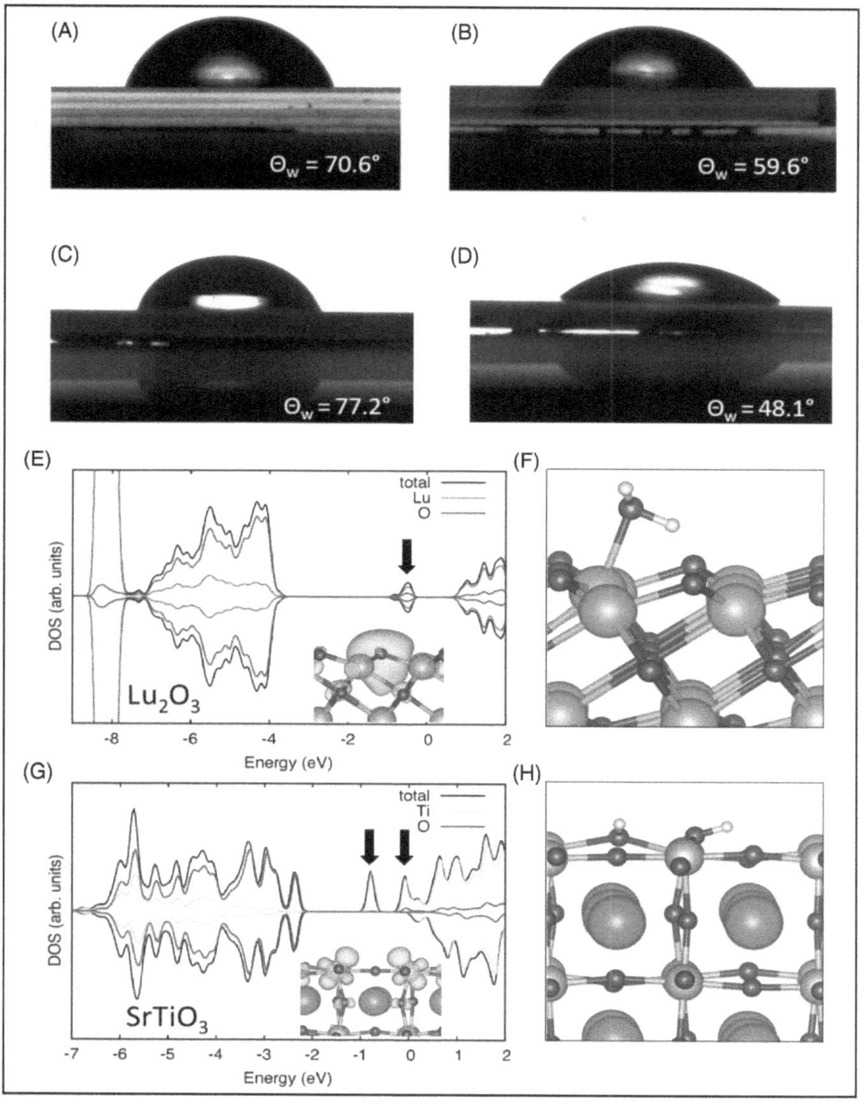

Figure 10.15 Water contact angle obtained on Lu_2O_3 (A) pristine, (B) annealed and $SrTiO_3$, (C) pristine, (D) annealed. Density of States (DOS) for a surface oxygen vacancy at the Lu_2O_3 (E) and $SrTiO_3$ (G). Most stable configurations for a water molecule at the surface of defective Lu_2O_3 (F) and $SrTiO_3$ (H). O atoms are represented in *red*, Lu in *light green*, Sr in *green*, Ti in *blue*, and H in *white*. *Reproduced from* (50) *with permission from The Royal Society of Chemistry.*

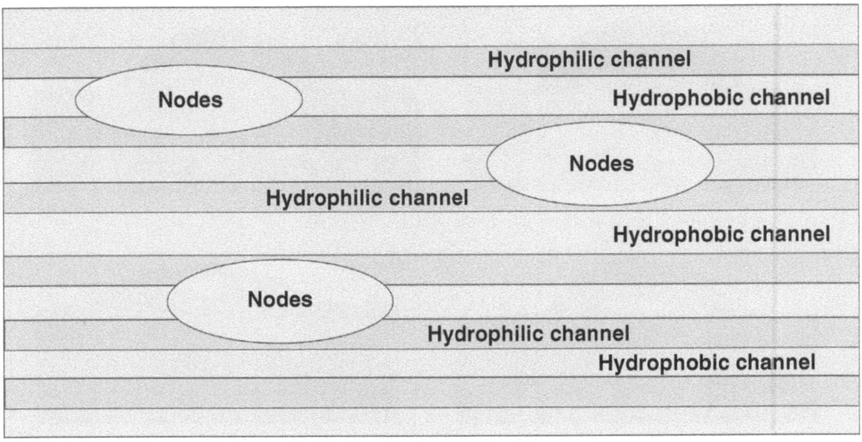

Figure 10.16 Liquid channels on a patterned substrate with alternating array of hydrophilic and hydrophobic channel: a possible pathway to surface tension-controlled microfluidics. Different hydrophilic channels are connected through "Nodes."

channel (Fig. 10.16) *(95,96)*. The flow of liquid on the channel patterns is driven by the curvature of expanding liquid *(97)*. The advancement of the liquid front along a hydrophilic channel is (in the case of a Newtonian fluid) well described by the standard lubrication equation derived from the Stokes equation, and shows a diffusion-like behavior *(98)*. Driving the liquid along the channels and making them merge at predefined locations offers a novel way to mix reactants or steer biochemical reactions, defining the concept of a "liquid microchip" *(95)* or "surface tension-confined microfluidics" *(96)*. One main advantage of the open structures over capillaries is their ease of cleaning coupled with another significant edge is that blocking of the capillary by unforeseen chemical reactions cannot occur in this particular case.

Finally, the ultimate frontier of novel application lies in the utilization of wettability research in understanding nano–bio interaction at fundamental level, i.e., how biological entities interact with single-crystalline, phase-pure, ultra-smooth flat surfaces, which is particularly important as numerous biological processes are regulated by surface topography and surface chemistry, such as protein adsorption, cell attachment, proliferation, and differentiation *(99,100)*. It has been shown decisively that R_{RMS} ranges between 20 and 50 nm are required for neural cell adhesion. Apart from surface roughness, grain size of the thin-film surfaces also has an important role to play—smaller grain size (~10–200 μm) has been extremely helpful for neural cell growth, whereas larger grain size (>250 μm) has an derogatory effect on cell growth and subsequent adhesion *(100)*. Moreover, it has been observed that cells tend to attach on edges of the edges of grains in polycrystalline thin-film surfaces *(68)*. Specialized neural cells like axons also show peculiar behavior of growing along the grooves on patterned solid surfaces rather than perpendicular to grooves. Previous studies have also reported the role of surface chemistry where the presence of graphene sheets has a positive influence on cell growth—to rule out the role of surface roughness, atomically flat surfaces (R_{rms} < 10 nm) were used to grow graphene on them *(101)*. Along with metal-oxide thin films, metal-oxide nanoparticles (N.P.s) can also act as novel biomaterials, like ZnO N.P. and microparticles, which can act as bio inhibitors with a negative effect on *E. coli* bacteria formation *(102)*. Bio inhibition effect also varies with size of ZnO N.P. where smaller N.P. size has more and more negative effect on biofilm formation of E. Coli. In terms of novel biomaterials in thin-film form, Bioglass® is one of the earliest—developed in 1985 by Hench; this incredible artificial bio-material has widespread use in implantology (as both soft and hard tissues can adhere to Bioglass surface seamlessly) *(59,103)*. More metal oxides are lately used in the bio-application such as $BaTiO_3$*(60)* and Niobium-doped titanates like TiO_2 or $SrTiO_3$ (with enhanced antioxidant properties) *(105)*. One very recent study has discussed in great detail the biofilm formation on various individual binary (CuO, ZrO_2, TiO_2 etc.) and ternary (i.e., $SrTiO_3$, $DyScO_3$, $NdGaO_3$, ZnO, Al_2O_3, LSAT, TiO_2, YSZ, etc.) metal-oxide surfaces grown by PLD technique. This was a systematic and thorough study where authors precisely quantified formation of biofilm of common human pathogen, *Salmonella enterica serovar Typhimurium (51)*. Along with study of the individual metal-oxide thin films, authors also turned their attention into growing binary solid solution of CuO and ZnO and ternary solid solution of ZnO, ZrO_2, and TiO_2 on a 4-inch Si wafer using the novel CCS technique (discussed and elaborated above in Fig. 10.12). Authors have thoroughly characterized the surface morphology using atomic force microscopy (AFM) and surface elemental composition using Rutherford back-scattering (RBS) techniques. After thorough surface characterization, they have exposed this entire PLD-CCS sample to the biological species for biofilm formation. Distinctly different amount of biofilms were formed in various parts of the 4-inch Si-wafer sample, which was also carefully quantified and reported in their article *(51)*. These types of studies open up a new direction of research where state-of-the art thin-film growth technique is brought into studying biofilm formation on metal oxide surfaces, which are widely used in healthcare industry. PLD technique can easily grow high-quality metal-oxide surfaces, which can be used to affordably create "microbiologically inert" surfaces urgently needed to combat rise of antibiotic resistance and nosocomial infections. Along with traditional PLD technique, novel PLD-CCS techniques can be further used to establish thin-film metal-oxide libraries for high-throughput screening of biofilm inhibitors *(105)*. Our ability to successfully identify affordable biofilm inhibitor surfaces can lead to successful elimination of many bacterial infections—formation of surface-attached consortium state (from earlier free-swimming organism state) is extremely crucial for pathogenic micro-organisms for successful establishment and maintenance of the disease state (Fig. 10.17).

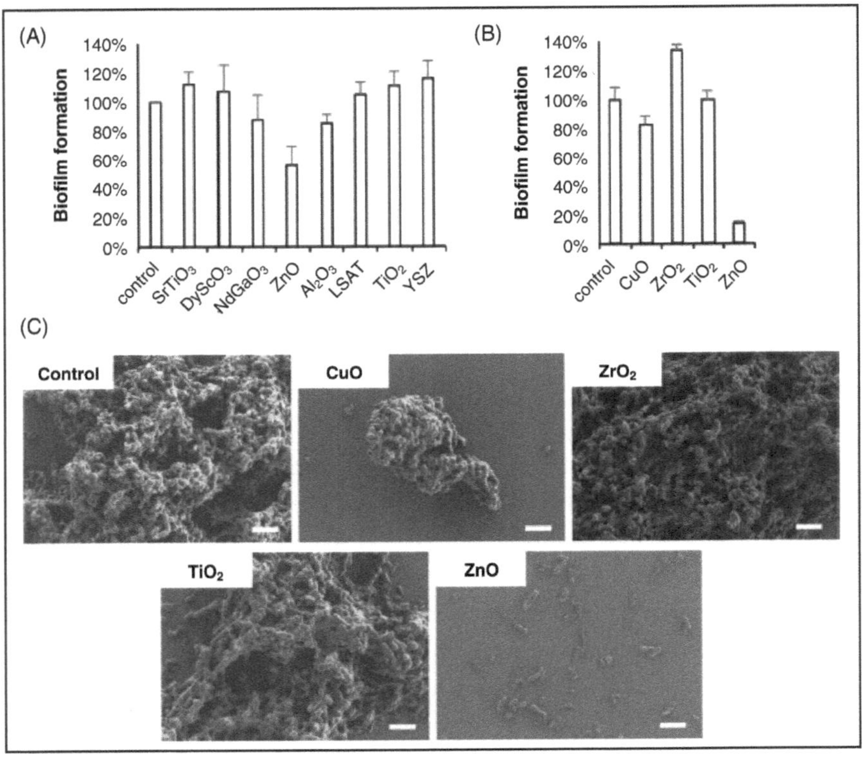

Figure 10.17 Biofilms formed when Salmonella was grown in the presence of (A) single metal oxide crystals of $SrTiO_3$, $DyScO_3$, $NdGaO_3$, etc. and (B) thin-film samples of binary metal oxides like CuO, ZrO_2, and TiO_2. All measurements are compared to a common glass slide used as a control baseline sample. (C) SEM images of control biofilms and biofilms grown on various oxide thin films (CuO, ZrO_2, TiO_2, and ZnO)—different morphological structures on individual metal oxide indicate distinct mechano-biological interaction at each interface. *Reprinted with permission from (7). Copyright (2017) American Chemical Society.*

8.5 Conclusion

In this chapter, we have discussed fundamentals of wettability starting from Young's equation, static and dynamic WCA like advancing and receding contact angle (ACA & RCA). ACA and RCA are two limiting stages of dynamic water contact angle, which are measured by either tilting the stages or by keeping the stage horizontal and changing the amount of liquid (with time) during contact angle measurement. Difference of ACA & RCA is described as contact angle hysteresis (CAH = ACA − RCA), which can be used for quantitative assessment of pinning force of any liquid droplet on a tilted solid surface. We have further discussed the two fundamental parameters that control wettability behavior, namely "surface chemistry" and "surface roughness." Role of surface chemistry has been straightforward where low energy surfaces bound to be hydrophobic for their lack of interaction with water molecule, whereas high energy surfaces will have better interaction with water molecule, which will make them subsequently more hydrophilic. Unlike the role of surface chemistry, the effect of surface roughness on wettability is more complex. There are two types of surface modifications—Wenzel type, where liquid droplet stays in full contact of solid surface and Cassie-Baxter type, where liquid droplet hangs on top of surface corrugations like a "fakirs' carpet." For Wenzel-type surface modulation, increasing roughness only increases hydrophilic/hydrophobic behavior (but cannot switch from hydrophilic to hydrophobic or vice versa), whereas in case of Cassie-Baxter type surface undulation, any surface can become progressively hydrophobic/super-hydrophobic irrespective of its chemical nature. Next part of this article focused on specific research direction of wettability studies on metal-oxide surfaces for potential self-cleaning and self-lubricating applications in microfluidics, healthcare industry, solar cell, and commercial household applications. In this part, we have initially discussed various recent studies undertaken on wettability studies of binary and ternary transition metal oxides (TMOs) like TiO_2, ZnO, ZrO_2, V_2O_5 and $SrTiO_3$, $NdGaO_3$, and $DyScO_3$. Subsequently, we went to discuss the controversial nature of rare-earth oxide surfaces (ranging from CeO_2 to Lu_2O_3), which is primarily reported to have hydrophobic/super-hydrophobic nature although various other research groups have reported them to be hydrophilic also. In this context, we have brought into the immense effect surface roughness has on the study of wettability and then justified the use of various *state-of-the-art* thin-film growth techniques (i.e., atomic layer deposition (ALD), pulsed laser deposition (PLD), RF & DC sputtering techniques, etc.) in this research. Use of these thin-film techniques (particularly PLD technique for its clean bottom-up growth process involving light matter interaction) enabled scientists to grow high-quality phase-pure, single crystalline, atomically flat metal-oxide surface free from nearly all surface imperfections, and hydro-carbon contaminations. It is extremely important to get rid of these topographic and chemical artefacts, which can introduce unwanted interfacial interactions leading toward unreliable experimental findings during contact angle measurements. Last part of this article focused on wettability study on REO & TMOs grown by primarily PLD technique. This chapter end with a brief discussion on how this wettability research can be expanded to understand and create affordable and novel bio-materials urgently needed for various niche applications in the vast and impactful health-care industry.

References

1. Wang, Y.; Runnerstrom, E. L.; Milliron, D. J. Switchable Materials for Smart Windows. *Annu. Rev. Chem. Biomol. Eng.* **2016**, *7*, 283–304.
2. Liu, T., et al. Smart Window Coating Based on F-TiO_2-$KxWO_3$ Nanocomposites with Heat Shielding, Ultraviolet Isolating, Hydrophilic and Photocatalytic Performance. *Sci. Rep.* **2016**, *6*, 27373.
3. Wei, Q.; Haag, R. Universal polymer coatings and their representative biomedical applications. *Mater. Horiz.* **2015**, *2*, 567–577.
4. Vasilev, K.; Cavallaro, A.; Zilm, P. Special Issue: Antibacterial Materials and Coatings. *Mol. Basel Switz* **2018**, *23*.
5. Tao Gao; Bjorn Petter Jelle. Nanoelectrochromics for Smart Windows: Materials and Methodologies, Materials for Energy, Efficiency and Sustainability. *Techconnect Briefs* **2016**, *2*, 279–282.
6. Kim, J.; Ong, G. K.; Wang, Y., et al. Nanocomposite Architecture for Rapid, Spectrally-Selective Electrochromic Modulation of Solar Transmittance. *Nano Lett.* **2015**, *15*, 5574–5579.
7. Baetens, R.; Jelle, B. P.; Gustavsen, A. Properties, requirements and possibilities of smart windows for dynamic daylight and solar energy control in buildings: A state-of-the-art review. *Sol. Energy Mater. Sol. Cells* **2010**, *94*, 87–105.
8. Xu, J., et al. Electrochromic-Tuned Plasmonics for Photothermal Sterile Window. *ACS Nano* **2018**, *12*, 6895–6903.
9. Wang, C.; Shim, M.; Guyot-sionnest, P. Electrochromic Nanocrystal Quantum Dots. *Sci. Rep.* **2001**, *291*, 2390–2392.
10. Garcia, G.; Buonsanti, R.; Liordes, A.; Runnerstorm, E. L.; Bergenud, A.; Milliron, D. J. Near-Infrared Spectrally Selective Plasmonic Electrochromic Thin Films. *Adv. Opt. Mater.* **2013**, *1*, 215–220.
11. Choy, J. -H.; Kim, Y. -I.; Kim, B. -W.; Park, N. -G.; Campet, G.; Grenier, J. -C. New Solution Route to Electrochromic Poly(acrylic acid)/WO3 Hybrid Film. *Chem. Mater.* **2000**, *12*, 2950–2956.
12. Llordés, A.; Garcia, G.; Gazquez, J.; Milliron, D. J. Tunable Near-Infrared and Visible-Light Transmittance in Nanocrystal-in-Glass Composites. *Nature* **2013**, *500*, 323–326.
13. Kostecki, R.; Richardson, T.; McLarnon, F. Photochemical and photoelectrochemical behavior of a novel TiO2/Ni(OH)2 electrode. *J. Electrochem. Soc.* **1998**, *145*, 2380–2385.

14. Schrauben, J. N.; Hayoun, R.; Valdez, C. N.; Braten, M.; Fridley, L.; Mayer, J. M. Titanium and Zinc Oxide Nanoparticles are Proton-Coupled Electron Transfer Agents. *Science* **2012**, *336*, 1298–1301.
15. Gavrilyuk, A. Photochromism in Transition Metal Oxides Induced by Photoinjection of Hydrogen. *App. Sci. innovations.pvt.ltd.* **2010**, 473–508.
16. Pan, L.; Wang, Y.; Wang, X.; Qu, H. Y.; Zhao, J. P.; Li, Y.; Gavrilyuk, A. Hydrogen Photochromism in Nb_2O_5 Powders. *Phys. Chem. Chem. Phys.* **2014**, *16*, 20828–20833.
17. Day, J. H. Thermochromism. *Chem. Rev.* **1963**, *63*, 65–80.
18. Day, J. H. Thermochromism of Inorganic Compounds. *Chem. Rev.* **1968**, *68*, 649–657.
19. Kiri, P.; Hyett, G.; Binions, R. Thermochromic Materials; Semiconductor; Solid-State; Thin Films. *Adv. Mater. Lett.* **2010**, *1*, 86–105.
20. Parkin, I. P.; Manning, T. D. Intelligent Thermochromic Windows. *J. Chem. Educ.* **2006**, *83*, 393–400.
21. Chang, T. C.; Cao, X.; Bao, S. H., et al. Review on Thermochromic Vanadium Dioxide Based Smart Coatings: From Lab to Commercial Application. *Adv. Manuf.* **2018**, *6*, 1–19.
22. Dalapati, G. K., et al. Color Tunable Low Cost Transparent Heat Reflector using Copper and Titanium Oxide for Energy Saving Application. *Sci. Rep.* **2016**, *6*, 20182.
23. Long, L.; Ye, H. How to be Smart and Energy Efficient: A General Discussion on Thermochromic Windows. *Sci. Rep.* **2014**, *4*, 6427.
24. Gao, Y.; Luo, H.; Zhang, Z.; Kang, L.; Chen, Z.; Du, J.; Kanehira, M.; Cao, C. Nanoceramic VO_2 Thermochromic Smart Glass: A Review on Progress in Solution Processing. *Nano Energ* **2012**, *1*, 221–246.
25. Li, Y.; Ji, S.; Gao, Y.; Luo, H.; Kanehira, M. Core-shell VO_2@TiO_2 Nanorods that Combine Thermochromic and Photocatalytic Properties for Application as Energy-Saving Smart Coatings. *Sci. Rep.* **2013**, *3*, 1370.
26. Bogdan, J.; Jackowska-Tracz, A.; Zarzyńska, J.; Pławińska-Czarnak, J. Chances and Limitations of Nanosized Titanium Dioxide Practical Application in View of its Physicochemical Properties. *Nanosc. Res. Lett.* **2015**, *10*, 57.
27. Wang, R.; Hashimoto, K.; Fujishima, A. Light-induced Amphiphilic Surfaces. *Nature* **1997**, *388*, 431–432.
28. Khataee, A.; Alireza, G.; Mansoori. Nanostructured Titanium Dioxide Materials. *World Scientific* **2011**.
29. Österlund, L.; Topalian, Z. Photocatalytic Oxide Films in the Built Environment. *J. Phys. Conf. Ser.* **2014**, *559*, 012009.
30. Tang Yixing, Self-cleaning Polyurethane and Polyester Coatings, Electronic thesis and Dissertation Repository.2013; p 1098.
31. Montali, A. Antibacterial Coating Systems. *Injury* **2006**, *37*, 81–86.
32. Sun, D.; Shahzad, M. B.; Li, M.; Wang, G.; Xu, D. Antimicrobial Materials with Medical Applications. *Mater. Technol.* **2015**, *30*, 90–95.
33. Haldar, J.; Weight, A. K.; Klibanov, A. M. Preparation, Application and Testing of Permanent Antibacterial and Antiviral Coatings. *Nat. Protoc.* **2007**, *2*, 2412–2417.
34. Cloutier, M.; Mantovani, D.; Rosei, F. Antibacterial Coatings: Challenges, Perspectives, and Opportunities. *Trends Biotechnol.* **2015**, *33*, 637–652.
35. Antimicrobial Paints & Coatings | Antimicrobial Additives | BioCote Ltd, United Kingdom.
36. Kumar, A.; Vemula, P. K.; Ajayan, P. M.; John, G. Silver-Nanoparticle-Embedded Antimicrobial Paints Based on Vegetable Oil. *Nat. Mater.* **2008**, 7, 236–247.
37. Krishnamoorthy, K.; Premanathan, M.; Veerapandian, M.; Kim, S. J. Nanostructured Molybdenum Oxide-Based Antibacterial Paint: Effective Growth Inhibition of Various Pathogenic Bacteria. *Nanotechnology* **2014**, *25*, 315101.
38. Natalio, F., et al. Vanadium Pentoxide Nanoparticles Mimic Vanadium Haloperoxidases and Thwart Biofilm Formation. *Nat. Nanotechnol.* **2012**, 7, 530–535.

39. Lee, D.; Cohen, R. E.; Rubner, M. F. Antibacterial Properties of Ag Nanoparticle Loaded Multilayers and Formation of Magnetically Directed Antibacterial Microparticles. *Langmuir ACS J. Surf. Colloids* **2005**, *21*, 9651–9659.
40. Feng, Q. L.; Wu, J.; Chen, G. Q.; Cui, F. Z.; Kim, T. N.; Kim, J. O. A Mechanistic Study of the Antibacterial Effect of Silver Ions on *Escherichia coli* and *Staphylococcus aureus*. *J. Biomed. Mater. Res.* **2000**, *52*, 662–668.
41. Rai, M.; Yadav, A.; Gade, A. Silver Nanoparticles as a New Generation of Antimicrobials. *Biotechnol. Adv.* **2009**, *27*, 76–83.
42. Park, S. -J.; Jang, Y. -S. Preparation and Characterization of Activated Carbon Fibers Supported with Silver Metal for Antibacterial Behavior. *J. Colloid Interface Sci.* **2003**, *261*, 238–243.
43. Yoon, K. -Y.; Hoon Byeon, J.; Park, J. -H.; Hwang, J. Susceptibility Constants of *Escherichia coli* and *Bacillus subtilis* to Silver and Copper Nanoparticles. *Sci. Total Environ.* **2007**, *373*, 572–575.
44. Ren, G.; Hu, D.; Cheng, E. W. C.; Vargas-Reus, M. A.; Reip, P.; Allaker, R. P. Characterisation of Copper Oxide Nanoparticles for Antimicrobial Applications. *Int. J. Antimicrobial. Agents* **2009**, *33*, 587–590.
45. Blossey, R. *Nat. Mater.* **2003**, *2*, 301.
46. Bocquet, L.; Lauga, E. *Nat. Mater.* **2011**, *10*, 334.
47. Cira, N. J.; Benusiglio, A.; Prakash, M. *Nature* **2015**, *519*, 446.
48. Rana, A., et al. *Nanoscale* **2016**, *8*, 15597.
49. BLossey, R. *Nat. Mater.* **2003**, *2*, 301.
50. Sarkar, T., et al. *RSC Adv.* **2016**, *6*, 109234.
51. Dykas, M. M.; Desai, S. K.; Patra, A., et al. *ACS Appl. Mater. Interfaces* **2018**, *10*, 12510.
52. Prakash, S., et al. *Nanoscale* **2018**, *10*, 3356.
53. Young, T. *Philos. Trans. Roy. Soc. London* **1804**, *65*.
54. Tadmor, R. *Langmuir* **2004**, *20*, 7659.
55. Ghanashyam Krishna, M.; Vinjanampati, M.; Dhar Purkayastha, D. *Eur. Phys. J. Appl. Phys.* **2013**, *62*.
56. Wenzel, R. N. *Ind. Eng. Chem.* **1936**, *28*, 988.
57. Cassie, A. B. D.; Baxter, S. *Trans. Faraday Soc.* **1944**, *40*, 546.
58. Irie, H.; Washizuka, S.; Yoshino, N.; Hashimoto, K. *Chem. Commun.* **2003**, 1298.
59. Wang, R.; Hashimoto, K.; Fujishima, A.; Chikuni, M.; Kojima, E.; Kitamura, A.; Shimohigoshi, M.; Watanabe, T. *Nature* **1997**, *388*, 431.
60. Frieder, M.; Jean-Christophe, B. *J. Phys.: Condens. Matter* **2005**, *17*, R705.
61. Zorba, V.; Chen, X.; Mao, S. S. *Appl. Phys. Lett.* **2010**, *96*, 093702.
62. Lu, G.; Linsebigler, A.; Yates, J. T. *J. Phys. Chem.* **1994**, *98*, 11733.
63. Bayati, M. R.; Joshi, S.; Molaei, R.; Narayan, R. J.; Narayan, J. *J. Appl. Phys.* **2013**, *113*, 063706.
64. Shibata, T.; Irie, H.; Ohmori, M.; Nakajima, A.; Watanabe, T.; Hashimoto, K. *Phys. Chem. Chem. Phys.* **2004**, *6*, 1359.
65. Neuheuser, T.; Hess, B. A.; Reutel, C.; Weber, E. *J. Phys. Chem.* **1994**, *98*, 6459.
66. Park, J. J.; Kim, D. Y.; Latthe, S. S.; Lee, J. G.; Swihart, M. T.; Yoon, S. S. *ACS Appl. Mater. Interf.* **2013**, *5*, 6155.
67. Lee, Y. C.; Hong, Y. P.; Lee, H. Y.; Kim, H.; Jung, Y. J.; Ko, K. H.; Jung, H. S.; Hong, K. S. *J. Colloid Interface Sci.* **2003**, *267*, 127.
68. Liu, K.; Cao, M.; Fujishima, A.; Jiang, L. *Chem. Rev.* **2014**, *114*, 10044.
69. Khranovskyy, V.; Ekblad, T.; Yakimova, R.; Hultman, L. *Appl. Surf. Sci.* **2012**, *258*, 8146.
70. Papadopoulou, E. L.; Barberoglou, M.; Zorba, V., et al. *J. Phys. Chem. C* **2009**, *113*, 2891.
71. Hu, H.; Ji, H. -F.; Sun, Y. *Phys. Chem. Chem. Phys.* **2013**, *15*, 16557.
72. Feng, X.; Zhai, J.; Jiang, L. *Angew. Chem. Int. Ed.* **2005**, *44*, 5115.
73. Zhu, W.; Feng, X.; Feng, L.; Jiang, L. *Chem. Commun.* **2006**, 2753.

74. Wang, S.; Feng, X.; Yao, J.; Jiang, L. *Angew. Chem. Int. Ed.* **2006**, *45*, 1264.
75. Feng, X.; Feng, L.; Jin, M.; Zhai, J.; Jiang, L.; Zhu, D. *J. Am. Chem. Soc.* **2004**, *126*, 62.
76. Lim, H. S.; Kwak, D.; Lee, D.Y.; Lee, S. G.; Cho, K. *J. Am. Chem. Soc.* **2007**, *129*, 4128.
77. Pang, C. L.; Sasahara, A.; Onishi, H.; Chen, Q.; Thornton, G. *Phys. Rev. B* **2006**, *74*, 073411.
78. Wendt, S.; Matthiesen, J.; Schaub, R.; Vestergaard, E. K.; Lægsgaard, E.; Besenbacher, F.; Hammer, B. *Phys. Rev. Lett.* **2006**, *96*, 066107.
79. Wendt, S., et al. *Surf. Sci.* **2005**, *598*, 226.
80. Bikondoa, O.; Pang, C. L.; Ithnin, R.; Muryn, C. A.; Onishi, H.; Thornton, G. *Nat. Mater.* **2006**, *5*, 189.
81. Miyauchi, M.; Nakajima, A.; Fujishima, A.; Hashimoto, K.; Watanabe, T. *Chem. Mater.* **2000**, *12*, 3.
82. Khan, S.; Azimi, G.; Yildiz, B.; Varanasi, K. K. *Appl. Phys. Lett.* **2015**, *106*, 061601.
83. Azimi, G.; Dhiman, R.; Kwon, H. -M.; Paxson, A. T.; Varanasi, K. K. *Nat. Mater.* **2013**, *12*, 315.
84. Azimi, G.; Kwon, H. -M.; Varanasi, K. K. *MRS Commun.* **2014**, *4*, 95.
85. Zenkin, S.; Kos, Š.; Musil, J. *J. Am. Ceram. Soc.* **2014**, *97*, 2713.
86. Preston, D. J.; Miljkovic, N.; Sack, J.; Enright, R.; Queeney, J.; Wang, E. N. *Appl. Phys. Lett.* **2014**, *105*, 011601.
87. Lundy, R.; Byrne, C.; Bogan, J., et al. *ACS Appl. Mater. Interfaces* **2017**, *9*, 13751.
88. Luo, Y.; Li, J.; Zhu, J.; Zhao, Y.; Gao, X. *Angew. Chem. Int. Ed.* **2015**, *54*, 4876.
89. Nakayama, K.; Hiraga, T.; Zhu, C.; Tsuji, E.; Aoki, Y.; Habazaki, H. *Appl. Surf. Sci.* **2017**, *423*, 968.
90. Oh, I. -K., et al. *Chem. Mater.* **2015**, *27*, 148.
91. Külah, E.; Marot, L.; Steiner, R.; Romanyuk, A.; Jung, T. A.; Wäckerlin, A.; Meyer, E. *Sci. Rep.* **2017**, 7, 43369.
92. Butt, H. -J.; Cappella, B.; Kappl, M. *Surf. Sci. Reports* **2005**, *59*, 1.
93. Mijatovic, D.; Eijkel, J. C. T.; van den Berg, A. *Lab Chip* **2005**, *5*, 492.
94. Manz, A.; Harrison, D. J.; Verpoorte, E. M. J.; Fettinger, J. C.; Paulus, A.; Lüdi, H.; Widmer, H. M. *J. Chromatogr. A* **1992**, *593*, 253.
95. Gau, H.; Herminghaus, S.; Lenz, P.; Lipowsky, R. *Science* **1999**, *283*, 46.
96. Lam, P.; Wynne, K. J.; Wnek, G. E. *Langmuir* **2002**, *18*, 948.
97. Grunze, M. *Science* **1999**, *283*, 41.
98. Darhuber, A. A.; Troian, S. M.; Reisner, W. W. *Phys. Rev. E* **2001**, *64*, 031603.
99. Fan, Y. W., et al. *J. Neurosci. Methods* **2002**, *120*, 17.
100. Anselme, K., et al. *Biomaterials* **2000**, *21*, 1567.
101. Nayak, T. R., et al. *ACS Nano* **2011**, *5*, 4670.
102. Nair, S., et al. *J. Mater. Sci.: Mater. Med.* **2009**, *20*, 235.
103. Hench, L. *J. Mater. Sci.: Mater. Med.* **2006**, *17*, 967.
104. Park, J. B., et al. *J. Biomed. Mater. Res.* **1981**, *15*, 103.
105. Contreras, R.; Sahlin, H.; Frangos, J. A. *J. Biomed. Mater. Res. Part A* **2007**, *80A*, 480.

CHAPTER 11

Upconverting nanoparticles: potential for a new heat regulating materials

V.G. Vasavi Dutt, Nimai Mishra
Department of Chemistry, SRM University, AP–Amaravati, Neerukonda, Guntur (Dt), Andhra Pradesh, India

1 Introduction

The optical properties are independent of the intensity of exciting light in the case of linear optics. The case in which the material's optical properties depend on the radiant flux density of excitation light is nonlinear optics. Such kind of phenomenon is the upconversion process *(1)*. The fluorescence follows the principle of Stokes' law in which the exciting photons are of higher energy than emitted photons (see Fig. 11.1A). Upconversion is an anti-Stokes process, which concerns with an excess of energies of emitted photons (see Fig. 11.1B). This phenomenon shows regulating the incident photon wavelength and thus could be used as a potential candidate for heat regulating material. It is usually characterized by the conversion of long-wavelength radiation, for example, infrared or NIR (near infrared) radiation to short-wavelength radiation in the visible range.

The principle of upconversion is depicted in Fig. 11.1B in which a luminescent center of ground state 1 absorbs energy and gets excited to state 2 via excitation photon or corresponding energy transfer (ET) process. Consequently, another excitation photon or ET process excites the luminescent center to state 3. A higher energy photon emission is observed as a result of the radiative emission from state 3 back to state 1 or any other lower energy levels. In the case of upconversion, the metastable intermediate excited state is responsible for the luminescence, which is intended to have relatively long lifetime in order to maintain high population in advance of the second excitation energy.

Under low energy excitation, two main mechanisms affect upconversion emission: the upconversion luminescence from lanthanide ions (e.g., Er^{3+}, Ho^{3+}, and Tm^{3+}) that depend on their abundant f-electrons and triplet–triplet annihilation (TTA)-based upconversion from the triplet states of

Energy Saving Coating Materials
http://dx.doi.org/10.1016/B978-0-12-822103-7.00011-X

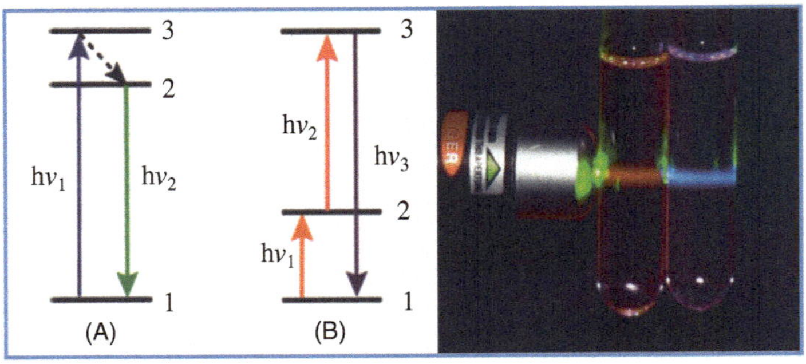

Figure 11.1 Schematic representation of (A) conventional photoluminescence and (B) upconversion luminescence process. Image to the right shows Stokes emission via fluorescence (red) and anti-Stokes emission (blue) by photon upconversion based on sensitized triplet–triplet annihilation, when the sample is excited with green light. *Copyright from Ref.* (3). *Image©Victor Gray from the Wikipedia Commons.*

molecular chromophores, which act as intermediate excited states *(2,4,5)*. Upconversion through the sequential absorption of multiple photons in trivalent lanthanide ions has been explored for producing higher anti-Stokes luminescence. In 1966 Auzel and coworkers first reported lanthanide-based upconversion luminescence in which the emission is observed from Er^{3+} by pumping Yb^{3+} in a glass matrix *(6)*. Due to the abundant energy levels with f-electron configurations of lanthanide materials, they are incorporated as sensitizers and emitters in many of the upconversion energy materials. Lanthanide-based upconversion materials exhibit distinctive emission properties like large anti-Stokes shifts in hundreds of nanometers, sharp emission lines, long upconversion lifetimes in ~ms and photostability *(7)*. On the other hand, for the first time, Parker and Hatchard studied organic chromophores—phenanthrene/naphthalene or proflavin/anthracene—that show both absorption and emission in the ultraviolet (UV) region *(8)*. Because of O_2-induced upconversion luminescence quenching and low photostability, this upconversion has been a challenge until recently when the interest on this area has enormously increased. In 2015 TTA-based upconversion emission can be observed by the naked eyes when excited with green laser at low power of <5 mW. Many metal complexes and organic dyes show TTA-based upconversion luminescence with spectral range from UV to NIR in organic solutions, polymer matrices, and nanoparticles have been studied *(3,9)*.

The growing interest of this topic has driven applications in photovoltaic devices *(10)*, bioimaging *(11)*, etc. The upconversion emission can be achieved by different mechanisms: energy transfer upconversion (ETU),

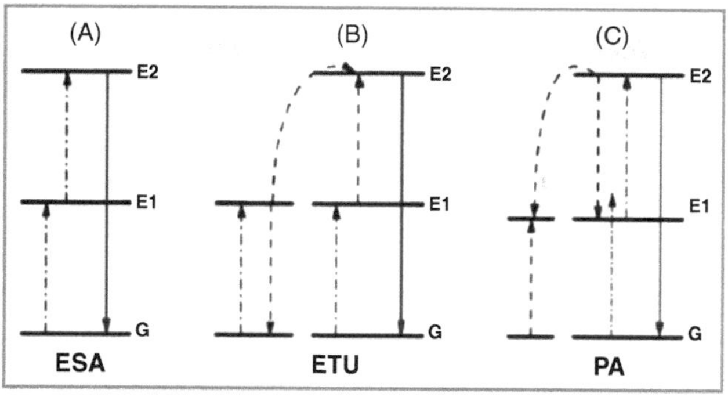

Figure 11.2 Schematic showing (A) excited-state absorption (ESA), (B) energy transfer upconversion (ETU), and (C) photon avalanche in lanthanide-doped crystals (dotted-dashed line: photon excitation, dashed line: energy transfer, and solid line: emission). *Copyright from Ref.* (12).

excited-state absorption (ESA), and photonic avalanche (PA). These mechanisms are based on sequential absorption of photons by long-lived metastable states. The upconversion emission thus occurs from the populated excited states as a result of this sequential absorption. In ESA, the emitting ions absorb a minimum of two photons to reach the emitting level, whereas in the case of ETU, there is a subsequent ET between the two ions in excited states, which lead to emission of short wavelength. The population of excited states of emitting ion occurs by the ET from the neighboring ion (see Fig. 11.2). The upconversion induced by PA starts with ground state excitation followed by resonant ESA to populate the upper emitting level. Cross-relaxation ET occurs between excited ion and neighboring ion ground state after the population of the metastable state. The two ions populate the levels to further initiate cross-relaxation followed by ESA to produce strong upconversion emission *(1,10)*.

Unlike the organic dyes and quantum dots that produce higher energy emission that needs an expensive pulsed laser, lanthanide-doped upconversion nanoparticles require low-cost diode laser as the excitation source. They can be synthesized in nanoscale and significantly show low cytotoxicity. Hence, they can be a good choice for bioimaging and biolabeling applications. Because of their unique optical properties, absorption and emission through forbidden 4f–4f transition provide long-lifetime and tunable emissions, they are explored in drug delivery applications as well. Photonic upconversion, the wavelength shifting technology signifies a sustainable route of converting low-energy terrestrial solar photons into the light of adequate energy in operational photovoltaics *(2,3,11)*.

2 Upconversion mechanisms

As discussed earlier, the mechanisms of lanthanide upconversion are categorized into three main classes: ESA, ETU, and PA. For doped upconversion materials, ESA is a suitable pumping mechanism. The general energy scheme of subsequent photon absorption is shown in Fig. 11.3. Under appropriate excitation, if the excited state energy is resonant to transition between ground state 1 to metastable state 2, it consecutively absorbs another photon, which is resonant to transition between state 2 and higher excited state 3. Upconversion emission occurs when the photon de-excites from higher excited state 3 to ground state 1. To elude the transfer losses through cross-section relaxation as shown in Fig. 11.3D and to enhance the gain from ESA, low concentrations of active ion doping in upconversion nanoparticles is imperative.

Different types of ETU process are illustrated in Fig. 11.3B–F. Firstly, the sensitizer absorbs photon and transits from ground state 1 to metastable state. Then energy transfer occurs from the sensitizer to activator, promoting to metastable state 2. Finally, the ESA process occurs to create a higher excited state 3. The sensitizer absorbs photons in the energy-transfer process

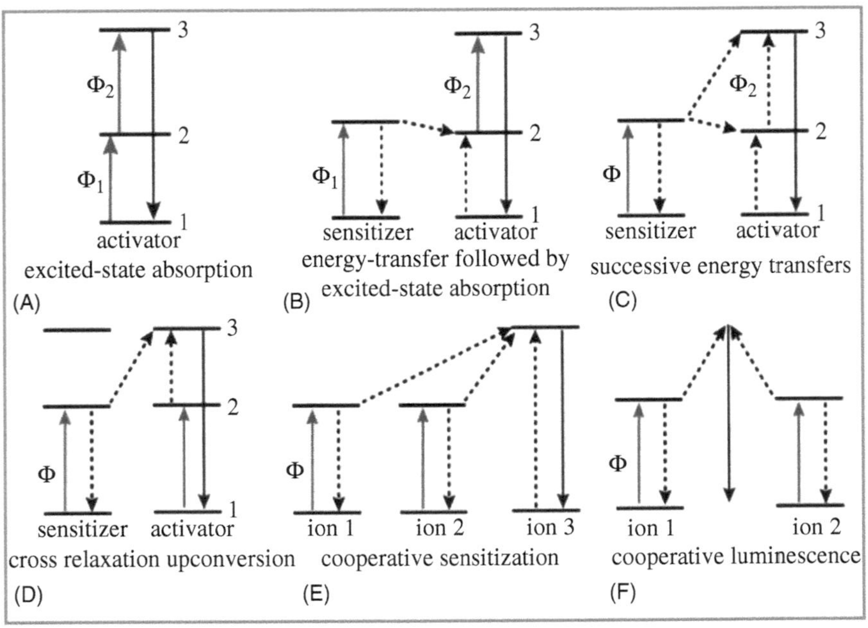

Figure 11.3 (A) General illustration of excited-state absorption (ESA). (B–F) Schemes related to energy-transfer upconversion (ETU). *Copyright from Ref.* (3).

while the activator is promoted to metastable state 2 and excited state 3 by two-photon energy process. Cross-relaxation (see Fig. 11.3C) might occur if the sensitizer and activator ions are similar in nature. Both the sensitizer and activator absorb excitation photons in their excited state 2. The energy-transfer process can transmit the activator to excited state 3 and revert the sensitizer to ground state. The last two processes (see Fig. 11.3E and F) involve cooperative effects because more than one center involves in sensitization or emission process. In cooperative sensitization, the two excited ions 1 and 2 absorb each of photon and transit to their respective excited states and then transfer the energy to another ion 3 so that it can elevate to its excited state 3. In the case of cooperative luminescence, the two excited ions 1 and 2 absorb each photon and cooperate in giving emission. The concentration of lanthanide ions should be considerably enough for ion–ion interactions to induce ETU process *(1,3)*.

2.1 Upconversion mechanism based on triplet–triplet annihilation

A bimolecular system is required for TTA upconversion, which consists of a sensitizer molecule to absorb the incident light and create triplet excitations and an emitter molecule to exhibit efficient TTA upconversion. This mechanism was first reported by Parker and Hatchard in 1962 *(13)*.

The general mechanism of upconversion is explained as follows (see Fig. 11.4): In step 1, the sensitizer molecule absorbs photons of low energy

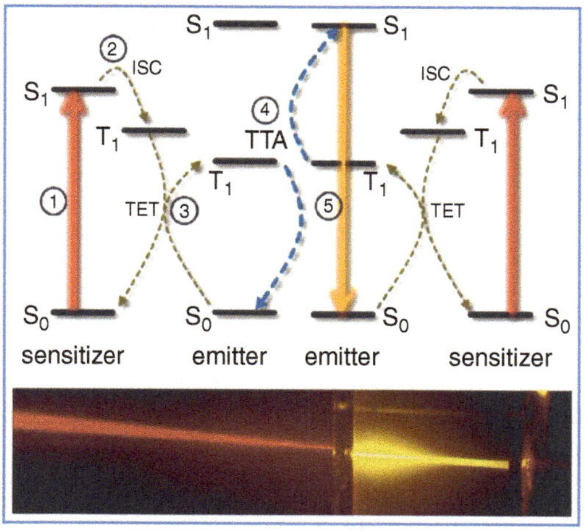

Figure 11.4 ***General illustration of triplet-triplet annihilation mechanism.*** The photograph at the bottom show TTA upconversion. *Copyright from Ref.* (14).

and go through the intersystem crossing to its lowest triplet state as shown in step 2 followed by triplet–triplet ET (Step 3). Step 4 indicates the TTA where the two emitters interact to create an excited singlet state in one molecule and the other is quenched to its ground state. This emitter which is in the excited state exhibits fluorescence (step 5) at higher energy than the absorbed photons in step 1. Essentially, the intermediate energy storage in triplet states provides a provision for efficient upconversion even under low-light conditions *(14)*. There are many examples reported for lanthanide-based upconversion under broadband, pulsed radiation *(15)*. Various reports have shown TTA-based upconversion under broadband white-light illumination even under sunlight as the excitation source *(16)*.

3 Upconversion nanoparticles

Since this unique anti-Stokes optical property has been explored in various applications, advancements on upconversion materials have been emphasized over the past few decades. In order to obtain high upconversion efficiency, it is crucial to co-dope sensitizer ions along with activator ions that possess similar intermediate excited state. The upconversion efficiency highly depends on the interactions between the activator ions and the network of sensitizer and the separation distance between the dopants. Hence, the optimal doping concentration is significant in leveraging the ET process and luminescence performance of the upconversion nanoparticle *(17,18)*.

However, concentration quenching has been a problem that hindered the advancements of highly luminescent materials. The concept of concentration quenching in inorganic phosphors was first explored by Dexter and Schulman in 1954 *(19)*. Later, different mechanisms of concentration quenching in organic dyes have been studied *(20)*. It inflicts a limitation to high luminescence intensity causing hindrance to further applications. To avoid luminescence quenching and for efficient upconversion, relatively low concentrations of sensitizers and activators are usually taken in hexagonal-phase alkaline rare-earth fluoride nanocrystal β-AREF$_4$ and are considered as the host material for upconversion. Low doping concentration is considered as the key point to obtain bright luminescent nanomaterials *(21)*.

Zhao et al. first developed highly doped homogeneous upconversion nanoparticles that exhibit excellent brightness. An increase in the excitation irradiance enhances the luminescence intensity for 0.5%, 4%, and 8% Tm^{3+}-doped nanocrystals as shown in Fig. 11.6A *(22)*. Generally, conventional upconversion materials achieve saturation in brightness with increasing the excitation power, but highly Er^{3+}-doped $NaYF_4$:Yb^{3+} nanocrystals exhibit much higher brightness in contrast with low Er^{3+}-doped materials as shown in Fig. 11.6B *(23)*. Meanwhile, high brightness of upconversion emission at doping concentrations of Er^{3+} is exhibited by $NaYF_4$:x%Er^{3+}@$NaLuF_4$ at 100% doping (see Fig. 11.6C) *(24)*. Highly Yb^{3+}-doped phase (α-) $NaYF_4$:Yb^{3+},Tm^{3+} nanocrystals show an emission enhancement by 43 times at NIR wavelengths compared to conventional Yb^{3+}-doped upconversion materials. Thus the optimal concentration of Yb^{3+} sensitizer ions has been investigated *(25)*. Using an inert shell doping, the material $NaYbF_4$:2%Er^{3+}@CaF_2 shows an intensity enhancement in visible light emission by 15 times (see Fig. 11.6D) *(26)*.

The controlled growth of heterogeneously doped core@shell upconversion nanoparticles can be widely studied to requisite for a specific application. High-performance upconversion can be achieved based on their excitation conditions for a desirable emission wavelength. The optimization of upconversion emission wavelengths through energy migration is first demonstrated using Gd^{3+} sublattice structure as a ET channel in core@shell structures as depicted in Fig. 11.6E. Capobianco and coworkers investigated on $NaGdF_4$:Yb^{3+},Er^{3+} upconversion nanoparticles with shell Yb^{3+}-doped $NaGdF_4$ and showed an improvement in green and red emission with respect to the shell (see Fig. 11.5A). The ET from Yb^{3+} from the shell to Er^{3+} ions in the core shows an enhancement in the overall upconversion efficiency *(1)*. In 2015 Zhang et al. developed high concentration lanthanide-doped $NaGdF_4$:Yb^{3+}, Tm^{3+},Er^{3+}@$NaGdF_4$:Eu^{3+}@$NaYF_4$ material that can show high emission brightness in the visible region *(27)*. Heterogeneous core@shell structures like $NaYbF_4$:Gd^{3+},Tm^{3+}@$NaGdF_4$@CaF_2:Ce^{3+} with a high-doping concentration of Ce^{3+} in the shell layer has been investigated to show an efficient upconversion. In order to match the Gd^{3+} energy level with the Ce^{3+} 4f–5d excitation frequency, the host CaF_2 has been employed *(28)*. Apart from the doping of lanthanide ions, organic dyes can be employed to the upconversion nanocrystals and have showed an improvement in sensitivity in FRET (Förster resonance energy transfer) limited measurements via Gd^{3+}-mediated ET. Later, it is also investigated that Mn^{2+} and Tb^{3+} ion exhibit similar functionality as Gd^{3+} in terms of energy migration at core@shell at high doping concentrations of activators *(29,30)*.

Yan and coworkers first investigated that high Nd^{3+} doping concentration in core@shell upconversion nanoparticles exhibits higher

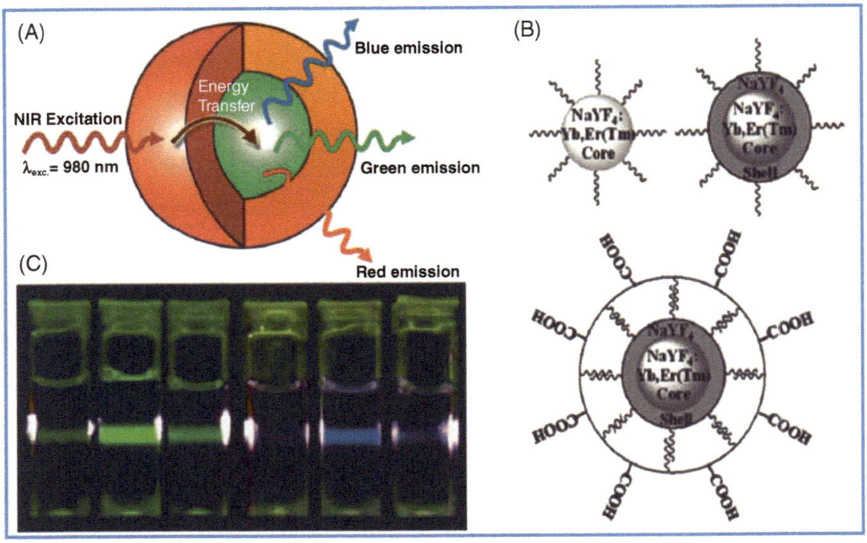

Figure 11.5 (A) Schematic representation of core@shell upconversion nanoparticles showing absorption of NIR light by Yb^{3+} rich shell (red) and energy transfer to Er^{3+},Yb^{3+}-doped core (green) giving upconverted blue, green, and red emissions *(37)*. (B) Structure of $NaYF_4$:Yb^{3+}, Er^{3+}/Tm^{3+} upconversion nanocrystals. (C) Photograph showing green emission from $NaYF_4$:Yb^{3+},Er^{3+} and blue emission from $NaYF_4$:Yb^{3+},Tm^{3+} respectively *(38)*.

luminescence compared to homogeneously doped conventional materials, and further studied its application in vivo imaging without the problem of tissue overheating (see Fig. 11.6F) *(31)*. Recently, a new class of upconversion nanocrystals β-$NaErF_4$:0.5%Tm^{3+}@$NaYF_4$ has been introduced by Chen et al. with bright red emission in which Tm^{3+} ions trap excitation energy to suppress the quenching effect *(32)*. High concentrations of Ce^{3+} into $NaYF_4$:Yb^{3+}/Ho^{3+} or $NaYF_4$:40%Gd^{3+} can improve the red-to-green emission via cross-relaxation between Ce^{3+} and Ho^{3+} *(33)*. Similarly, a single band red upconversion emission has been reported from $NaGdF_4$:Yb^{3+},Ho^{3+},Ce^{3+}@$NaYF_4$:Yb^{3+},Nd^{3+} core@shell nanocrystals with highly Nd^{3+}-doped shell *(34)*. Usually, lanthanide-doped nanocrystals show narrow FWHM (full width at half maximum) and low absorption coefficients. However the organic dyes show broader absorption than Yb^{3+} sensitizer ions that are used in upconversion nanocrystals. Even with the photostability issues, the hybrid nanoparticles provide many possibilities as explained by Chen et al. Dye-sensitized upconversion has been investigated via efficient ET dynamics of cyanine derivatives that are functionalized on the surface of $NaYbF_4$:Tm^{3+}@$NaYF_4$:Nd^{3+} nanocrystals *(35)*.

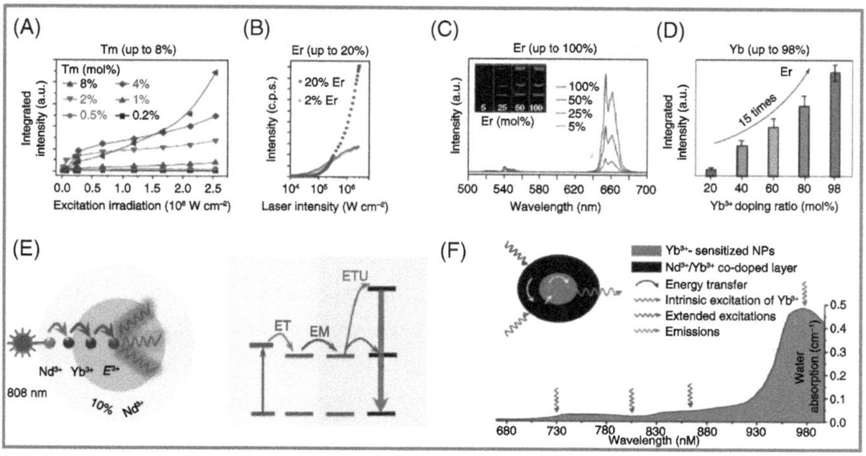

Figure 11.6 (A) Trend showing upconversion luminescence intensity with respect to excitation for Tm^{3+}-doped upconversion nanocrystals *(22)*. (B) Graph showing upconversion luminescence intensity against excitation for upconversion nanocrystals with 20% and 2% Er^{3+}, each with 20% Yb^{3+} *(23)*. (C) Emission spectra of colloidal $NaYF_4$:*x*%Er@$NaLuF_4$ nanocrystals (*x* = 5, 25, 50, 100) with inset of luminescence image of the nanocrystals in cyclohexane when excited with 980 nm laser *(24)*. (D) Upconversion luminescence intensity plotted against doping ratio of α-$NaY_{0.98-x}Yb_xF_4$:2%Er@CaF_2 (*x* = 0.2, 0.4, 0.6, 0.8, 0.98) *(26)*. (E) Schematic showing the core@shell upconversion nanoparticle with energy migration at the excitation part (where E^{3+} = Er^{3+}, Tm^{3+}, or Ho^{3+}). (F) Graph showing the absorption of water in NIR and depiction of Nd^{3+} Yb^{3+} ET process by introducing Nd^{3+}/Yb^{3+} codoped shell. It represents lower water absorption where ET process could extend effective excitation bands that can minimize the over-heating of tissues with NIR laser exposure *(31)*.

Apparently, Zheng and coworkers reported a promising approach of fine tuning of upconversion luminescence of inorganic lead halide perovskite quantum dots by sensitization of lanthanide-doped nanoparticles. Tailoring of the band gap of perovskite quantum dots can allow full-color emission beyond the availability of lanthanides alongside the high conversion efficiency of ETU under low excitation. A prolonged luminescent lifetime of excitons can be achieved based on the lifetimes of lanthanide ions. For sensitizing the upconversion in $CsPbBr_3$ quantum dots, they are treated with $LiYbF_4$:0.5%Tm^{3+}@$LiYF_4$ core/shell nanoparticles in 2:1 weight ratio where the nanoparticles can be used as internal UV or blue source to irradiate perovskite quantum dots. This provides a versatile approach for fine tuning of NIR based photon upconversion, which may be explored in several applications for high-resolution bioimaging and full-color tunability for solar cells *(36)*.

4 Applications

As already discussed above, the upconversion mechanisms work differently from other Stokes' mechanisms as the upconversion depends on multiple low energy pump photons for high energy emission. The common upconversion nanoparticles, which use Yb^{3+} as sensitizer and $Er^{3+}/Tm^{3+}/Ho^{3+}$ as emitter utilize 980 nm excitation wavelength because of large absorption cross-section of sensitizer and highly efficient ET between them. Later, Nd^{3+} has also been used in the shell to understand its utilization of 800 nm excitation in core@shell upconversion materials *(1,12,18)*. In that aspect, lanthanide-doped upconversion nanoparticles exhibit broad emission bands from UV to NIR wavelengths *(3,21,25,27)*. Because of such unique features of upconversion nanoparticles as discussed earlier, they can be employed in a wide range of applications like bioimaging, biosensing, solar cells, photodetectors, lasers, light-emitting diodes, displays, etc. *(3,10,11,50)*. Due to their high biocompatibility, upconversion nanoparticles find their use in biology, medicine, and nanotechnology because of their good emission penetration depth and controlled ET process. They can uplift the application of existing biological imaging techniques because most of the biological samples show a narrow window of optical transparency in the infrared region. The infrared wavelengths penetrate through the tissues facilitating optical imaging where the light is upconverted to provide easy detection by photodetectors. Since 1990, there have been significant contributions in the advancements of biological applications based on the concept of upconversion *(11)*.

The upconversion photoluminescence imaging is a powerful tool in biodetection, therapy, and screening *(39)*. Due to their high photostability and nonphotoblinking character, upconversion nanoparticles are best suitable for biological applications. However the only limitation is their relatively low upconversion quantum yield. Hence, these upconversion nanoparticles are being explored to obtain an enhancement in their emission yield. PEG (poly-ethylene glycol)-capped $NaYbF_4{:}Er^{3+},Gd^{3+}$ upconversion nanoparticles have been investigated, which can effectively internalize by HeLa cells and give good contrast images of a rat's heart than a clinical contrasting agent *(40)*.

Nanocomposite materials containing upconversion lanthanide nanoparticles have been extensively used in therapeutics from photodynamic therapy *(41,42)*, photothermal therapy *(43)*, and targeted delivery to multifunctional cancer therapies *(44)*. They are widely used as therapeutic agents because they can be operated in NIR in vivo. The wavelength converting ability of lanthanide upconversion materials can provide platform for operating them as therapeutic agents from UV or visible to NIR region, which can be extensively employed for photodynamic therapy or drug carriers. On the other hand, upconversion nanoparticles are used as photoluminescent probes to study the distribution and metabolism of drugs. The amount of drug can also be monitored on the basis of efficient ET between upconversion nanoparticles and drug molecules. Such materials have been designed for controlled drug release in photodynamic therapy and chemotherapy. Idris et al. used mesoporous silica- coated upconversion nanoparticles as nanotransducer to inhibit tumor growth, which is directly injected into melanoma tumors conjugated with tumor-targeting agent in conventional photodynamic therapy. According to the results shown in Fig. 11.8, the upconversion nanoparticles-based photodynamic therapy has been used for the inactivation of tumor cells and can be extended for future noninvasive cancer therapy *(42)*. Due to its unique optical properties, these materials have been extensively studied in terms of their usage in optoelectronic devices.

4.1 Optoelectronic devices

Solar energy is the most abundantly available source of energy that comprises infrared, visible, and UV light. The broad spectrum of filtered sunlight through the Earth's atmosphere poses a challenge for an efficient power conversion using a finite band gap semiconductor material photovoltaic cell. In the case of the c-Si solar cell, only 31% of incident light is harvested and the rest of the light is transmitted through the cell. The unused IR sunlight offers a promising stage for enhancement of efficiency of solar cell and other electronic devices. Upconversion mechanism facilitates an efficient solar energy conversion by combining two or more IR photons and converting into high energy single visible or UV photon *(45)*.

4.2 In solar cells

An efficient photonic upconversion with sub-band gap energies that create electron–hole pair in semiconductor would surpass the limitation in single-junction photovoltaic devices by using IR light. On illumination with 0.86 W 980 nm laser, a photocurrent of 5 nA was obtained in a photoelectronic device comprising CdSe quantum dots bound to $NaYF_4$:Yb,Er nanoparticles of 22 nm *(46)*. Initially, Gibart et al. used Yb^{3+},Er^{3+} codoped ceramic upconversion layer on the back contact of a solar cell *(47)*. The external efficiency is improved to 3.4% under 1523 nm laser illumination when the silver is coated on the back of the upconversion material layer to reflect all the photons back to the active layer *(48)*. The photoconversion efficiency of a quantum dot-based solar cell with TiO_2/CdS/ZnS configuration is enhanced by incorporating Yb^{3+}–Er^{3+} codoped ZrO_2 nanoparticles layer. Electrochemical impedance measurements revealed reduced charge recombination after coating with upconversion nanoparticle *(49)*.

Crystalline silicon being a promising material for photovoltaics, several efforts have been dedicated for the enhancement of photovoltaic efficiency beyond the theoretical limit. In this regard, upconversion has been one of the possible routes, which can harness even sub-band gap photons, that is, to increase absorption in 800–1100 nm range. Er^{3+}-doped upconversion materials are of special importance because they absorb in 1450–1580 nm, emit at 980 nm, 540 nm, and 650 nm, and generate hot excitons that contribute to the improvement of photocurrent *(11)*.

In the case quantum dot-based solar cell, the whole process of excitation, emission, upconversion, and electron transportation is depicted in Fig. 11.7. When the photoanode is irradiated with sub-band gap energy levels (NIR), the upconversion nanoparticles emit green and red light based on the transitions of Er^{3+}. The visible light emission as a result of upconversion followed by simultaneous absorption by quantum dots present an efficient configuration of photoanode which exhibit high optical gain in photovoltaic device module *(50)*.

4.3 Infrared photodetectors

Infrared photodetectors have gained attention because of its high penetration depth of IR light. Low-cost and flexibility in processing make nanocrystal-based IR detector further benefit in its application. Control of the size to obtain an appropriate band gap plays a key role for attaining high efficiency of IR photodetectors. Near-infrared photodetectors are commercialized devices with a wide range of applications in atmospheric sounding, bioimaging, night vision, etc. Narrow band gap semiconductors like InAs, Ge, organic semiconductors, etc. can function as photoactive materials in IR photodetector applications because they can create photo-induced charge carriers that can be separated at the interface of a heterojunction *(11,51,52)*.

Upconversion nanoparticles are decorated to the surface of graphene and the schematic is shown in Fig. 11.8A. From *I*–*V* characteristics of graphene/GaAs with and without upconversion nanoparticles (see Fig. 11.8B). It is observed that I_{sc} is enhanced by 2.6 times after using upconversion

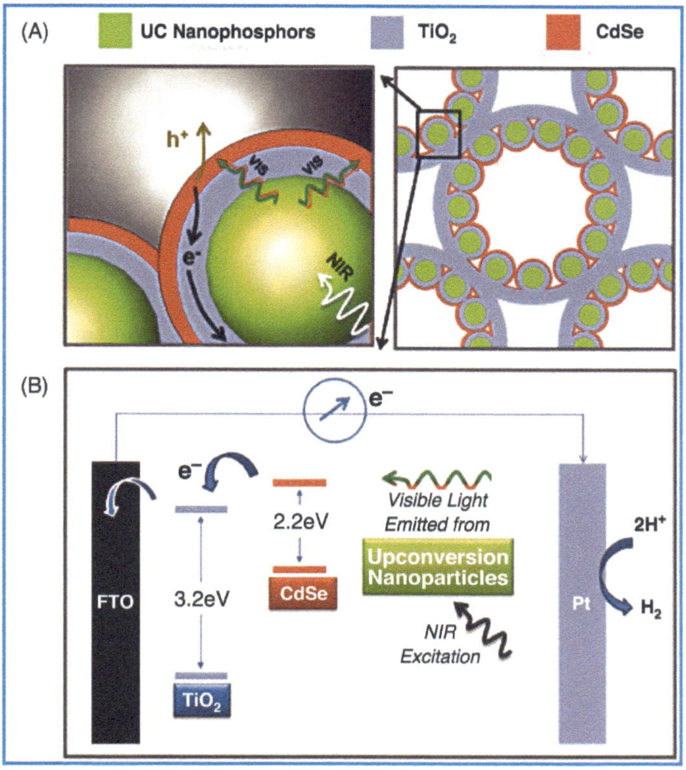

Figure 11.7 (A) Schematic illustration representing energy flow in as-synthesized hetero nanostructures from upconversion nanoparticle to CdSe layer and toward TiO_2 nanoparticles. (B) Band energy diagram showing charge transport from excited CdSe to TiO_2 and toward electrode for hydrogen. *Copyright from Ref.* (50).

nanoparticles, which concludes that they help in enhancing the photocurrent of graphene/GaAs photodetector. The charge carrier transfer across the heterojunction is depicted in Fig. 11.8C. The fermi level of graphene is moved up after introducing upconversion nanoparticles in device configuration, which eventually lead to an effective charge separation resulting in larger photocurrent. It is also interpreted that the reflectance is reduced relatively with the introduction of upconversion nanoparticles for which the photocurrent is increased (see Fig. 11.10D) *(52)*.

4.4 In lasing

The upconversion mechanism found potential application in fiber lasers. An efficient generation and propagation of upconversion luminescence in $NaYF_4$:Yb^{3+}/Er^{3+} microtubes under NIR laser excitation is demonstrated

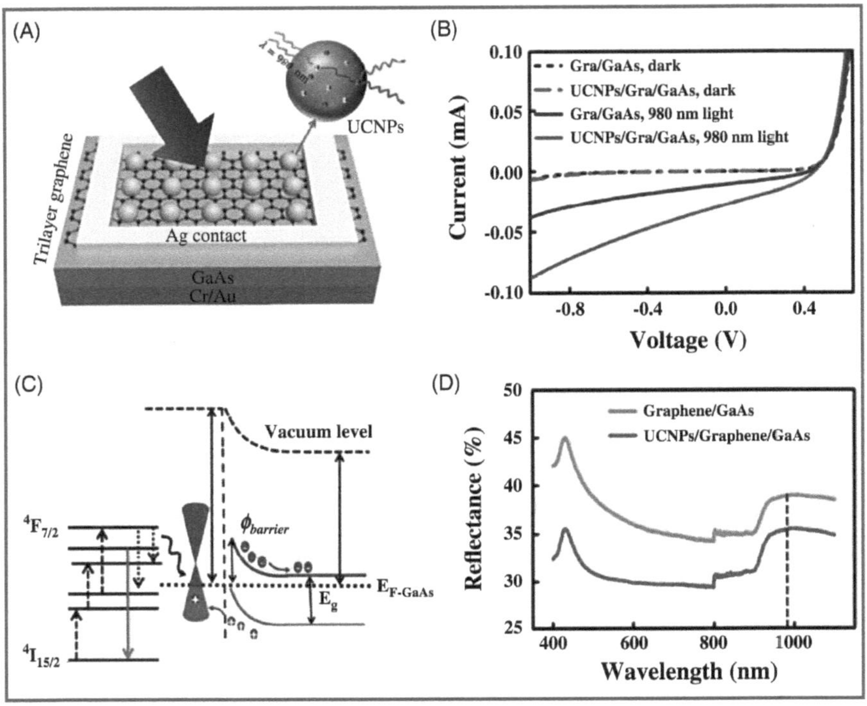

Figure 11.8 (A) Schematic illustration of graphene/GaAs NIR photodetector applied with upconversion nanoparticles. (B) *I–V* curves of the device in 980 nm illumination and dark current. (C) Representation of band alignment of upconversion nanoparticles/graphene/GaAs heterostructure. (D) Graph showing reflectance of the device with and without upconversion nanoparticles. *Copyright from Ref.* (52).

by Gao et al. Coupling of the incident pump laser and remote propagation of upconversion luminescence are crucial factors to get longer propagation distances. Since the material used is of high refractive index with low phonon energy, it is best suited for applications based on upconversion. Different colors like blue and pink luminescence from these microtubes are acquired by codoping with Yb^{3+}/Tm^{3+} and Yb^{3+}/Ho^{3+} *(53)*.

The upconversion laser is explored by Lin and Jin as shown in Fig. 11.9. With relatively low absorption from the ground state, the upconversion process in energy-looping upconversion nanoparticles has started and this is accelerated by intermediate ESA at 1064 nm. A higher energy excited-state of Tm^{3+} cross-relax giving energy to ion in its ground state. This ultimately results in population inversion in Tm^{3+}-doped energy looping upconversion nanoparticles in $^3H_4$ state, emission at 800 nm and $^1G_4$ state (blue emission) *(54)*.

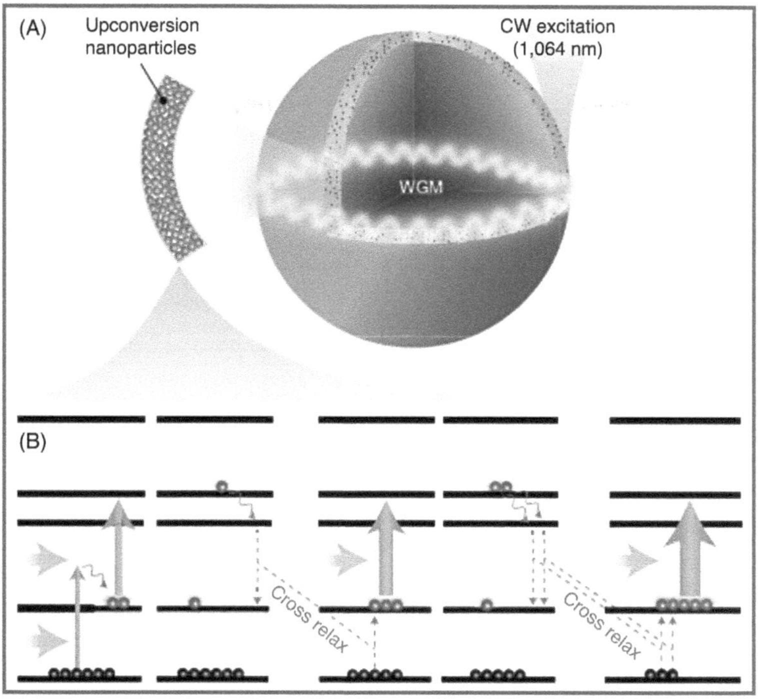

Figure 11.9 (A) Schematic representation of polystyrene bead of 5 μm used as optical cavity. Tm^{3+}-doped upconversion nanoparticles are incorporated onto the system's surface. (B) Band energy diagrams showing absorptions, energy upconversion followed by cross-relaxation-assisted amplification of excited states resulting population inversion. *Copyright from Ref.* (54).

4.5 Heat regulation

The advancement in luminescence thermometry has inclined the research toward the thermo bioimaging and exploring the properties of thermometers to determine the heat transport properties at extremely small scales. $KLu(WO_4)_2:Ho^{3+},Tm^{3+}$ nanoparticles have the capability to calibrate the thermal resistance, a key factor to model the heat transfer at small scale by controlled heat release. Relative thermal sensitivity of 2.8% K^{-1} and photothermal efficiency of 34 ± 2% are achieved by these nanoparticles as a new multifunctional thermometer of unprecedented performance (see Fig. 11.10E) *(55)*. To investigate the change in luminescence ratio with respect to excitation intensity, Pickel et al. used a technique based on the lifetime study of luminescence exhibited by the nanoparticle. Fig. 11.10 shows the temperature-dependent lifetime study of five different nanoparticles.

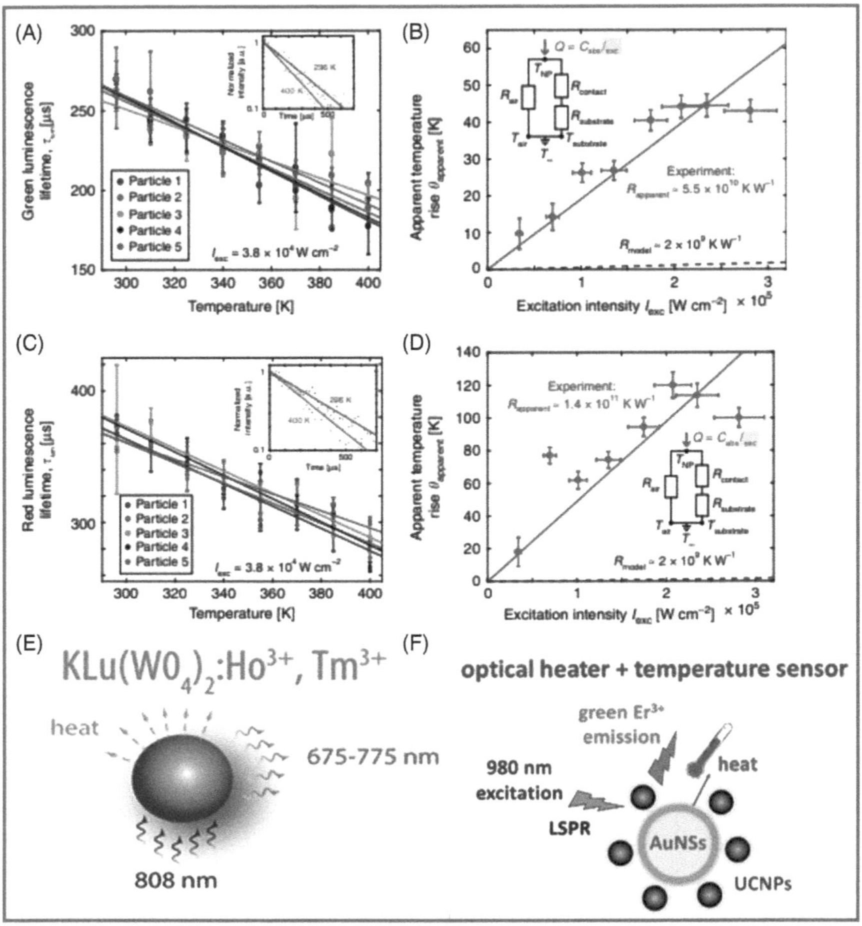

Figure 11.10 ***The study of luminescence lifetime thermometry and self-heating.*** (A) Time-dependent luminescence lifetime study for green emission band. (B) Temperature rise against excitation intensity of the green spectral band for single nanoparticle. (C, D) Similar results as (A) and (B) for green band emission *(56)*. (E) Luminescence thermometry of $KLu(WO_4)_2$: Ho^{3+},Tm^{3+} nanoparticles *(55)*. (F) Induced heating and thermometry by upconversion nanoparticle-decorated gold shell *(57)*.

The luminescence lifetime clearly shows a decrease with increasing temperature, which may be attributed to phonon-mediated decay. Such changes in luminescence lifetime spectra of Yb^{3+}/Er^{3+} codoped materials implicate impeded thermalization of nonequilibrium phonon distribution, which is referred to as "phonon bottleneck." The effect of laser excitation on temperature rise is also deliberated in ratiometric temperature measurements *(56)*.

A multifunctional system of lanthanide-doped $NaGdF_4$ upconversion nanoparticles with gold nanoshells is explored as optical heater and temperature sensors. Due to their excellent linearity and sensitivity, $NaGdF_4$:Yb^{3+}:Er^{3+} (oleate- or silica-coated) upconversion nanoparticles are suitable a temperature probes because of their high thermal sensitivities. The heating capacity of these materials was investigated on the basis of Er^{3+} emission to explore applications as hyperthermia agent controlled with a nanothermometer functionality *(57)*.

From the above studies it is evident that this material can be used as a new type of heat regulating material where it can absorb higher wavelength photons and convert it into lower wavelength.

References

1. Haase, H.; Schäfer, H. Upconverting Nanoparticles. *Angew. Chem.* **2011**, *50*, 5808–5829.
2. Singh-Rachford, T. N.; Castellano, F. N. Photon Upconversion Based on Sensitized Triplet–Triplet Annihilation. *Coord. Chem. Rev.* **2010**, *254*, 2560–2573.
3. Zhou, J.; Liu, Q.; Feng, W.; Sun, Y.; Li, F. Upconversion Luminescent Materials: Advances and Applications. *Chem. Rev.* **2015**, *115*, 395–465.
4. Nadort, A.; Zhao, J.; Goldys, E. M. Lanthanide Upconversion Luminescence at the Nanoscale: Fundamentals and Optical Properties. *Nanoscale* **2016**, *8* (27), 13099–13130.
5. Cheng, X.; Ge, H.; Wei, Y., et al. Design for Brighter Photon Upconversion Emissions via Energy Level Overlap of Lanthanide Ions. *ACS Nano* **2018**, *12* (11), 10992–10999.
6. Auzel, F. Quantum Counter by Energy Transfer Between Two Rare Earth Ions in a Mixed Tungstate and in a Glass. *Acad. Sci.* **1966**, *262*, 1016.
7. Auzel, F. Upconversion and Anti-Stokes Processes with *f* and *d* Ions in Solids. *Chem. Rev.* **2004**, *104*, 139–173.
8. Parker, C. A.; Hatchard, C. G. Delayed Fluorescence from Solutions of Anthracene and Phenanthrene. *Proc. Chem. Soc.: London* **1962**, *14*, 386.
9. Zhao, J.; Ji, S.; Guo, H. Triplet–Triplet Annihilation Based Upconversion: From Triplet Sensitizers and Triplet Acceptors to Upconversion Quantum Yields. *RSC Adv.* **2011**, *1*, 937–950.
10. Green, M. A.; Bremner, S. P. Energy Conversion Approaches and Materials for High-efficiency Photovoltaics. *Nat. Mater.* **2017**, *16*, 23–34.
11. Sun, Q. -C.; Ding, Y. C.; Sagar, D. M., et al. Photon Upconversion Towards Applications in Energy Conversion and Bioimaging. *Progr. Surf. Sci.* **2017**, *92* (4), 281–316.
12. Wang, F.; Liu, X. Recent Advances in the Chemistry of Lanthanide-doped Upconversion Nanocrystals. *Chem. Soc. Rev.* **2009**, *38*, 976–989.
13. Parker, C. A.; Hatchard, C. G. . *Proc. Chem. Soc.* **1962**, 386–387.
14. Schulze, T. F.; Schmidt, T. W. Photochemical Upconversion: Present Status and Prospects for its Application to Solar Energy Conversion. *Energy Environ. Sci.* **2015**, *8* (1), 103–125.
15. MacDougall, S. K. W.; Ivaturi, A.; Marques-Hueso, J., et al. Ultra-high Photoluminescent Quantum Yield of β-$NaYF_4$: 10% Er^{3+} via Broadband Excitation of Upconversion for Photovoltaic Devices. *Opt. Express* **2012**, *20*, A879–A887.
16. Borjesson, K.; Dzebo, D.; Albinsson, B., et al. Photon Upconversion Facilitated Molecular Solar Energy Storage. *J. Mater. Chem. A* **2013**, *1* (30), 8521–8524.
17. Yang, P.; Deng, P.; Yin, Z. Concentration quenching in Yb:YAG. *J. Lumin.* **2002**, *97*, 51–54.

18. Wen, S.; Zhou, J.; Zheng, K., et al. Advances in Highly Doped Upconversion Nanoparticles. *Nat. Commun.* **2018**, *9* (1), 2415.
19. Dexter, D. L.; Schulman, J. H. Theory of Concentration Quenching in Inorganic Phosphors. *J. Chem. Phys.* **1954**, *22*, 1063–1070.
20. Zhao, G. -J.; Liu, J. -Y.; Zhou, L. -C., et al. Site-selective Photoinduced Electron Transfer from Alcoholic Solvents to the Chromophore Facilitated by Hydrogen Bonding: A New Fluorescence Quenching Mechanism. *J. Phys. Chem. B* **2007**, *111*, 8940–8945.
21. Wang, F.; Liu, X. Upconversion Multicolor Fine-tuning: Visible to Near Infrared Emission from Lanthanide-doped $NaYF_4$ Nanoparticles. *J. Am. Chem. Soc.* **2008**, *130*, 5642–5643.
22. Zhao, J.; Jin, D.; Schartner, E. P., et al. Single-nanocrystal Sensitivity Achieved by Enhanced Upconversion Luminescence. *Nat. Nanotechnol.* **2013**, *8*, 729–734.
23. Gargas, D. J.; Chan, E. M.; Ostrowski, A. D., et al. Engineering Bright Sub-10-nm Upconverting Nanocrystals for Single-molecule Imaging. *Nat. Nanotechnol.* **2014**, *9*, 300–305.
24. Johnson, N. J. J.; He, S.; Diao, S., et al. Direct Evidence for Coupled Surface and Concentration Quenching Dynamics in Lanthanide-doped Nanocrystals. *J. Am. Chem. Soc.* **2017**, *139*, 3275–3282.
25. Chen, G.; Ohulchanskyy, T. Y.; Kumar, R., et al. Ultrasmall Monodisperse $NaYF_4$:Yb^{3+}/Tm^{3+} Nanocrystals with Enhanced Near-infrared to Near-infrared Upconversion Photoluminescence. *ACS Nano* **2010**, *4*, 3163–3168.
26. Shen, B.; Cheng, S.; Gu, Y., et al. Revisiting the Optimized Doping Ratio in Core/Shell Nanostructured Upconversion Particles. *Nanoscale* **2017**, *9*, 1964–1971.
27. Zhang, C.; Yang, L.; Zhao, J., et al. White-light Emission from an Integrated Upconversion Nanostructure: Toward Multicolor Displays Modulated by Laser Power. *Angew. Chem. Int. Ed.* **2015**, *54*, 11531–11535.
28. Chen, X.; Jin, L.; Sun, T., et al. Energy Migration Upconversion in Ce(III)-doped Heterogeneous Core–Shell–Shell Nanoparticles. *Small* **2017**, *13*, 1701479.
29. Li, X.; Liu, X.; Chevrier, D. M., et al. Energy Migration Upconversion in Manganese(II)-doped Nanoparticles. *Angew. Chem. Int. Ed.* **2015**, *54*, 13312–13317.
30. Deng, R.; Wang, J.; Chen, R., et al. Enabling Förster Resonance Energy Transfer from Large Nanocrystals through Energy Migration. *J. Am. Chem. Soc.* **2016**, *138*, 15972–15979.
31. Wang, Y. -F.; Liu, G. -Y.; Sun, L. -D., et al. Nd^{3+}-sensitized Upconversion Nanophosphors: Efficient in vivo Bioimaging Probes with Minimized Heating Effect. *ACS Nano* **2013**, 7, 7200–7206.
32. Chen, Q.; Xie, X.; Huang, B., et al. Confining Excitation Energy in Er^{3+}-sensitized Upconversion Nanocrystals through Tm^{3+}-mediated Transient Energy Trapping. *Angew. Chem. Int. Ed.* **2017**, *56*, 7605–7609.
33. Gao, W.; Zheng, H.; Han, Q., et al. Enhanced Red Upconversion Luminescence by Codoping Ce^{3+}in β-NaY(Gd0.4)F_4:Yb^{3+}/Ho^{3+} Nanocrystals. *J. Mater. Chem.* **2014**, *C 2*, 5327–5334.
34. Chen, D.; Liu, L.; Huang, P., et al. Nd^{3+}-sensitized Ho^{3+} Single-band Red Upconversion Luminescence in Core–Shell Nanoarchitecture. *J. Phys. Chem. Lett.* **2015**, *6*, 2833–2840.
35. Chen, G.; Damasco, J.; Qiu, H., et al. Energy-cascaded Upconversion in an Organic Dye-sensitized Core/Shell Fluoride Nanocrystal. *Nano Lett.* **2015**, *15*, 7400–7407.
36. Zheng, W.; Huang, P.; Gong, Z., et al. Near-infrared-triggered Photon Upconversion Tuning in All-inorganic Cesium Lead Halide Perovskite Quantum Dots. *Nat. Commun.* **2018**, *9*, 3462.
37. Vetrone, F.; Naccache, R.; Mahalingam, V., et al. The Active-core/Active-shell Approach: A Strategy to Enhance the Upconversion Luminescence in Lanthanide-doped Nanoparticles. *Adv. Funct. Mater.* **2009**, *19*, 2924.

38. Yi, G. -S.; Chow, G. -M. Water-Soluble $NaYF_4$:Yb,Er(Tm)/$NaYF_4$/Polymer Core/Shell/Shell Nanoparticles with Significant Enhancement of Upconversion Fluorescence. *Chem. Mater.* **2007**, *19* (3), 341–343.
39. Chen, G.; Qiu, H.; Prasad, P. N., et al. Upconversion Nanoparticles: Design, Nanochemistry, and Applications in Theranostics. *Chem. Rev.* **2014**, *114*, 5161.
40. Liu, Y.; Ai, K.; Liu, J., et al. A High-performance Ytterbium-based Nanoparticulate Contrast Agent for in vivo X-ray Computed Tomography Imaging. *Angew. Chem. Int. Ed.* **2012**, *51*, 1437–1442.
41. Wang, C.; Cheng, L.; Liu, Z. Upconversion Nanoparticles for Photodynamic Therapy and Other Cancer Therapeutics. *Theranostics* **2013**, *3*, 317.
42. Idris, N. M.; Gnanasammandhan, M. K.; Zhang, J., et al. In Vivo Photodynamic Therapy using Upconversion Nanoparticles as Remote-controlled Nanotransducers. *Nat. Med.* **2012**, *18*, 1580.
43. Cheng, L.; Yang, K.; Li, Y. G., et al. Multifunctional Nanoparticles for Upconversion Luminescence/MR Multimodal Imaging and Magnetically Targeted Photothermal Therapy. *Biomaterials* **2012**, *33*, 2215.
44. Tian, G.; Zhang, X.; Gu, Z., et al. Recent Advances in Upconversion Nanoparticles-based Multifunctional Nanocomposites for Combined Cancer Therapy. *Adv. Mater.* **2015**, *27* (47), 7692–7712.
45. Goldschmidt, J. C. Upconversion for Photovoltaics – A Review of Materials, Devices and Concepts for Performance Enhancement. *Adv. Opt. Mater.* **2015**, *3*, 510.
46. Yan, C.; Dadvand, A.; Rosei, F., et al. Near-IR Photoresponse in New Up-converting CdSe/$NaYF_4$:Yb,Er Nanoheterostructures. *J. Am. Chem. Soc.* **2010**, *132*, 8868.
47. Gibart, P.; Auzel, F.; Guillaume, J. C., et al. Below Band-gap IR Response of Substrate-free GaAs Solar Cells using Two-photon Up-conversion. *J. Appl. Phys.* **1996**, *35*, 4401.
48. Richards, B. S.; Shalav, A. Enhancing the Near-infrared Spectral Response of Silicon Optoelectronic Devices via Up-conversion. *IEEE Trans. Electron Devices* **2007**, *54*, 2679.
49. Ramachari, D.; Esparza, D.; Lopez-Luke, T., et al. Synthesis of Co-doped Yb^{3+}-Er^{3+}:ZrO_2 Upconversion Nanoparticles and their Applications in Enhanced Photovoltaic Properties of Quantum Dot Sensitized Solar Cells. *J. Alloys Compd.* **2017**, *698*, 433–441.
50. Su, L. T.; Karuturi, S. K.; Luo, J., et al. Photon Upconversion in Hetero-nanostructured Photoanodes for Enhanced Near-Infrared Light Harvesting. *Adv. Mater.* **2013**, *25*, 1603–1607.
51. Sun, C. -J.; Xu, Z.; Hu, B., et al. Application of $NaYF_4$:Yb,Er Upconversion Fluorescence Nanocrystals for Solution-processed Near Infrared Photodetectors. *Appl. Phys. Lett.* **2007**, *91*, 191113.
52. Wu, J.; Yang, Z.; Qiu, C., et al. Enhanced Performance of a Graphene/GaAs Self-driven Near-infrared Photodetector with Upconversion Nanoparticles. *Nanoscale* **2018**, *10*, 8023–8030.
53. Gao, D.; Tian, D.; Zhang, X., et al. Simultaneous Quasi-one-dimensional Propagation and Tuning of Upconversion Luminescence through Waveguide Effect. *Sci. Rep.* **2016**, *6*, 22433.
54. Lin, G.; Jin, D. Taking Upconversion to Lase in Microcavity. *Nat. Nanotechnol.* **2018**, *13*, 534–536.
55. Savchuk, O. A.; Carvajal, J. J.; Brites, C. D. S., et al. Upconversion Thermometry: A New Tool to Measure the Thermal Resistance of Nanoparticles. *Nanoscale* **2018**, *10*, 6602–6610.
56. Pickel, A. D.; Teitelboim, A.; Chan, E. M., et al. Apparent Self-heating of Individual Upconverting Nanoparticle Thermometers. *Nat. Commun.* **2018**, *9*, 4907.
57. Nigoghossian, K.; Ouellet, S.; Plain, J., et al. Upconversion Nanoparticle-decorated Gold Nanoshells for Near-infrared Induced Heating and Thermometry. *J. Mater. Chem. B* **2017**, *5*, 7109–7117.

CHAPTER 12

Machine learning approach for materials technologies

Jatindra Kumar Dash[a], Mohit Sharma[b], Goutam Kumar Dalapati[c,d]
[a]Department of Computer Science and Engineering, SRM University-AP, Amaravati, Andhra Pradesh, India
[b]Agency for Science, Technology and Research (A*STAR), Institute of Materials Research and Engineering, Singapore
[c]Department of Physics, SRM University-AP, Amaravati, Andhra Pradesh, India
[d]School of Engineering & Innovation, The Open University, Walton Hall, Milton Keynes, United Kingdom

1 Introduction

Machine learning (ML) techniques have potential to accelerate the material development and process optimization through accurate prediction. Recently, machine learning techniques have been used for materials designing *(1–4)*, performance optimization of the p-type transparent conductors *(5)*, optical properties of materials *(6)*, design of new glass *(7)*, thermal behavior of building façade and prediction of building energy consumption *(8–12)*. The accurate implementation of machine learning with materials processing knowledge can reduce the experimental cost, material consumption, and timeline required for the scaling-up of the technologies. In literature, computer modeling or machine learning is successfully utilized to swift development and design of nanomaterials more precisely *(13)*. Machine learning algorithms can be used to create computational model that can examine the complex structure of nanoparticles and simulate its scattering properties along with different colors of light. This utilization of computational approaches can reduce the development cost and improve process accuracy. For example, ML techniques have been used to optimize the p-type transparent conducting materials synthesized by using chemical bath deposition (CBD) process. By using ML and strategic design of experiment (DOE), the number of experimental process required for process optimization has been reduced. An optimal process for transparent film with average visible transmittance of ~80% and conductive of ~430 S cm^{-1} obtained for the transparent film after two rounds of optimization *(5)*.

A conventional programmatic approach to find a solution for a given problem is to write a computer program that takes input from user, process these inputs using some definite steps and produce the output. This process

Energy Saving Coating Materials
http://dx.doi.org/10.1016/B978-0-12-822103-7.00012-1

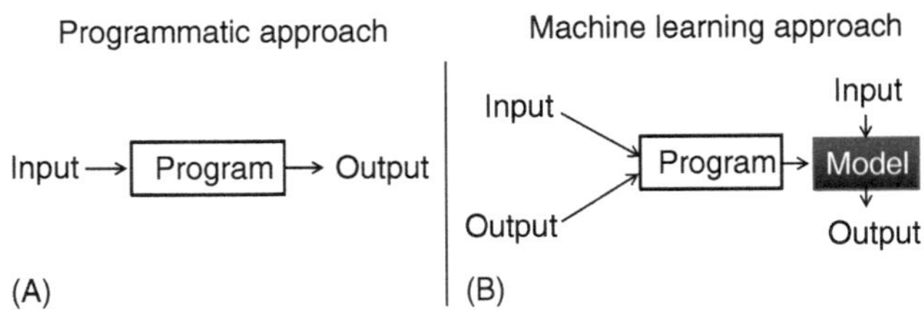

Figure 12.1 (A) Programmatic approach. (B) Machine learning approach.

is depicted in Fig. 12.1.A This approach usually applied for problems whose solutions can be derived using a fixed set of mathematical equations. In contrast to this, a machine learning program learns the concept or hypothesis from existing data (obtained through experimentation or mathematical simulations) and creates a model that can compute (more precisely predict) the output for unseen data. The machine learning algorithm or program helps to develop a model that learns the input and output mapping from the existing data. The schematic for the machine learning approach is depicted in Fig. 12.1B. The machine learning approach is followed for the problems whose solutions cannot be derived using a fixed set of equations.

The formal definition for machine learning is "A computer program is said to learn from Experience E with respect to some class of Tasks T and a Performance measure P, if its performance on Tasks in T, as measured by P, improves with experience E" *(14)*. The existing data (experimental or simulation) in any domain serves as Experience (E). The machine learning algorithms are broadly categorized into mainly supervised and unsupervised (Fig. 12.2).

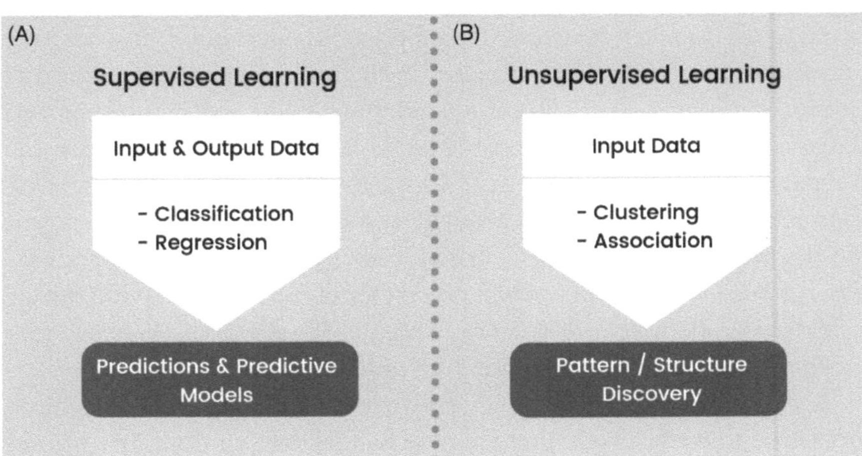

Figure 12.2 ***Types of machine learning algorithm.*** *Image Courtesy: https://medium.com/datadriveninvestor/supervised-vs-unsupervised-machine-learning-5200ffa7301a.*

1.1 Supervised learning

As the name indicates, these types of algorithm make use of supervisor as a teacher to train or teach the machine using data. In such cases, the data are properly labeled. In other words, in supervised learning, each input set of data is tagged with the correct answer. The algorithm in this category tries to learn the input output relationship (usually referred to as concept) from the labeled datasets (set of training examples). The model generated through the training process is then used to predict the output for the unseen input data. Two different types of task are usually performed using supervised learning namely Classification and Regression. In case of classification, the task is to classify a given input to an output class or category. In regression, the task is to predict an output that is continuous in nature. Both classification and regression are used for predicting output for a given input. Classification and Regression both share the same concept of utilizing known datasets (referred to as training datasets) to make predictions.

1.2 Unsupervised learning

In unsupervised learning there is no teacher present during the process of learning. In other words, there will not be any labeled prior knowledge in case of unsupervised learning. In this case, the model learns the concept by analyzing the (dis)similarity the concept from the input data itself. Data clustering is an example of unsupervised learning (Table 12.1)

2 Machine learning model in materials science

There are basically three important phases in building a machine learning model for any application.

Table 12.1 Comparison of different supervised and unsupervised algorithms 1

Supervised Learning	Unsupervised Learning
Regression	Clustering
Linear: simple, multiple regression	K-means clustering
Nonlinear: simple, multiple	Agglomerative clustering
Classification:	Fuzzy C-means clustering
Naïve Bayes	Graph-based clustering
ANN	
SVM	
Decision tree	
kNN	
Logistic regression	

ANN, artificial neural network; *SVM*, support vector machine; *kNN*, K-nearest neighbor.

2.1 Data collection and preprocessing

In this phase, the necessary data are collected from previous experiments or can be generated through some simulations. These data work as experience for the model to learn the concept. Usually the data collected are not in a suitable structure and thus need to be structured before it can be used to train the model. The methods used for making the data suitable for application falls under feature engineering or data preprocessing. Data preprocessing includes several activities such as normalization, standardization, data cleansing, missing data discovery, data reduction for reducing complexity, etc.

2.2 Model creation

In this phase, the processed data are fed to one of the machine learning algorithm selected suitably to create a model. This phase is also called training phase. The model generated through training phase is then put to work on unseen data.

2.3 Evaluation

Any model created should be evaluated thoroughly for its performance to examine the quality of learning. In this phase, the model is evaluated critically and a performance score is computed that helps to take a decision on applicability of that model.

To develop energy saving smart building, thermal behavior of a wall has been modeled by using deep learning techniques to predict accurate behavior of a specific enclosure *(12)*. The connexionist model has been used for the modeling. For the training of connexionist model, data acquired from several measurements in the specific enclosure *(12)*. The connexionist model is more effective and can be useful to predict energy consumption of the building. In another work, Zheng et al. *(6)*, have performed numerical study on the thermal and optical properties of an aerogel glazing system using advanced machine learning algorithm and multivariable optimization to form a translucent window with super-insulating performance. The results show artificial neural network modeling is more accurate and computational-efficient than the traditional fitting methodology *(6)*. It is also worth noting that the teaching-learning-based optimization is more robust and competitive.

3 Conclusion

Materials innovation process and thermal behavior of an enclosed environment can be predicted efficiently and accurately by data driven machine learning technique. Strategic design of experiment and ML optimization is essential in finding experimental conditions which attempts to reduce experimental process and time significantly.

References

1. Schmidt, J.; Marques, M. R. G.; Botti, S.; Marques, M. A. L. Recent Advances and Applications of Machine Learning in Solid State Materials Science. *NPJ Comput. Mater.* **2019**, *83*, 1–36.
2. Pilania, G.; Wang, C.; Jiang, X.; Rajasekaran, S.; Ramprasad1, R. Accelerating Materials Property Predictions Using Machine Learning. *Sci. Rep.* **2013**, *3*, 2810.
3. Liu, Y.; Zhao, T.; Ju, W.; Shi, S. Materials Discovery and Design Using Machine Learning. *J. Materiomics* **2017**, *3*, 159–177.
4. Brunin, G.; Ricci, F.; Ha, V.; Rignanese, G.; Hautier, G. Transparent Conducting Materials Discovery Using High-Throughput Computing. *Comp. Mater.* **2019**, *63*, 1–13.
5. Wei, L.; Xu, X.; Gurudayal, J.; Bullock, J. W.; Ager. Machine Learning Optimization of p-Type Transparent Conducting Films. *Chem. Mater.* **2019**, *31*, 7340–7350.
6. Zheng, S.; Zhou, Y. Numerical Study on the Thermal and Optical Performances of an Aerogel Glazing System with the Multivariable Optimization Using an Advanced Machine Learning Algorithm. *Adv. Theory Simul.* **2019**, *2*, 1900092.
7. Stein, H. S.; Guevarra, D.; Newhouse, P. F.; Soedamadji, E.; Gregoire, J. M. Machine Learning of Optical Properties of Materials-Predicting Spectra from Images and Images from Spectra. *Chem. Sci.* **2019**, *10*, 47.
8. Liu, H.; Fu, Z.; Yang, K.; Xu, X.; Bauchy, M. Machine Learning for Glass Science and Engineering: A Review. *J. Non-Crystalline Solids: X* **2019**, *4*, 100036.
9. Ngarambe, J.; Yun, G. Y.; Kim, G. Prediction of Indoor Clothing Insulation Levels: A Deep Learning Approach. *Energy Build.* **2019**, *202*, 109402.
10. Zhou, Y.; Zheng, S.; Zhang, G. Artificial Neural Network Based Multivariable Optimization of a Hybrid System Integrated with Phase Change Materials, Active Cooling, and Hybrid Ventilations. *Energy Convers. Manag.* **2019**, *197*, 111859.
11. Seyedzadeh, S.; Rahimian, F.; Glesk, I.; Roper, M. Machine Learning for Estimation of Building Energy Consumption and Performance: A Review. *Visualiz. Eng.* **2018**, *6*, 5.
12. Aznar, F.; Echarri, V.; Rizo, C.; Rizo, R. Modelling the Thermal Behaviour of a Building Facade Using Deep Learning. *PLoS ONE* **2018**, *13* (12), e0207616.
13. AI-based method could speed development of specialized nanoparticles, http://news.mit.edu/2018/.
14. Mitchell, T. M. Machine Learning and Data Mining. *Commun. ACM* **1999**, *42*, 31–36.

CHAPTER 13

Roadmap for materials selection and energy saving coatings

Mohit Sharma[a], Goutam Kumar Dalapati[a,b,c]
[a]Agency for Science, Technology and Research (A★STAR), Institute of Materials Research and Engineering, Singapore
[b]Department of Physics, SRM University-AP, Amaravati, Andhra Pradesh, India
[c]School of Engineering & Innovation, The Open University, Walton Hall, Milton Keynes, United Kingdom

1 Progression of low-emission coatings for energy technology

The state of the art for industrial development of low-emission coating materials technology and its advancement has been demonstrated in Table 13.1. Up to 1975, the energy saving films was grown by evaporation and pyrolitic techniques, subsequently the new sputtering technologies were adopted to reduce the heat gain and save energy with affordability. Interface engineering and metal–dielectric optimization is required to enhance durability and optical performance of the coating.

The deposition methods and parameters considerably influence the performance of the energy saving coatings. The two key parameters: visible transmittance and IR reflectance are essential for energy saving utility of transparent heat reflectors. However, in multilayered coatings the metal oxide dielectric materials is necessary to improve the visible transmittance through the coating layer and to protect thin metal layer from environmental effects. Interface in between the metal layer and the metal oxide dielectric reduces visible transmittance and IR reflectance. The feeble interfacial effects can be reduced by inserting a passivation layer within the metal and dielectric. The passivation can be employed by utilizing the metal oxide target and oxygen ambient for dielectric deposition. The thickness of the metal layer, metal oxide dielectric quality, growth process, passivation, and interfacial layer formation are the key parameters for effective heat regulating multilayered performance. The presence of oxygen and nitrogen change surface morphology and the growth mechanism of ultra-thin film metal. The refractive index, optical band gap and microstructural properties of the coatings can be controlled by optimization of the coating process and

Energy Saving Coating Materials
http://dx.doi.org/10.1016/B978-0-12-822103-7.00013-3

Table 13.1 Low-e transparent coatings. Industrial growth and advancement of technology since 1975[a] *(1)*.

Description	**Specification of commercial product in the coatings technology**				**Ongoing research on metals based multilayer films**
Year	**1975–1990**	**1990–2010**	**2010–2018**		
Coating materials	BiO_x/Au, SnO_2, ZnS/Au/ZnS	Oxide/Ag/oxide	Multi-layered (double glaze, triple glaze) Dielectric and Dielectric/Ag/Dielectric	Advanced interface engineering for Metal-based multilayered	Metal-based multifunctional coatings with enhanced stability and durability
Technology	Heat insulating glasses/window shielding for building and automotive	Asymmetric and double layered silver system with light to solar gain (LSG) range 2–2.3	Multilayer silver system	Multilayer silver and bronze system	Multilayered Metal Cu, Ag, Sn, Au oxide transparent coating systems
Visible transmission % (EN-410)	40–60	50–60	50–70	47–89	45–85
Shading coefficient (EN 410/0,80)	–	–	0.25–0.49	0.28–0.76	0.55–0.66
U Value W/m²K (EN 673)	–	–	0.6–1.0	2.0–6.0	5.0–6.0
Solar factor SHGC (EN-410)	–	–	0.17–0.47	0.44–0.65	0.55–0.57
Deposition method	Sputtering pyrolysis, evaporation, APCVD	Sputtering	Sputtering	Sputtering	Sputtering

[a] Data collected from product specifications: 3M, AGC interpane Flashglas/DELOG, Thermoplus Cardinal, Double silver, Solarban glass.

parameters during dielectric deposition. The visible transmittance of the dielectric increases after thermal annealing process.

2 Prospective on heat regulating materials for energy technologies

Chapter 1 elucidates the basic technologies to extract solar radiation from earth's atmosphere and conversion into energy needs. The right nanostructured multi-layered system, for example, dielectric/metal/dielectric coatings, with suitable chemical and physical properties can effectively improve the heat regulating and cooling properties. The advanced functional coating on the glass/plastic substrate effectively reduces the energy consumption and help to reduce carbon footprints globally. Further, strategically synchronization of transparent solar cells integrated with multi-layered coatings are effectively improve properties of smart windows applications. Chapter 2 includes the interaction of infrared radiation with the energy saving coatings and their role to tune the optical performance. Chapter 3 includes the state of the art of the transparent heat regulating materials, including metal-dielectric THR, heat mirror coatings for tropical conditions, and energy generating transparent heat regulating window glazing systems. Chapter 4 comprises the details of dielectric-metal-dielectric THR multilayers coatings; architectures and design. In the subsequent section of the chapter the development of Metal-dielectric nanocomposite materials as high-stability dielectric/metal/dielectric multilayer coatings has been included. Chapter 5 includes heat regulating radiative cooling and cool paint applications. The paint industry is strategically aligned with recent technologies to achieve around 10°C reduction in temperature after employing cool radiative and paint coatings. The materials prospective and theoretical consideration and experimental demonstration of the cool radiative cooling process and cool paint additives has in discussed in the subsequent section of the chapter.

Chapter 6 includes the properties of the solar absorbers and their influence solar selectivity on photo thermal conversion efficiency. The future potential of solar thermal systems was discussed using low dimensional ultra high temperature ceramics (UHTCs) films. Chapter 7 consists of atomic layer deposition (ALD), liquid-ALD, and successive ionic layer adsorption and reaction (SILAR) techniques to develop thin films that as transparent heat regulations (THR). The detail processing and optimization of the techniques and surface morphological, electronic properties of these materials were discussed. Chapter 8 includes new coatings materials and efficient thermoelectric technologies. A smart window with anti-bacterial activity has been discussed in Chapter 9. Chapter 10 discusses the development in smart window engineering toward antibacterial biomaterials for healthcare industry, further it highlights antibacterial activities of smart coating technology wettability and stability of heat regulating metal oxide thin-film coatings. The chapter included the wettability research to expand the progress of affordable. Innovative technology and up-conversion nano-particles have been included in Chapter 11 for environmental friendly heat regulation technology. Chapter 12 comprises the basic discussions on the artificial intelligence for materials selection of energy saving materials, followed by the overall summary of the topics discussed in the book in this chapter. Smart-window technology has recently developed and its assessments on viable commercial potential are in progress. The precision implementation of coating technologies has yet to deliver a larger prospective in terms of environment friendly heating management services. The smart technologies, for example, thermo chromic windows can help to modulate solar transmittance in near infrared and visible region, depending on personal preference and climatic conditions. The engineered multilayer nano-structures materials considered as new opportunities for future smart window and energy efficient coating technology. The strategic insight of the energy saving marker landscape has suggested that the energy saving coating based technologies will observe strong rise by 2024. The key forecast areas for this prediction includes aerospace, construction and defense applications [2]. In current scenario, the majority of energy demands are fulfilled through non-renewable energy sources, which in-turns create a huge carbon footprints, and thus environment pollution as well as contribute to global warming. It is essential to address the key issues of greenhouse gas emissions mitigation accorded to United Nations Framework Convention on Climate Change (UNFCCC). In thoughtfulness to address the adverse and abrupt climatic changes globally, it is obligatory to control worldwide temperature to well below 2°C above preindustrial levels. The framework countries required to adopt essential actions to mitigate global warming and carbon footprints through compulsory regularities. In coming future, owing to the rigorous regulatory for emission control, it will be essential to implement environmental friendly/biodegradable and energy efficient and smart technologies to drive the regional product demand. The development of innovative technologies comprising energy saving materials and coatings is a key consideration to achieve the vital energy saving milestones worldwide. Henceforth, reduction of global warming is much sought-after for the society.

References

1. Dalapati, G. K., et al. Transparent Heat Regulating (THR) Materials and Coatings for Energy Saving Window Applications: Impact of Materials Design, Micro-Structural, and Interface Quality on the THR Performance. *Progr. Mater. Sci.* **2018**, *95*, 42–131.
2. https://www.gminsights.com/industry-analysis/energy-efficient-coatings-market.

Index

Note: Page numbers followed by "f" indicate figures, "t" indicate tables.

Q

R

S

Kolibri 360 GmbH
Ferdinand-Jühlke-Straße 7
99095 Erfurt, Deutschland
produktsicherheit@kolibri360.de